Geometry and Computing

Jörg Peters • Ulrich Reif

Subdivision Surfaces

With 52 Figures

Springer

Jörg Peters
University of Florida
Dept. Computer & Information Science
Gainesville FL 32611-6120
USA
jorg@cise.ufl.edu

Ulrich Reif
TU Darmstadt
FB Mathematik
Schloßgartenstr. 7
64289 Darmstadt
Germany
reif@mathematik.tu-darmstadt.de

ISBN 978-3-642-09527-6 e-ISBN 978-3-540-76406-9

Springer Series in Geometry and Computing

Mathematics Subjects Classification (2000): 65D17, 65D07, 53A05, 68U05

Softcover reprint of the hardcover 1st edition 2008

Cover design: deblik, Berlin
Printed on acid-free paper

9 8 7 6 5 4 3 2 1

springer.com

for

Anousha
Daniel
Julian
Kavina
Tobias

Preface

Akin to B-splines, the appeal of subdivision surfaces reaches across disciplines from mathematics to computer science and engineering. In particular, subdivision surfaces have had a dramatic impact on computer graphics and animation over the last 10 years: the results of a development that started three decades ago can be viewed today at movie theaters, where feature length movies cast synthetic characters 'skinned' with subdivision surfaces. Correspondingly, there is a rich, ever-growing literature on its fundamentals and applications.

Yet, as with every vibrant new field, the lack of a uniform notation and standard analysis tools has added unnecessary, at times inconsistent, repetition that obscures the simplicity and beauty of the underlying structures. One goal in writing this book is to help shorten introductory sections and simplify proofs by proposing a standard set of concepts and notation.

When we started writing this book in 2001, we felt that the field had sufficiently settled for standardization. After all, Cavaretta, Dahmen, and Micchelli's monograph [CDM91] had appeared 10 years earlier and we could build on the habilitation of the second author as well as a number of joint papers. But it was only in the process of writing and seeing the issues in conjunction, that structures and notation became clearer. In fact, the length of the book repeatedly increased and decreased, as key concepts and structures emerged.

Chapter 2/15, for example, was a late addition, as it became clear that the differential geometry for singular parameterizations, of continuity, smoothness, curvature, and injectivity, must be established upfront and in generality to simplify the exposition and later proofs. By contrast, the key definition of subdivision surfaces as splines with singularities, in Chap. 3/39, was a part of the foundations from the outset. This point of view implies a radical departure from any focus on control nets and instead places the main emphasis on nested surface rings, as explained in Chap. 4/57. Careful examination of existing proofs led to the explicit formulation of a number of assumptions, in Chap. 5/83, that must hold when discussing subdivision surfaces in generality. Conversely, placing these key assumptions upfront, shortened the presentation considerably. Therefore, the standard examples of subdivision algorithms reviewed in Chap. 6/109 are presented with a new, shorter and simpler analysis than in earlier publications. Chapters 7/125 and 8/157 were triggered by very recent, new insights and results and partly contain unpublished material. The suitability of the major known subdivision algorithms for engineering design was at the heart of the investigations into the shape of subdivision surfaces in Chap. 7/125. The shortcomings of the standard subdivision algorithms discovered in the process forced a renewed search for an approach to subdivision capable of meeting shape and higher-order continuity requirements. Guided subdivision was devised in response. The second part of Chap. 7/125 recasts this class of subdivision algorithms in a more abstract form that may be used as a prototype for a number of new curvature continuous subdivision algorithms. The first part of Chap. 8/157 received a renewed impetus from a recent stream of publications aimed at predicting the distance of a subdivision surface from their geometric control structures after some m refinement steps. The introduction of proxy surfaces and the distance to the corresponding subdivision surface subsumes this set of questions and provides a framework for algorithm-specific optimal estimates. The second part of Chap. 8/157 grew out of the surprising observation that the Catmull–Clark subdivision can represent the same sphere-like object starting from any member of a whole family of initial control configurations. The final chapter, Chap. 9/175, shows that a large variety of subdivision algorithms is fully covered by the exposition in the book. But it also outlines the limits of our current knowledge and opens a window to the fascinating forms of subdivision currently beyond the canonical theory and to the many approaches still awaiting discovery.

As a monograph, the book is primarily targeted at the subdivision community, including not only researchers in academia, but also practitioners in industry with an interest in the theoretical foundations of their tools. It is not intended as a course text book and contains no exercises, but a number of worked out examples. However, we aimed at an exposition that is as self-contained as possible, requiring, we think, only basic knowledge of linear algebra, analysis or elementary differential geometry. The book should therefore allow for independent reading by graduate students in mathematics, computer science, and engineering looking for a deeper understanding of subdivision surfaces or starting research in the field.

Two valuable sources that complement the formal analysis of this book are the SIGGRAPH course notes [ZS00] compiled by Schröder and Zorin, and the book 'Subdivision Methods for Geometric design' by Warren and Weimer [WW02]. The notes offer the graphics practitioner a quick introduction to algorithms and their implementation and the book covers a variety of interesting aspects outside our focus; for example, a connection to fractals, details of the analysis of univariate algorithms, variational algorithms based on differential operators and observations that can simplify implementation.

We aimed at unifying the presentation, placing for example *bibliographical notes* at the end of each core chapter to point out relevant and original references. In addition to these, we included a large number of publications on subdivision surfaces in the reference section. Of course, given the ongoing growth of the field, these notes cannot claim completeness. We therefore reserved the internet site

`www.subdivision-surface.org`

for future pointers and additions to the literature and theme of the book, and, just possibly, to mitigate any damage of insufficient proof reading on our part.

It is our pleasure to thank at this point our colleagues and students for their support: *Jianhua Fan, Ingo Ginkel, Jan Hakenberg, René Hartmann, Kęstutis Karčiauskas, Minho Kim, Ashish Myles, Tianyun Lisa Ni, Andy LeJeng Shiue, Georg Umlauf*, and *Xiaobin Wu* who worked with us on subdivision surfaces over many years. *Jan Hakenberg, René Hartmann, Malcolm Sabin, Neil Stewart*, and *Georg Umlauf* helped to enhance the manuscript by careful proof-reading and providing constructive feedback. *Malcolm Sabin* and *Georg Umlauf* added valuable material for the bibliographical notes. *Nira Dyn* and *Malcolm Sabin* willingly contributed two sections to the introductory chapter, and it was again *Malcolm Sabin* who shared his extensive list of references on subdivision which formed the starting point of our bibliography. *Chandrajit Bajaj* generously hosted a retreat of the authors that brought about the final structure of the book. Many thanks to you all! The work was supported by the NSF grants CCF-0430891 and DMI-0400214.

Our final thanks are reserved for our families for their support and their patience. You kept us inspired.

Contents

List of Definitions, Theorems, Examples and Lemmas

Chapter 1
Introduction and Overview

Subdivision surfaces can be viewed from at least three different vantage points. A designer may focus on the increasingly smooth shape of *refined polyhedra*. The programmer sees *local operators applied to a graph data structure*. This book views subdivision surfaces as *spline surfaces with singularities* and it will focus on these singularities to reveal the analytic nature of subdivision surfaces. Leveraging the rich interplay of linear algebra, analysis and differential geometry that the spline approach affords, we will, in particular, be able to clarify the necessary and sufficient constraints on subdivision algorithms to generate smooth surfaces. Viewing subdivision surfaces as spline surfaces with singularities is, at present, an unconventional point of view. Visualizing a sequence of polyhedra or tracking a sequence of control nets appears to be more intuitive. Ultimately, however, both views fail to capture the properties of subdivision surfaces due to their discrete nature and lack of attention to the underlying function space. In Sects. 1.1/1 and 1.2/2, we now briefly discuss the two points of view not taken in this book while in Sect. 1.3/4 the analytic view of subdivision surfaces as splines with singularities is sketched out. Section 1.4/6 delineates the focus and scope and Sect. 1.5/7 gives an overview over the topics covered in the book. A useful section to read is Sect. 1.6/7 on notation.

The trailing two sections are special. We felt a need to recall the state of the art in subdivision in the regular, shift invariant setting, and to give an overview on the historical development of the topic discussed in this book. In view of our own, limited expertise in these fields, we decided to seek prominent help. Nira Dyn and Malcolm Sabin, two pioneers and leading researchers in the subdivision community agreed to contribute, and their insightful overviews form Sects. 1.7/8 and 1.8/11.

1.1 Refined Polyhedra

For a graphics designer, subdivision is a tool for automatically cutting off sharp edges from a carefully crafted polyhedral object. The goal is to obtain a finer and

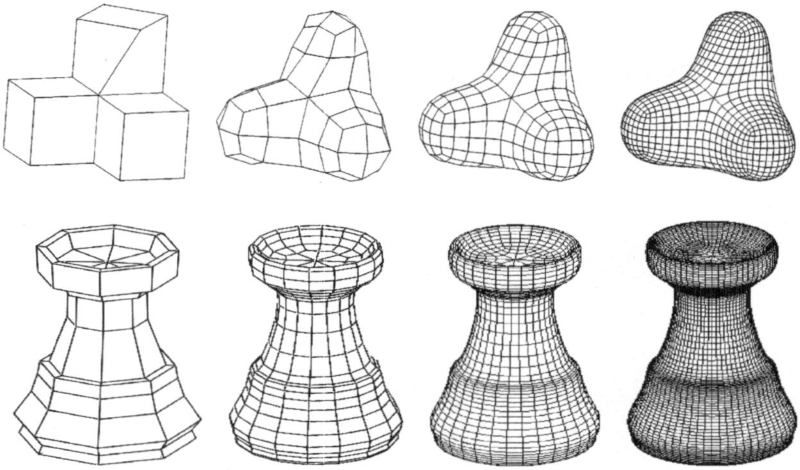

Fig. 1.1 Catmull–Clark algorithm: Starting from a given input mesh, iterated mesh refinement yields a sequence of control nets converging to a smooth limit surface. Vertices with $n \neq 4$ neighbors require *extraordinary* subdivision rules.

finer faceted representation that converges to a visually smooth limit surface (see Fig. 1.1/2). In effect, subdivision is viewed here as *geometric refinement and smoothing*. This intuitive view of subdivision has made it popular for a host of applications. This book could be faulted for failing to celebrate the rich content that can be generated with such faceted representations that have taken, for example, movie animation by storm. Indeed, we neglect the graphics designer's faceted control polyhedron until Sect. 8.1/157. This is due to the fact that a number of restrictions and assumptions have to be placed on subdivision algorithms before the notion of a control polyhedron even makes geometric sense. The cases where the control polyhedron is well-defined are therefore justifiably famous and popular.

The actual relationship between the properties of the finite control polyhedron and those of the limit subdivision surface is not straightforward, already for position and more so for higher-order differential geometric quantities. Moreover, in many design packages, the control polyhedron is ultimately projected onto the limit surface.

1.2 Control Nets

For the computer scientist, subdivision is primarily a set of operations on a graph data structure. While the vertices still carry geometric meaning, the edges serve to encode connectivity. Facets play a subordinate role, relevant only for rendering. This point of view, subtly different from faceted approximation, was also taken by the early literature on subdivision surfaces. Subdivision surfaces were correctly

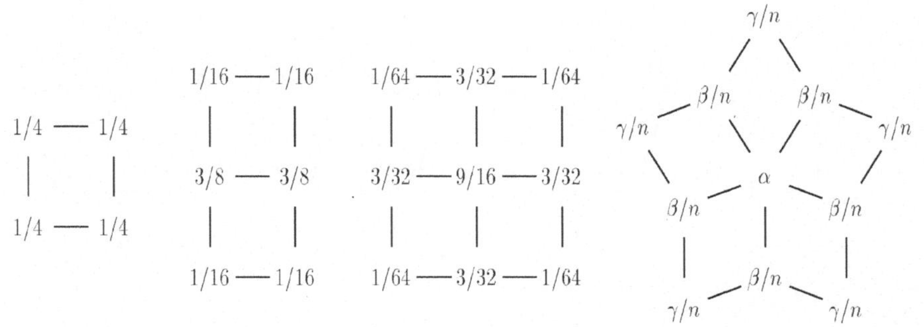

Fig. 1.2 Stencils for Catmull–Clark algorithm: (*left three*) Rules for determining new B-spline control points of a uniform bicubic spline from old ones after uniform knot insertion. The numbers placed at the grid points give the averaging weights: for example (*left*), a new point is generated as $1/4$ of each of four control points of a quadrilateral. The rules establish a new control point for each face, edge and vertex respectively. (*right*) Vertices with $n \neq 4$ neighbors require a rule generalizing the regular case $n = 4$.

characterized as generalizing a property of tensor-product B-splines: where the vertices connected by edges form a regular grid, they are interpreted as the B-spline control net of a uniform tensor-product spline. Representing such splines on a subdivided domain, by a standard technique called ‘uniform knot insertion’, yields a finer regular grid. Figure 1.2/3, left three, illustrates this process for bicubic splines.

When the regular grid of control points is replaced by an irregular configuration, the rules of regular grid refinement can obviously no longer be applied. The contribution of the seminal papers [DS78,CC78] are ‘extraordinary subdivision rules’ that mimic the regular rules and apply to irregular networks of points. The vertices of the input are taken to be control points and the edges determine how a *mesh refinement operator* is applied (Fig. 1.2/3, right).

To analyze these extraordinary rules, the early subdivision literature viewed subdivision surfaces as the limit of a sequence of ever finer control nets. The rules of refinement correspond to smoothing operators that map a neighborhood of the control point to an equivalent neighborhood of the corresponding control point in the refined control net. To track the mesh near any given control point, all smoothing operators are placed into the rows of a subdivision matrix. Repeated refinement can then locally be viewed as repeated application of the subdivision matrix to a vector of control points of the neighborhood. This *discrete, linear algebraic* view immediately yields important guidelines for constructing extraordinary rules. In particular, it provides *necessary conditions* for a smooth limit surface that take the form of restrictions on the eigenvalues of the subdivision matrix.

However, the discrete, linear-algebraic point of view fails to provide sufficient conditions since it neglects the functions associated with the control points. The splines defined by the ever-increasing regular parts of the control net give a foothold to tools of analysis and differential geometry.

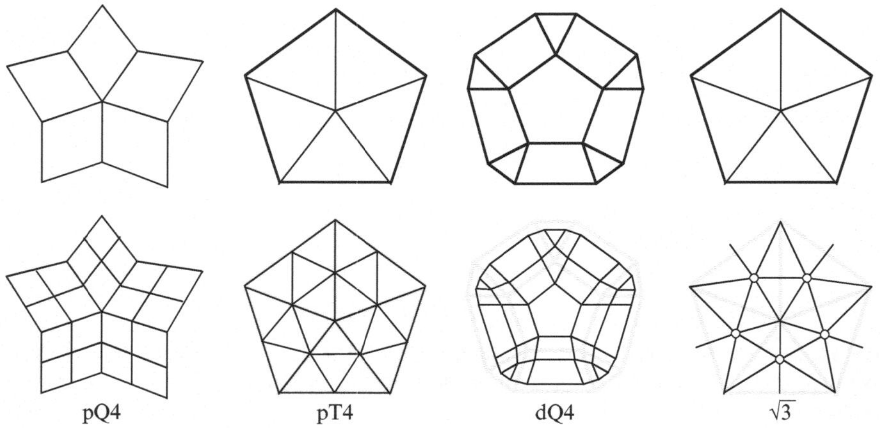

Fig. 1.3 Types of mesh refinement: *top* Initial mesh and *bottom* refined mesh. We focus on algorithms of type pQ4 and dQ4 that result in quadrilateral patches; the analysis and structure of other subdivision algorithms is analogous (see Chap. 9/175).

1.3 Splines with Singularities

To adequately characterize the continuity properties of subdivision surfaces, this book emphasizes a third view. As before, where the connectivity allows, the points of the control net may be interpreted as, e.g., spline coefficients. Refinement isolates pieces of the surface where such an interpretation is not possible and extraordinary rules have to be applied. These pieces of the surface are defined as the union of *nested sequences of surface rings and their limit points*. This approach supplies a concrete parametrization that allows us to leverage tools of *analysis and differential geometry* to expose the structure of subdivision surfaces. We contrast the concepts as follows.

- *Mesh refinement* generates a sequence of finer and finer control nets, converging to a limit surface. This is the appropriate setup for data structures and implementation.
- *Subdivision* generates a sequence of nested rings, whose union forms a spline in the generalized sense. This is the appropriate setup for analytic purposes.

Typically, the resulting objects coincide: the union of rings defines the same surface as the limit of control nets. Because this book investigates analytic properties, it focuses on subdivision in the sense of the second item. The exposition will use the following concepts.

Splines in a generalized sense. The common attribute of the numerous variants of splines appearing in the literature[1] is a segmentation of the domain. Hence, we use

[1] The 'zoo of splines' is a crowded place: A large part of the latin and greek alphabets is already reserved for one-letter prefixes such as for 'B-spline' or 'G-spline'. In addition, there are arc splines, box splines, Chebyshev splines, discrete splines, exponential splines, trigonometric splines, rational splines, simplex splines, perfect splines, monosplines, Euler splines, Whittaker splines . . .

the term ‘spline’ in the following, much generalized sense. A *spline* is a function defined on a domain which consists of indexed copies of a standard domain, such as the unit square for quadrilateral bivariate splines (cf. Definition 3.1/43). Beforehand, we make no assumptions on the particular type of functions to be used. Therefore our use of the word ‘spline’ covers not only linear combinations of B-splines or box-splines,[2] but also a host of non-polynomial cases like piecewise exponentials or even wavelet-type functions. The key observation is that we can regard subdivision surfaces as a special case of these general splines. Because, ultimately, we are aiming at the representation of smooth surfaces, we assume throughout that splines are at least continuous.

Quadrilateral splines. We focus on splines defined on a union of indexed *unit squares*, called *cells*, and subdivision that iterates *binary refinement* of these cells. The analysis of subdivision surfaces based on a triangular domain partition (see Fig. 1.3/4) is analogous; as is ternary or finer tessellation rather than dyadic refinement (see for example [IDS02, Ale02, ZS01] for various classifications of mesh refinement patterns). Even vector-valued subdivision does not require new concepts but is fully covered by the theory to be developed. Chapter 9/175 lists classes of subdivision algorithms that share the structure of subdivision based on quadrilateral splines and that therefore need not be developed separately.

Splines as union of rings. To properly characterize continuity, the spline domain is given the topological structure of a two-dimensional manifold. This avoids a more involved characterization by means of matching smoothness conditions for abutting patches. The key to understanding subdivision surfaces are the isolated singularities of splines on a topological domain. That is, we focus on the neighborhood of extraordinary domain points where $n \neq 4$ quadrilateral cells join.

In the language of control points and meshes, each refinement enlarges the ‘regular parts’ of the control mesh, i.e., the submeshes where standard subdivision rules apply. At the same time, the region governed by extraordinary rules shrinks. As this process proceeds, a nested sequence of smaller and smaller ring-shaped surface pieces is well-defined, corresponding to the newly created regular region. Eventually, these rings, together with a central limit point, cover all of the surface (see Fig. 4.3/61).

In the language of splines,

a spline in subdivision form is a nested sequence of rings.

Since all rings are mappings from the same annular domain to $\mathbb{R}^d$, where typically $d = 3$, we can consider spaces of rings spanned by a single, finite-dimensional system of *generating rings*.

A subdivision algorithm is a recursion that generates a sequence of rings

within the span of such a system of generating rings. The word ‘ring’ will not lead to confusion since no rings in the algebraic sense will be considered in this context.

Smoothness at singularities. With rings contracting ad infinitum towards a singularity of the parametrization, it is necessary to use, in the limit, a differential geometric

[2] See e.g. [dHR93] or [PBP02,Chap. 17].

characterization of smoothness. Smoothness is measured in a natural local coordinate system. Injectivity with respect to this coordinate system is crucial but not always guaranteed by subdivision algorithms; and the lack of second-order differentiability with respect to the coordinate system presents a challenge for characterizing shape. We therefore devote Chap. 2/15 of this book to a review of concepts of differential geometry specifically of surfaces with isolated singularities. This differential geometry of singularities is rarely discussed in the classical literature and is crucial for understanding subdivision surfaces.

1.4 Focus and Scope

The analysis of subdivision on regular grids has been well-documented and we can point to a rich literature (see Sect. 1.7/8) on the subject. In particular, [CDM91, p. 18] gives a general technique for evaluating functions in subdivision form, polynomial or otherwise, at any rational parameter value. Differentiability of such functions can typically be established by proving contraction of difference sequences of the coefficients[3] The resulting surfaces are splines in the generalized sense discussed above.

The continuity and shape analysis in this book will therefore focus on the singularities corresponding to 'extraordinary rules'. These singularities are assumed to be isolated so that a *local analysis*, based the union of rings, suffices to establish *necessary and sufficient conditions for* C^1 *and* C^2 *continuity*.

We focus on at *stationary linear algorithms*. The analysis then combines the discrete, linear-algebraic view with the analytic differential geometric view, i.e., considers both the subdivision matrix and the surface parametrization.

This analysis develops *simple recipes* for checking properties of subdivision algorithms and their limit surfaces. Such recipes are needed to verify the correctness of newly proposed algorithms and to assess their strengths and deficiencies. An important component in deriving these recipes is to make assumptions explicit. For example, if we fail to check for 'ineffective eigenvectors' (Definition 4.19/76), we cannot conclude that the subdivision matrix ought to have a single leading eigenvalue of 1, a property that is often taken for granted. Or, to conclude that a C^1-subdivision algorithm generates a C^1-surface, we need to check that the input control points are 'generic' (Definition 5.1/84). Once such prerequisites have been established, even the 'injectivity-test' becomes simple (see, e.g., Theorem 5.24/105). We illustrate this process for three well-known subdivision algorithms and provide a framework for constructing new algorithms, in particular for generating C^2-surfaces.

---

[3] The technique relies on the following observation [CDM91, DGL91, Kob98b] [PBP02, p.117]: Let $q^m := [\ldots, q_i^m, \ldots]$ be a sequence with $2^{mk}\nabla^{k+1}q^m$ converging uniformly to zero as m tends to infinity. Then the limit q^c is a C^k function and for $j = 0 : k$, the sequences $2^{mj}\nabla^j q^m$ converge uniformly to the derivatives $\partial^j q^c$.

1.5 Overview

Chapter 2/15 reviews some little known material on the differential geometry of surfaces in the presence of singularities, and lays the groundwork for most of the proofs in Chaps. 3/39–7/125. Chapter 3/39 formally defines the objects of the investigation: splines on topological domains and their forced singularities. Chapter 4/57 introduces the refinement aspect for these splines and defines the resulting class of surfaces obtained by subdivision. We now narrow the focus to stationary algorithms, i.e., algorithms where the same rules are applied at each step.

Chapter 5/83 characterizes stationary subdivision algorithms that generate smooth surfaces, that is, at least C^1-manifolds. While a very general class of algorithms is covered here, particular scrutiny is given to 'standard algorithms' which are characterized by subdivision matrices with a double subdominant eigenvalue. In Chap. 6/109, the resulting powerful analysis techniques are applied to three well-known subdivision algorithms. In Chap. 7/125 we derive constraints that further restrict the class of admissible subdivision algorithms to those that are able to represent the full spectrum of second order shapes. A further restriction of this class finally yields C^2-subdivision algorithms, and we present a new framework for constructing such algorithms. Finally, in Chap. 8/157, we determine bounds on the distance of a subdivision to a proxy surface, and in particular to its control polyhedron. Further, the question of local and global linear independence of systems of generating splines is discussed. Chapter 9/175 then summarizes what schemes fall in the scope of the book and points to algorithms outside.

For a *quick tour* through the material, one may proceed as follows. Not skipping the notational conventions in Sect. 1.6/7 below, Sect. 2.1/16 is *indispensable* for understanding whatever follows; also Sect. 2.3/23 should not be missed. In Chap. 3/39, Sects. 3.2/41–3.4/47 are fundamental, as well as the whole of Chap. 4/57. In Chap. 5/83, Sect. 5.3/89 may be skipped on first reading. Chapter 6/109 provides examples by applying the techniques to specific algorithms; its content is not prerequisite to understanding the remaining chapters. Parts of the material in Chaps. 7/125 and 8/157 are brand-new. Here, the exposition is less tutorial, but rather intended to prepare the ground for new research in the field.

1.6 Notation

As a mnemonic help, in particular to discern objects and maps into the range $\mathbb{R}^d$ from objects and maps into the bivariate domain, we use bold greek letters for objects and for maps into $\mathbb{R}^2$. For example, planar curves and reparametrizations, such as the 'characteristic' reparametrization, will be represented by bold greek letters. We use plain roman letters for real or complex-valued functions and constants. The constants, λ for eigenvalues and κ for curvature, are an exception to conform to well-established usage. Bold roman font is used, in particular, for points and functions in the embedding space $\mathbb{R}^d$. For example, the subdivision surface $\mathbf{x}$, its normal $\mathbf{n}$ and the control points $\mathbf{q}_i$ are so identified.

Points and functions in $\mathbb{R}^2$ and $\mathbb{R}^d$ are always understood as row-vectors, e.g.

$$\boldsymbol{\xi} = [\boldsymbol{\xi}_1, \boldsymbol{\xi}_2] \in \mathbb{R}^2, \quad \mathbf{x} = [\mathbf{x}_1, \dots, \mathbf{x}_d] \in \mathbb{R}^d.$$

Consequently, linear maps in $\mathbb{R}^2$ and $\mathbb{R}^d$ are represented by matrix multiplication *from the right*. For example,

$$\tilde{\boldsymbol{\xi}} := \boldsymbol{\xi} R, \quad R := \begin{bmatrix} \cos t & \sin t \\ -\sin t & \cos t \end{bmatrix},$$

is a counter-clockwise rotation about the origin by the angle t. We summarize:

Bold greek – point or map into $\mathbb{R}^2$ – row vector
Bold roman – point or map into $\mathbb{R}^d$ – row vector

As in Matlab, elements in a row of a matrix or vector are separated by a comma, while rows are separated by a semicolon. For example,

$$[1,\ 2,\ 3;\ 4,\ 5,\ 6] = \begin{bmatrix} 1 & 2 & 3 \\ 4 & 5 & 6 \end{bmatrix}.$$

We have made an effort to clarify concepts by a consistent use of names. For example, what appears currently in the literature as 'characteristic map' is called characteristic ring when we want to emphasize its structure as a map over a topological ring and distinguish it from the characteristic spline that is defined as a union of rings and their limit point (cf. Fig. 4.3/61). Replicated from the Index, here are the key variables:

$\mathbf{x}$	spline	$\mathbf{x}^m$	m-th ring of $\mathbf{x}$
b_ℓ	generating spline	g_ℓ	generating ring
e_ℓ	eigenspline	f_ℓ	eigenring
χ	characteristic spline	ψ	characteristic ring

Generating splines span the space of subdivision surfaces. They have no relationship with the formal power series of the z-transform that is sometimes called generating function (see also the footnote on p. 10).

1.7 Analysis in the Shift-Invariant Setting

Contributed by Nira Dyn

A 'classical' subdivision scheme on a *regular mesh* generates a limit object, such as function, curve, surface, from initial data consisting of discrete points (control points), parametrized by the vertices of the mesh. The limit object is obtained by two processes; first by recursive refinements of the control points, based on a fixed local refinement rule, and then by a limiting process on the sequence of control points generated by the recursive refinements.

The theory of subdivision schemes on regular meshes is quite different from the analysis presented in this book. It can be traced back to two papers by de Rham [dR47, dR56], who designed schemes for generating univariate functions with certain unusual smoothness properties.

The use of subdivision schemes in geometric modeling started with the efficient rendering of B-spline curves [Cha74, For74]. This method is directly extendable to tensor-product B-spline surfaces. The topology of such surfaces is rather limited, and in order to design surfaces of general topology, it was necessary to introduce irregular points and faces in the initial net of control points, together with special valence-dependent refinement rules. (The valence of points and faces in the regular mesh on which tensor-product B-spline surfaces are defined is 4. Points and faces with valence different from 4 are termed irregular or extraordinary). The necessity to refine such general nets lead to the design and analysis of the Doo–Sabin scheme and the Catmull–Clark scheme, which extend to arbitrary meshes the tensor-product quadratic and cubic B-spline schemes respectively [CC78, DS78]. At a later stage the Loop scheme extended a certain box-spline subdivision scheme defined on regular triangulations (all vertices of valence 6), to general triangulations [Loo87].

In all these cases the limit surface and its properties were known away from a finite number of irregular points, and the analysis was concentrated at these points. This book presents the state-of-the-art theory about subdivision surfaces in the vicinity of irregular points.

The analysis of convergence and smoothness of subdivision schemes on regular meshes became important when *interpolatory schemes* were introduced by Dubuc and Deslauriers for univariate functions [DD89, Dub86] and independently by Dyn, Gregory and Levin, in [DGL87] for curves and in [DGL90] for surfaces. The limit objects of these schemes are no more piecewise analytic functions, as in the spline cases, but are of *fractal* nature. These limits have only a procedural definition in terms of the refinement rule of the corresponding scheme. The existence of limit objects and their properties had to be deduced from a finite number of coefficients, called the mask of the scheme, which define the subdivision refinement rule.

The need to analyze both the convergence of subdivision schemes defined on regular meshes, and the smoothness properties of the limit objects generated by these schemes, gave rise to the development of analysis tools of several kinds. In parallel to the developments in the geometric-modeling literature, there were many independent developments in the wavelets literature, since a very important class of wavelets is defined by subdivision.

In [DGL91], Dyn, Gregory and Levin presented the analysis of a general univariate subdivision scheme for curves, in terms of *derived subdivision schemes* for the differences and the divided differences of the data generated by the investigated scheme. This analysis extends the one used for the four-point scheme [DGL87] to any order of smoothness.

In [CDM91], Cavaretta, Dahmen and Micchelli considered subdivision schemes on regular meshes in any space dimension. They introduced the important notion of a *symbol* of a scheme, which replaces the finite set of mask coefficients by a *Laurent polynomial*. With this notion, algebraic methods became relevant to the analysis of subdivision schemes. Also in [CDM91], the principle of 'contractivity relative to a positive function' was introduced, for checking convergence. This lead to simple sufficient conditions for the convergence of schemes with positive mask coefficients. The seminal work [CDM91] dealt with many other aspects of subdivision, in particular with the *refinement equation* satisfied by the 'B-spline-like' function defined by a convergent subdivision scheme. This observation related subdivision with that part of *wavelets* theory which is based on a refinement equation (see [HD03] for further developments on this relation).

While the analysis tools in [DGL91, CDM91] dealt with subdivision schemes as refinement operators on control points, the analysis of the interpolatory schemes in [DD89, Dub86] was done in the Fourier domain. Also, most of the analysis of solutions of refinement equations in the wavelets literature was done in the Fourier domain [Dau88, Dau92, DL92a]. Yet in [DL92b], Daubechies and Lagarias devised methods which follow the development of the control points during the subdivision process, and obtained in particular interesting observations about the fractal nature of limits of the four-point interpolatory scheme of [DD89].

The analysis tools based on derived schemes in [DGL91] were simplified and extended to the multivariate setting by Dyn, Levin and Hed [Hed90]. Using the symbol and the z-transform[4] of the control points, it was possible to investigate convergence and smoothness by simple algebraic operations (see [Dyn92] and references therein). The analysis of convergence and smoothness of multivariate subdivision schemes was based on the existence of derived 'non-degenerate' (called now 'full rank') *matrix subdivision* schemes for differences and divided differences of the data generated by the investigated scheme. (A matrix subdivision scheme refines sequences of vectors using a matrix-valued mask, and generates a limit vector-function). The convergence part of this analysis, was first presented in [CDM91].

In a series of papers, Sauer with co-authors developed algebraic methods for the derivation of matrix-Laurent-polynomial symbols of the matrix subdivision schemes, used in the analysis of convergence and smoothness of multivariate subdivision schemes (see [MS04] and references therein). They also studied the analysis of matrix subdivision schemes in general [CCS05].

In the wavelet literature, there was an intense study of matrix subdivision schemes, mainly for the construction of multiwavelets from a refinable vector of functions [Str96]. In this context rank 1 matrix schemes are of interest in contrast to the full rank case above. B-splines with equidistant multiple knots are limits of such schemes [Plo97].

Matrix subdivision schemes are not affine invariant, and therefore not adequate for geometric modeling from control points. Yet *Hermite subdivision* schemes,

[4] The analysis of sequences is aided by the z-transform [Jur64] or 'generating function method'. The z-transform converts a sequence (of coefficients) into a Laurent series i.e. a summation of rational and polynomial terms such that the j-th summand is the j-th element of the sequence weighted by z^{-j}. The sequence of coefficients can then be analyzed as a continuous functions, called *symbol*.

which comprise a special class of matrix subdivision schemes, are of interest in the functional setting. These schemes refine function values and derivatives values. Hermite schemes can be used for the design of curves from control points and tangent vectors attached to them. The first Hermite schemes were introduced and investigated by Merrien in [Mer92, Mer94b]. General analysis tools for univariate, interpolatory Hermite schemes were presented by Dyn and Levin in [DL99], as extension of the tools for the 'classical' case.

The analysis of scalar or matrix schemes, defined on regular meshes, and based on the same refinement rule, operating at all locations and at all refinement levels, can be done in the Fourier domain. Yet, the analysis based on derived schemes can be easily adapted to *non-uniform* schemes, as was done by Daubechies, Guskov and Sweldens [DGS01, DGS99].

For the analysis of *non-stationary subdivision* schemes, where a fixed refinement rule operates in each refinement level, but is changing with the refinement level, new tools were introduced by Dyn and Levin in [DL95]. The analysis of convergence is done by comparison with converging stationary schemes, and the analysis of smoothness by extension of the notion of a smoothing factor, which is related to the derived first divided-difference scheme in the stationary case. Non-stationary schemes can generate exponential splines, which are piecewise analytic functions, as well as compactly supported infinitely smooth functions [DL95].

In recent years, much of the research effort is invested in the design and analysis of *non-linear subdivision* schemes. Such schemes can operate on other types of data, as manifold-valued data [RDS$^+$05, WD05], or can be data dependent [MDL05, DY00, KvD98]. The available analysis tools for linear schemes are not applicable for non-linear schemes, and new methods of analysis have to be developed.

1.8 Historical Notes on Subdivision on Irregular Meshes

Contributed by Malcolm Sabin

The beginnings of the subdivision story can be dated back to the papers of de Rham [dR56, dR47], over fifty years ago, but the relevance to the modeling of shape started with the proposal of Chaikin [Cha74], who devised a method of generating smooth curves for plotting. This was soon analyzed by Forrest [For74] and by Riesenfeld [Rie75] and linked with the burgeoning theory of B-spline curves. It became clear that equal-interval B-spline curves of any degree would have such a subdivision construction.

The extension to surfaces took just a few years, until 1978, when Catmull and Clark [CC78] published their descriptions of both quadratic and cubic subdivision surfaces, the exciting new point being that a surface could be described which was not forced to have a regular rectangular grid in the way that the tensor product B-spline surfaces were. The definition of a specific surface in terms of a control mesh could follow the needs of the boundaries and the curvature of the surface. This was made possible by the extension of the subdivision rules to allow for 'extraordinary vertices' where other than four faces come together and 'extraordinary faces' where a face has other than four sides.

At about the same time Doo [DS78] and Sabin, who had also been working on quadratic subdivision, showed a way of analyzing the behavior of these algorithms at the extraordinary points, treating the refinement process in terms of matrix multiplication, and using eigenanalysis of the spectrum of this matrix [DS78], a technique also used in the univariate case by de Rham. This aspect was followed up by Ball and Storry [BS84, BS86] who made this analysis process more formal and succeeded in making some improvements to the coefficients used around the extraordinary points in the Catmull–Clark algorithm. Storry identified that in the limit, the configuration around an extraordinary point was always an affine transform (dependent on the original polyhedron) of a point distribution which was completely defined by the eigenvectors of the subdivision matrix. He called this the *natural configuration.*

The next two big ideas emerged in 1987. Loop, in his Masters' thesis [Loo87], described a subdivision algorithm defined over a grid of triangles. This not only gave a new domain over which subdivisions could be defined, but also showed that the eigenanalysis could be used explicitly in the original design of an algorithm, in the choice of coefficients which should be used around extraordinary points.

The other significant publication that year was the description by Dyn, Levin and Gregory [DGL87] of their four-point curve scheme. This was new in two ways: it was an interpolating scheme, rather than smoothing, and the limit curve did not consist of parametric polynomial pieces. The analysis of its continuity and differentiability therefore required new tools.

The first tool was provided in [DGL87] and tools of a greater generality were provided in [CDM91] and [DGL91]. The method in the later paper together with the idea of the *symbol* of a subdivision scheme, presented in [CDM91], was later expressed in terms of z-transforms, which turn convolution of sequences of numbers into multiplication of Laurent polynomials. Algebraic manipulation of these polynomials allows such processes as the taking of differences to be expressed very simply, and it has turned out that many of the arguments we need to deploy can be expressed very elegantly in this notation. It also provides sufficient conditions for a scheme to have a certain level of derivative continuity, whereas the eigenanalysis approach provides only necessary conditions.

The generalization of the four-point ideas to an interpolating surface scheme came in 1990, with the description by Dyn, Gregory and Levin [DGL90] of the Butterfly scheme, an interpolating surface scheme defined over a triangular grid.

In 1995 Reif [Rei95c] showed that there was rather more to continuity than had been dreamt of. He identified that the natural configuration implies a parametrization of the rings of regular pieces which surround each extraordinary point and that it is essential, in order to obtain a scheme which generates well-behaved surfaces at the extraordinary points, to ensure that this parametrization is injective. Later, Peters and Reif went further, and in [PR98] they constructed a scheme (a variant of the quadratic) for which the injectivity test fails, resulting in severe folding of the limit surface in every ring.

The following year Reif [Rei96a] showed that the attempts to make a C^2-variant of Catmull–Clark were not going to succeed, because a surface C^2 at the extraordinary points would need to have regular pieces at least bi-sextic.

Thus as we passed the mid-1990s, subdivision theory stood like this:

- A surface subdivision scheme takes a manifold mesh of vertices joined by faces, usually called the polyhedron, and creates a new, finer, polyhedron by constructing new vertices as linear combinations of the old ones, in groups defined by the connectivity of the polyhedron, and joining them up by new faces in a way related to the old connectivity.
- This refinement can be repeated as often as desired, and there are conditions on the scheme guaranteeing the existence of a well-defined limit surface to which the sequence of finer and finer polyhedra converges. During the refinement process the number of extraordinary points remains constant, and they become separated by regular mesh of a kind which is dependent on the topological rules of the scheme.
- The regular mesh is often well-described by box-spline theory (the Butterfly scheme was almost alone in not being describable in those terms) but the z-transform analysis can always be applied to determine the smoothness of the limit surface in the regular regions. The extraordinary points are surrounded by rings of regular mesh, and close to the extraordinary point these are just affine transforms of the natural configuration, and can be parametrized by the characteristic map.
- Because every box-spline has a generating subdivision scheme [DM84], we had a way in principle of creating as many different subdivision schemes as we might want. Each such scheme would have to have its extraordinary point rules invented, of course, but nobody had bothered to go through the exercise. We also had a sequence of interpolating curve schemes, generated by letting an increasing number $(2n)$ of points influence the new vertex in the middle of each span [DD89], but this had not led to a sequence of interpolating triangular surface schemes. In fact Catmull–Clark, Loop and Butterfly were regarded as the significant surface schemes, and the cubic B-spline subdivision and the four-point scheme as the significant curve schemes, any others being only of academic interest.
- The question of the behavior of the limit surface in the immediate vicinity of the extraordinary points was still of interest. Indeed, the papers [DS78,BS88,Sab91a] before Reif's key result[Rei96a] on the lower bound of the polynomial order of patches surrounding an extraordinary point have been more than balanced by those after [PU98b, PU98a, QW99, PU00a, PU00b, GBDS05].

Chapter 2
Geometry Near Singularities

Subdivision surfaces have to be analyzed in the terms of differential geometry. This chapter summarizes well-known concepts, such as the Gauss map, the principal curvatures and the fundamental forms, but also develops material that is not found in standard text books, such as the embedded Weingarten map, that is crucial to understanding subdivision surfaces.

Parametric singularities in the form of isolated 'extraordinary points' are a key feature of subdivision surfaces. The analysis of such singularities requires a separate assessment of parametric and geometric continuity. Accordingly, we will define function spaces C^k_r where k indicates the smoothness of the parametrization, except at isolated points, and r measures the smoothness of the resulting surface in the geometric sense.

After providing special notations for dot and cross products in Sect. 2.1/16, we consider basic concepts from the differential geometry of regularly parametrized surfaces in Sect. 2.2/17. In particular, the *embedded Weingarten map*, which is given by a (3×3)-matrix, is introduced as a geometric invariant for the study of curvature properties. Unlike the principal directions, it is uniquely defined and continuous even at umbillic points. This property is crucial for our subsequent considerations of limit properties of subdivision surfaces at singular points.

In Sect. 2.3/23, the standard requirement on the regularity of the parametrization is suspended at an isolated point to allow for the structural conditions of subdivision surfaces. To establish geometric continuity, we first introduce the concept of 'normal continuity'. That is, we require that the normal map can be continuously extended from the regular neighborhood to the singular point. This unique normal is used to define a differential-geometric notion of smoothness. If and only if the projection of the surface to the tangent plane is injective, the surface is single-sheeted and meets the requirements of a two-dimensional manifold. Then, the surface can be viewed as the graph of a scalar-valued function in a local coordinate system: the parameters are associated with the tangent plane, and function values are measured in the normal direction. To capture both analytic and geometric smoothness, we call a single-sheeted surface C^k_r if its parametrization is C^k and the local height function is C^r. In case of single-sheetedness, we can use continuity of the Gauss map and the embedded Weingarten map to decide membership in C_1^k and C_2^k, respectively. This approach circumvents an explicit construction of the local height function. Using the embedded Weingarten map avoids having to select consistent coordinate systems in the set of tangent planes, as is necessary when working with the standard Weingarten map.

2.1 Dot and Cross Products

In this section, we introduce notations for dot and cross products of vectors and matrices, that should be familiar to the reader before reading on. They are carefully designed to reduce the notational complexity throughout the book.

We denote the transpose of a matrix B by B^{t}, and AB^{t} is the standard matrix product of two matrices A and B^{t} of suitable dimensions. Then the *dot product* of A and B is defined by

$$A \cdot B := AB^{\mathrm{t}}.$$

In particular, for two row-vectors $\mathbf{a}, \mathbf{b} \in \mathbb{R}^d$ with components a_i, b_i

$$\mathbf{a} \cdot \mathbf{b} := \mathbf{a}\mathbf{b}^{\mathrm{t}} = \sum_{i=1}^{d} a_i b_i$$

is the Euclidean inner product. The associated norm is

$$\|\mathbf{a}\| := \sqrt{\mathbf{a} \cdot \mathbf{a}}.$$

The *cross product* of two vectors $\mathbf{a} = [a_1, a_2, a_3], \mathbf{b} = [b_1, b_2, b_3]$ in $\mathbb{R}^3$ is the vector

$$\mathbf{a} \times \mathbf{b} := [a_2 b_3 - a_3 b_2,\ a_3 b_1 - a_1 b_3,\ a_1 b_2 - a_2 b_1],$$

while the cross product of two vectors $\boldsymbol{\alpha} := [a_1, a_2], \boldsymbol{\beta} := [b_1, b_2]$ in $\mathbb{R}^2$ is the real number

$$\boldsymbol{\alpha} \times \boldsymbol{\beta} := \det[\boldsymbol{\alpha}; \boldsymbol{\beta}] = a_1 b_2 - a_2 b_1.$$

For differentiable bivariate functions, partial differentiation is denoted by the operators D_1 and D_2. If the differentiable function $\mathbf{x} := \mathbb{R}^2 \to \mathbb{R}^d$ depends on s and t, we also write

$$D_1 \mathbf{x} = \mathbf{x}_s, \quad D_2 \mathbf{x} = \mathbf{x}_t.$$

Combining both partial derivatives, we obtain the operator $D := [D_1; D_2]$. That is,

$$D\mathbf{x} = \begin{bmatrix} D_1 \mathbf{x} \\ D_2 \mathbf{x} \end{bmatrix} = \begin{bmatrix} \mathbf{x}_s \\ \mathbf{x}_t \end{bmatrix}$$

is a $(2 \times d)$-matrix because $\mathbf{x}$ is always written as a row-vector. Further, we define the *partial cross product operator* ${}^{\times}\!D := D_1 \times D_2$. We distinguish the following two cases:

- If applied to a differentiable surface $\mathbf{x} : \mathbb{R}^2 \to \mathbb{R}^3$ then, as detailed in Sect. 2.2/17,

$$ {}^{\times}\!D\mathbf{x} = D_1\mathbf{x} \times D_2\mathbf{x} \tag{2.1}$$

  is a field of *vectors* perpendicular to this surface.
- If applied to a differentiable function $\boldsymbol{\xi} : \mathbb{R}^2 \to \mathbb{R}^2$ then

$$ {}^{\times}\!D\boldsymbol{\xi} = D_1\boldsymbol{\xi} \times D_2\boldsymbol{\xi} = \xi_{1,s}\xi_{2,t} - \xi_{1,t}\xi_{2,s}$$

  is a *real-valued function*, also known as the *Jacobian determinant* of $\boldsymbol{\xi}$.

For later use, we define the following rules, which are easily verified by inspection:

- If $\mathbf{p}, \mathbf{q} \in \mathbb{R}^3$ are constant vectors, and $f, g : \mathbb{R}^2 \to \mathbb{R}$ are differentiable functions, then

$$ {}^{\times}\!D(f\mathbf{p} + g\mathbf{q}) = {}^{\times}\!D[f, g]\,(\mathbf{p} \times \mathbf{q}). \tag{2.2}$$

  We recall that ${}^{\times}\!D[f, g] = f_s g_t - f_t g_s$ is real-valued since $[f, g]$ is a function with two coordinates. For general sums of that type,

$$ {}^{\times}\!D\Big(\sum_k f_k\mathbf{p}_k\Big) = \sum_{k<\ell} {}^{\times}\!D[f_k, f_\ell]\,(\mathbf{p}_k \times \mathbf{p}_\ell). \tag{2.3}$$

- If $\mathbf{x} : \mathbb{R}^2 \to \mathbb{R}^3$ is a surface and $\mathbf{F}$ is an orthogonal (3×3)-matrix, i.e., $\mathbf{F}\cdot\mathbf{F} = \mathbb{1}$ is the identity, then

$$ {}^{\times}\!D(\mathbf{x}\cdot\mathbf{F}) = ({}^{\times}\!D\mathbf{x})\cdot\mathbf{F}. \tag{2.4}$$

- If $\boldsymbol{\xi} : \mathbb{R}^2 \to \mathbb{R}^2$ is a function with two coordinates and L is a (2×2)-matrix, then

$$ {}^{\times}\!D(\boldsymbol{\xi}L) = ({}^{\times}\!D\boldsymbol{\xi})\ \det L. \tag{2.5}$$

2.2 Regular Surfaces

In differential geometry, domains of surfaces are typically assumed to be open subsets of $\mathbb{R}^2$. In our context however, domains are always required to be closed. Thus, let $\boldsymbol{\Sigma}$ be a non-empty closed subset of $\mathbb{R}^2$ that allows for differentiation up to the boundary.[1] Points in the *domain* $\boldsymbol{\Sigma}$, later on also called *parameters*, are denoted by

$$\boldsymbol{\sigma} = (s, t) \in \boldsymbol{\Sigma}.$$

[1] For now, $\boldsymbol{\Sigma}$ can be visualized as the unit square. More generally, $\boldsymbol{\Sigma}$ needs only satisfy the following cone property: for every point $\boldsymbol{\sigma}$ on the boundary of $\boldsymbol{\Sigma}$, there exists $\varepsilon > 0$ and a vector $\rho \in \mathbb{R}^2\backslash\{\mathbf{0}\}$ such that $\boldsymbol{\sigma} + \rho' \in \boldsymbol{\Sigma}$ for all ρ' with $\|\rho'\| \le \|\rho\|$ and $\big\|\rho'/\|\rho'\| - \rho/\|\rho\|\big\| \le \varepsilon$. For instance, the L-shaped set $\boldsymbol{\Sigma} := [0, 2]^2\backslash[0, 1)^2$ has this property.

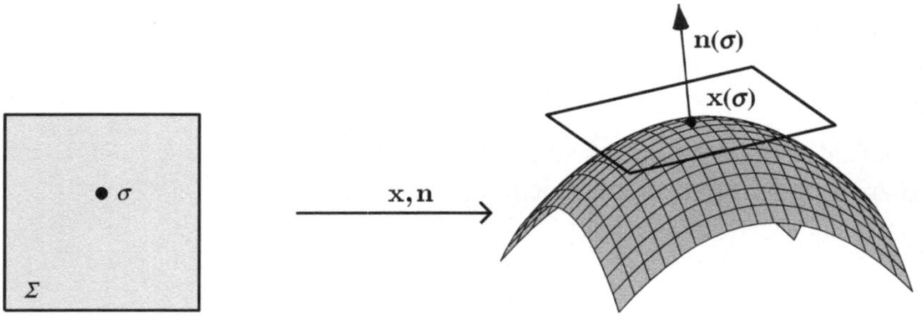

Fig. 2.1 Illustration of Definition 2.1/18: Regular surface $\mathbf{x}$ and its Gauss map $\mathbf{n}$ at a parameter $\boldsymbol{\sigma} \in \boldsymbol{\Sigma}$.

A C^k*-surface* $\mathbf{x}$ on the domain $\boldsymbol{\Sigma}$ is a k-times continuously differentiable function

$$\mathbf{x} : \boldsymbol{\Sigma} \ni \boldsymbol{\sigma} \mapsto \mathbf{x}(\boldsymbol{\sigma}) \in \mathbb{R}^3,$$

and the space of all such functions is denoted by $C^k(\boldsymbol{\Sigma}, \mathbb{R}^3)$. Unless otherwise stated, we assume $k \geq 1$, throughout. The *image* of $\mathbf{x}$ is the set

$$\mathbf{x}(\boldsymbol{\Sigma}) := \{\mathbf{x}(\boldsymbol{\sigma}) : \boldsymbol{\sigma} \in \boldsymbol{\Sigma}\}.$$

Further, we say that $\mathbf{x}$ is *embedded* in $\mathbb{R}^3$, if the map $\mathbf{x}$ is injective. Using the notation ${}^{\times}\!D\mathbf{x}$ for the cross product of partial derivatives, as introduced in the preceding section, we define regular surfaces as usual (see also Fig. 2.1/18).

Definition 2.1 (Regular surface, Gauss map). A C^1-surface $\mathbf{x}$ is called *regular*, if ${}^{\times}\!D\mathbf{x}(\boldsymbol{\sigma}) \neq \mathbf{0}$ for all $\boldsymbol{\sigma} \in \boldsymbol{\Sigma}$. For a regular surface $\mathbf{x}$, the *Gauss map* is defined by

$$\mathbf{n} : \boldsymbol{\Sigma} \ni \boldsymbol{\sigma} \mapsto \frac{{}^{\times}\!D\mathbf{x}(\boldsymbol{\sigma})}{\|{}^{\times}\!D\mathbf{x}(\boldsymbol{\sigma})\|} \in \mathbb{R}^3.$$

The Gauss map assigns to each point $\mathbf{x}(\boldsymbol{\sigma}_0)$ a normalized *normal vector* $\mathbf{n}(\boldsymbol{\sigma}_0)$ that is asymptotically perpendicular to the surface,

$$\lim_{\boldsymbol{\sigma} \to \boldsymbol{\sigma}_0} \frac{\big(\mathbf{x}(\boldsymbol{\sigma}) - \mathbf{x}(\boldsymbol{\sigma}_0)\big) \cdot \mathbf{n}(\boldsymbol{\sigma}_0)}{\|\mathbf{x}(\boldsymbol{\sigma}) - \mathbf{x}(\boldsymbol{\sigma}_0)\|} = 0. \tag{2.6}$$

The plane which is orthogonal to $\mathbf{n}(\boldsymbol{\sigma}_0)$ and passes through $\mathbf{x}(\boldsymbol{\sigma}_0)$ is called the *tangent plane* at the point $\mathbf{x}(\boldsymbol{\sigma}_0)$. Property (2.6/18) can be used to show that the Gauss map is a *geometric invariant* in the following sense.

Theorem 2.2 (Invariance of n). Let $\mathbf{x}$ and $\tilde{\mathbf{x}}$ be two regular embedded C^1-surfaces with equal image. Then, for any pair $\boldsymbol{\sigma}, \tilde{\boldsymbol{\sigma}}$ of parameters with $\mathbf{x}(\boldsymbol{\sigma}) = \tilde{\mathbf{x}}(\tilde{\boldsymbol{\sigma}})$, the normal vectors are equal up to sign. That is,

$$\mathbf{n}(\boldsymbol{\sigma}) = r\,\tilde{\mathbf{n}}(\tilde{\boldsymbol{\sigma}})$$

either for $r = 1$ or for $r = -1$.

Now, in order to study curvature properties, we assume $\mathbf{x} \in C^k$, $k \geq 2$, for the rest of this chapter. Differentiating the identity $\mathbf{n} \cdot \mathbf{n} = 1$, we obtain $D\mathbf{n} \cdot \mathbf{n} = 0$. That means that the row vectors of $D\mathbf{n}$ lie in the tangent plane of $\mathbf{x}$ at the corresponding point. Hence, by regularity, there exists a (2×2)-matrix W, called the *Weingarten map,*[2] with

$$-D\mathbf{n} = W D\mathbf{x}.$$

Multiplication with the transpose of $D\mathbf{x}$ yields

$$II = WI,$$

where

$$I := D\mathbf{x} \cdot D\mathbf{x} = \begin{bmatrix} \mathbf{x}_s \cdot \mathbf{x}_s & \mathbf{x}_t \cdot \mathbf{x}_s \\ \mathbf{x}_s \cdot \mathbf{x}_t & \mathbf{x}_t \cdot \mathbf{x}_t \end{bmatrix}$$

and, as we will show below,

$$II := -D\mathbf{n} \cdot D\mathbf{x} = -\begin{bmatrix} \mathbf{x}_s \cdot \mathbf{n}_s & \mathbf{x}_t \cdot \mathbf{n}_s \\ \mathbf{x}_s \cdot \mathbf{n}_t & \mathbf{x}_t \cdot \mathbf{n}_t \end{bmatrix} \tag{2.7}$$

are symmetric (2×2)-matrices, called the *first and second fundamental form* of $\mathbf{x}$, respectively. It is easily verified by inspection that

$$\det I = \|{}^{\times}D\mathbf{x}\|^2.$$

Thus, I is invertible since the parametrization is assumed to be regular. By the product rule,

$$0 = D_k(D_i\mathbf{x} \cdot \mathbf{n}) = D_iD_k\mathbf{x} \cdot \mathbf{n} - II_{i,k}.$$

Hence, the components of the second fundamental form are given by

$$II_{i,k} = D_iD_k\mathbf{x} \cdot \mathbf{n} = \frac{\det[D_iD_k\mathbf{x}; D\mathbf{x}]}{\|{}^{\times}D\mathbf{x}\|} = \frac{\det[D_iD_k\mathbf{x}; D\mathbf{x}]}{\sqrt{\det I}}, \tag{2.8}$$

showing that II is indeed symmetric. Because I is invertible, we obtain

$$W = II\, I^{-1}.$$

Given a smooth curve $\gamma(t) = \boldsymbol{\sigma} + t\boldsymbol{\sigma}' + o(t)$ in the domain $\boldsymbol{\Sigma}$ of $\mathbf{x}$, we define the related curves

$$\mathbf{c}_{\mathbf{x}}(t) := \mathbf{x}(\gamma(t))$$

on the surface, and

$$\mathbf{c}_{\mathbf{n}}(t) := \mathbf{n}(\gamma(t))$$

on the unit sphere. Dropping, as usual, the parameter $\boldsymbol{\sigma} = \gamma(0)$, we obtain by the chain rule

$$\mathbf{c}'_{\mathbf{x}}(0) = \boldsymbol{\sigma}' D\mathbf{x}, \quad \mathbf{c}'_{\mathbf{n}}(0) = -\boldsymbol{\sigma}' W D\mathbf{x}.$$

---

[2] In the literature, the Weingarten map is also referred to as the *shape operator.*

Now, we are looking for curves γ with the property that

- $\mathbf{r} := \mathbf{c}'_{\mathbf{x}}(0)$ has unit length;
- $\mathbf{c}'_{\mathbf{n}}(0) = -\kappa\mathbf{r}$ for some $\kappa \in \mathbb{R}$, i.e., $\mathbf{c}'_{\mathbf{x}}(0)$ and $\mathbf{c}'_{\mathbf{n}}(0)$ are parallel.

Then the vector $\mathbf{r} \in \mathbb{R}^3$ is called a *principal direction*, and κ is the corresponding *principal curvature* of $\mathbf{x}$ at the point $\mathbf{x}(\boldsymbol{\sigma})$. The condition $\mathbf{c}'_{\mathbf{n}}(0) = -\kappa\mathbf{r}$ is equivalent to

$$\boldsymbol{\sigma}' W = \kappa \boldsymbol{\sigma}', \quad \mathbf{r} = \boldsymbol{\sigma}' D\mathbf{x}.$$

In other words, $\boldsymbol{\sigma}'$ is a left eigenvector of W to the eigenvalue κ. We will show that W has always two real eigenvalues κ_1, κ_2, and that the corresponding pair $\mathbf{r}_1, \mathbf{r}_2$ of principal directions can be chosen orthonormal. It is well known that the principal curvatures and directions are *geometric invariants* in the sense that they do not depend on the parametrization, but only on the shape and the orientation of the surface.

Theorem 2.3 (Invariance of principal curvatures and directions). *Let* $\mathbf{x}$ *and* $\tilde{\mathbf{x}}$ *be two regular embedded* C^2*-surfaces with equal image. Then, for any pair* $\boldsymbol{\sigma}, \tilde{\boldsymbol{\sigma}}$ *of parameters with* $\mathbf{x}(\boldsymbol{\sigma}) = \tilde{\mathbf{x}}(\tilde{\boldsymbol{\sigma}})$ *and* $\mathbf{n}(\boldsymbol{\sigma}) = r\tilde{\mathbf{n}}(\tilde{\boldsymbol{\sigma}})$ *according to Theorem (2.2/18), the following holds: if* $\mathbf{r}$ *is a principal direction of* $\mathbf{x}$ *at* $\mathbf{x}(\boldsymbol{\sigma})$ *to the principal curvature* κ*, then it also a principal direction of* $\tilde{\mathbf{x}}$ *at* $\tilde{\mathbf{x}}(\tilde{\boldsymbol{\sigma}})$ *to the principal curvature* $r\kappa$.

The theory developed so far is well established, but does not suffice for analyzing subdivision surfaces.

The principal curvatures depend continuously on the parameter $\boldsymbol{\sigma} \in \Sigma$ if we fix the order $\kappa_1 \leq \kappa_2$, but the principal directions have discontinuities at *umbilic points*, characterized by $\kappa_1 = \kappa_2$. Here, any direction in the tangent plane is a principal direction, and $\mathbf{r}_1, \mathbf{r}_2$ do not converge when approaching such a point. Example 2.15/29 illustrates this fact with a paraboloid of revolution: here the principal directions diverge near the vertex. In standard textbooks on differential geometry, the phenomenon of diverging principal directions is described, but then shrugged off as a degenerate situation. In our context, however, convergence properties will play a most important role so that we resort to a less common, yet natural approach.

Definition 2.4 (Embedded Weingarten map W). For a regular C^2-surface $\mathbf{x}$, let

$$D\mathbf{x}^+ := (I^{-1} D\mathbf{x})^{\mathrm{t}}$$

denote the pseudo-inverse of $D\mathbf{x}$, which is a (3×2)-matrix. Then we define the *embedded Weingarten map* of $\mathbf{x}$ as the symmetric (3×3)-matrix

$$\mathbf{W} := D\mathbf{x}^+ \, II \cdot D\mathbf{x}^+.$$

We claim that this object is a geometric invariant that contains the complete curvature information in a continuous way. Because $\mathbf{W} \cdot \mathbf{n} = \mathbf{0}$, the normal vector is always an eigenvector of $\mathbf{W}$ to the eigenvalue 0. The two other eigenvectors can be chosen orthonormal (mutually and with respect to $\mathbf{n}$), and are collected in a

(2×3)-matrix $\mathbf{R}$. The diagonal matrix of the corresponding pair of eigenvalues is denoted by $K := \mathrm{diag}(\kappa_1, \kappa_2)$. We obtain the factorization

$$\mathbf{W} = [\mathbf{R}^{\mathrm{t}}\, \mathbf{n}^{\mathrm{t}}] \begin{bmatrix} K & 0 \\ 0 & 0 \end{bmatrix} \begin{bmatrix} \mathbf{R} \\ \mathbf{n} \end{bmatrix} = \mathbf{R}^{\mathrm{t}} K \mathbf{R},$$

and therefore

$$\mathbf{R}\mathbf{W} = K\mathbf{R}.$$

By orthogonality of the eigenvectors, we have $\mathbf{R} \cdot \mathbf{n} = \mathbf{0}$. Hence, there exists a (2×2)-matrix Σ with $\mathbf{R} = \Sigma D\mathbf{x}$. Together with the definitions $\mathbf{W} = D\mathbf{x}^{\mathrm{t}} I^{-1} I\!I\, I^{-1} D\mathbf{x}$ and $W = I\!I\, I^{-1}$, we conclude from the last display $\Sigma W D\mathbf{x} = W \Sigma D\mathbf{x}$, and eventually

$$\Sigma W = K \Sigma.$$

That is, the diagonal entries of $K = \mathrm{diag}(\kappa_1, \kappa_2)$ are the eigenvalues of W, and the rows of $\Sigma = [\boldsymbol{\sigma}_1'; \boldsymbol{\sigma}_2']$ are the corresponding left eigenvectors which, via $[\mathbf{r}_1; \mathbf{r}_2] := \Sigma D\mathbf{x} = \mathbf{R}$, yield the principal directions. Another useful identity is obtained by multiplying $\mathbf{W}$ from both sides by $D\mathbf{x}$,

$$D\mathbf{x}\, \mathbf{W} \cdot D\mathbf{x} = (D\mathbf{x} \cdot D\mathbf{x}) I^{-1} I\!I I^{-1} (D\mathbf{x} \cdot D\mathbf{x}) = I\!I.$$

Substituting in the definition (2.7/19) of $I\!I$, we find $D\mathbf{x}\mathbf{W} \cdot D\mathbf{x} = -D\mathbf{n} \cdot D\mathbf{x}$ and

$$D\mathbf{x}\, \mathbf{W} = -D\mathbf{n}.$$

Hence, just as W, the matrix $\mathbf{W}$ describes the connection between the differentials $D\mathbf{x}$ and $D\mathbf{n}$. But, formally speaking, this connection is expressed in the dual of the tangent space. The resulting advantage of $\mathbf{W}$ over W is that $\mathbf{W}$ refers to the coordinates of the embedding space. By contrast, W refers to coordinates of the tangent space, and there is no distinguished choice for them. This ambiguity becomes a substantial problem for segmented surfaces such as subdivision surfaces, because W may then be discontinuous even in case of geometric smoothness (see Example 2.15/29).

Together with the last display, the condition $\mathbf{n}\mathbf{W} = \mathbf{0}$ uniquely defines $\mathbf{W}$,

$$\begin{bmatrix} D\mathbf{x} \\ \mathbf{n} \end{bmatrix} \mathbf{W} = \begin{bmatrix} -D\mathbf{n} \\ \mathbf{0} \end{bmatrix}.$$

Let $\mathbf{F}$ be any orthogonal (3×3)-matrix, and $\mathbf{x}_0$ a point in $\mathbb{R}^3$. If we define $\bar{\mathbf{x}} := (\mathbf{x} - \mathbf{x}_0) \cdot \mathbf{F}$ then $D\bar{\mathbf{x}} = D\mathbf{x} \cdot \mathbf{F}$. By (2.4/17), this yields

$$\bar{\mathbf{n}} = \mathbf{n} \cdot \mathbf{F}, \tag{2.9}$$

and $D\bar{\mathbf{n}} = D\mathbf{n} \cdot \mathbf{F}$. Thus, the penultimate display easily verifies that the embedded Weingarten maps are related by

$$\bar{\mathbf{W}} = \mathbf{F}\,\mathbf{W} \cdot \mathbf{F}. \tag{2.10}$$

So far, we have found the following: the embedded Weingarten map $\mathbf{W}$ is a symmetric (3×3)-matrix with a trivial eigenvalue 0 corresponding to the surface normal $\mathbf{n}$. The other two eigenvalues κ_1, κ_2 are the principal curvatures, and the corresponding eigenvectors $\mathbf{r}_1, \mathbf{r}_2$ are the principal directions of the surface. But unlike these directions, the matrix $\mathbf{W}$ depends continuously on the parameter $\boldsymbol{\sigma}$. The following theorem establishes $\mathbf{W}$ as a geometric invariant:

Theorem 2.5 (Invariance of W). Let $\mathbf{x}$ and $\tilde{\mathbf{x}}$ be two regular embedded C^2-surfaces with equal image, and let $\mathbf{n}(\boldsymbol{\sigma}) = r\tilde{\mathbf{n}}(\tilde{\boldsymbol{\sigma}})$ for $\mathbf{x}(\boldsymbol{\sigma}) = \tilde{\mathbf{x}}(\tilde{\boldsymbol{\sigma}})$ as in Theorem 2.3/20. Then the corresponding embedded Weingarten maps are equal up to sign,

$$\mathbf{W}(\boldsymbol{\sigma}) = r\tilde{\mathbf{W}}(\tilde{\boldsymbol{\sigma}}).$$

Proof. Let $r = 1$. If $\mathbf{r}$ is a principal direction of $\mathbf{x}$ at $\mathbf{x}(\boldsymbol{\sigma})$ to the principal curvature κ, then, by Theorem 2.3/20, it is also a principal direction of $\tilde{\mathbf{x}}$ at $\tilde{\mathbf{x}}(\tilde{\boldsymbol{\sigma}})$ to κ. Further, $\mathbf{W}(\boldsymbol{\sigma}) \cdot \mathbf{n}(\boldsymbol{\sigma}) = \tilde{\mathbf{W}}(\tilde{\boldsymbol{\sigma}}) \cdot \tilde{\mathbf{n}}(\tilde{\boldsymbol{\sigma}}) = \mathbf{0}$. Hence, the eigenspaces and eigenvalues of $\mathbf{W}(\boldsymbol{\sigma})$ and $\tilde{\mathbf{W}}(\tilde{\boldsymbol{\sigma}})$ coincide so that $\mathbf{W}(\boldsymbol{\sigma}) = \tilde{\mathbf{W}}(\tilde{\boldsymbol{\sigma}})$. If $r = -1$, then again, the corresponding eigenspaces coincide. However, the eigenvalues and hence the matrices have opposite sign. □

The *mean curvature* κ_{M} and the *Gaussian curvature* κ_{G} of $\mathbf{x}$ are defined by

$$\kappa_{\mathrm{M}} := \frac{\kappa_1 + \kappa_2}{2}, \quad \kappa_{\mathrm{G}} := \kappa_1 \kappa_2.$$

They can be computed from W,

$$\kappa_{\mathrm{M}} = \frac{1}{2} \operatorname{trace} W, \quad \kappa_{\mathrm{G}} = \det W = \det I\!I / \det I,$$

or, equally well, from $\mathbf{W}$,

$$\kappa_{\mathrm{M}} = \frac{1}{2} \operatorname{trace} \mathbf{W}, \quad \kappa_{\mathrm{G}} = \frac{1}{2} \operatorname{trace}^2 \mathbf{W} - \frac{1}{2} \|\mathbf{W}\|_{\mathrm{F}}^2, \tag{2.11}$$

where $\|\mathbf{W}\|_{\mathrm{F}}^2 := \sum_{i,j} \mathbf{W}_{i,j}^2$ is the squared Frobenius norm of $\mathbf{W}$. A point $\mathbf{x}(\boldsymbol{\sigma})$ is called

- *elliptic* if $\kappa_{\mathrm{G}}(\boldsymbol{\sigma}) > 0$,
- *hyperbolic* if $\kappa_{\mathrm{G}}(\boldsymbol{\sigma}) < 0$,
- *parabolic* if $\kappa_{\mathrm{G}}(\boldsymbol{\sigma}) = 0$.

For later use, and as an important special case, we consider a surface in Euler form.

Example 2.6 (Euler form). Consider a surface in *Euler form*

$$\bar{\mathbf{x}}(u, v) := [u,\ v,\ h(u, v)], \quad (u, v) \in \bar{\boldsymbol{\Sigma}}, \tag{2.12}$$

where $\bar{\Sigma}$ is a closed domain and h is a C^2-function. Obviously, the parametrization is always regular, and with

$$R := \begin{bmatrix} r_v & -h_u h_v \\ -h_u h_v & r_u \\ h_u & h_v \end{bmatrix}, \quad \begin{aligned} r_u &:= 1 + h_u^2 \\ r_v &:= 1 + h_v^2 \\ w &:= (1 + h_u^2 + h_v^2)^{-1/2}, \end{aligned}$$

we obtain

$$D\bar{\mathbf{x}} = \begin{bmatrix} 1 & 0 & h_u \\ 0 & 1 & h_v \end{bmatrix}, \quad \bar{I} = \begin{bmatrix} r_u & h_u h_v \\ h_u h_v & r_v \end{bmatrix}, \quad (D\bar{\mathbf{x}})^+ = w^2 R.$$

Using

$$\bar{\mathbf{n}} = w\,[-h_u,\, -h_v,\, 1], \tag{2.13}$$

the second fundamental form turns out to be a multiple of the Hessian matrix H of h,

$$\bar{II} = w \begin{bmatrix} h_{uu} & h_{uv} \\ h_{uv} & h_{vv} \end{bmatrix} =: w\,H.$$

Together, the embedded Weingarten map is

$$\bar{\mathbf{W}} = w^5\, RH \cdot R, \tag{2.14}$$

In particular, at a point with $h_u = h_v = 0$, we have

$$\bar{\mathbf{W}} = \begin{bmatrix} H & 0 \\ 0 & 0 \end{bmatrix},$$

and the principal curvatures are the eigenvalues of H. □

2.3 Surfaces with a Singular Point

As the next chapter will demonstrate, subdivision surfaces have isolated points where the natural parametrization of such surfaces is either non-regular or fails to be differentiable at all. There the theory developed above does not apply. However, the surfaces can still be smooth from a geometric point of view. The following example illustrates this difference between analytic and geometric smoothness.

Example 2.7 (Geometric and analytic smoothness). Consider the two surfaces with $(s,t) \in [-1,1]^2$ (see Fig. 2.2/24),

$$\begin{aligned} (s,t) &\mapsto [2s + |s|, t, 0] \\ (s,t) &\mapsto [s^3,\, t^3,\, s^2 + t^2]. \end{aligned} \tag{2.15}$$

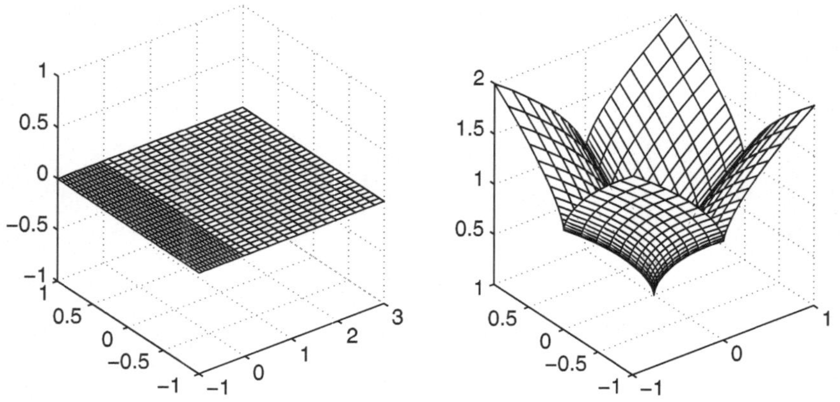

Fig. 2.2 Illustration of Example 2.7/23: (*left*) C^0_∞-surface and (*right*) C^∞_0-surface.

The first parametrization is not C^1, but generates a perfectly smooth plane. By contrast, the second parametrization is C^∞, but generates a surface with a cusp at the origin. Its partial derivatives vanish at the origin, i.e., they are linearly dependent. □

This section is devoted to a study of geometric smoothness properties of surfaces that are not differentiable at a single parameter, say $\boldsymbol{\sigma} = \mathbf{0}$. We start with the following general definition that covers functions with values in $\mathbb{R}^d$.

Definition 2.8 (C^k_0-function, almost regular). Let $\boldsymbol{\Sigma} \subset \mathbb{R}^2$ be a domain containing the origin $\boldsymbol{\sigma} = \mathbf{0}$. A continuous function $\mathbf{x} : \boldsymbol{\Sigma} \to \mathbb{R}^d$ is called C^k_0, if it is C^k everywhere except at the origin. The space of all such functions is denoted by $C^k_0(\boldsymbol{\Sigma}, \mathbb{R}^d)$. The image

$$\mathbf{x}^{\mathrm{c}} := \mathbf{x}(\mathbf{0})$$

is called the *central point* of $\mathbf{x}$. For $d = 2$ or $d = 3$, the function $\mathbf{x} \in C^k_0(\boldsymbol{\Sigma}, \mathbb{R}^d)$ is called *almost regular* if ${}^{\times}\!D\mathbf{x}(\boldsymbol{\sigma}) \neq 0$ for all $\boldsymbol{\sigma} \in \boldsymbol{\Sigma}\backslash\{\mathbf{0}\}$.

The subscript of C^k_0 does not refer to $\boldsymbol{\sigma} = \mathbf{0}$ but to $\mathbf{x}$ being continuous there. Typically, it is impossible to define a normal vector of a C^k_0-surface at the central point in a meaningful way. However, in special situations it may still be possible. The following definition addresses such a case.

Definition 2.9 (Normal continuity). An almost regular surface $\mathbf{x} \in C^1_0(\boldsymbol{\Sigma}, \mathbb{R}^3)$ is *normal continuous,*[3] if the limit

$$\mathbf{n}^{\mathrm{c}} := \lim_{\boldsymbol{\sigma} \to \mathbf{0}} \mathbf{n}(\boldsymbol{\sigma}),$$

called the *central normal*, exists.

[3] In the literature, normal continuity is also called *tangent plane continuity*.

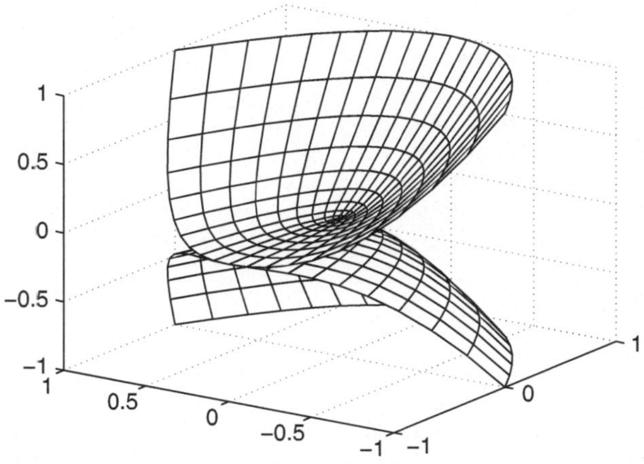

Fig. 2.3 Illustration of Example 2.10/25: Normal continuous surface with local self-intersection.

In other words, normal continuity requires that the Gauss map can be extended in a continuous way to all of the domain. For instance, if we consider the first surface in (2.15/23), we find $\mathbf{n}(\boldsymbol{\sigma}) = [0,\ 0,\ 1]$ for $\boldsymbol{\sigma} \neq \mathbf{0}$ so that, trivially, the surface is normal continuous with $\mathbf{n}^c = [0,\ 0,\ 1]$. By contrast, the surface $\mathbf{x}(s,t) := \left[s, t, \sqrt{s^2+t^2}\right]$ is almost regular, but *not* normal continuous.

Normal continuity is a weak notion of smoothness as the following example shows.

Example 2.10 (Multi-sheeted surface). Consider the almost regular C_0^∞-surface shown in Fig. 2.3/25

$$\mathbf{x}(s,t) := \left[s^2 - t^2,\ st,\ s^3\right], \quad (s,t) \in [-1,1]^2.$$

It is easily verified by inspection that $\mathbf{x}$ is normal continuous with $\mathbf{n}^c = [0,\ 0,\ 1]$. However, the image of $\mathbf{x}$ is not a 2-manifold in the sense of differential geometry because it is not homeomorphic to a subset of $\mathbb{R}^2$, not even if restricted to an arbitrarily small neighborhood of the central point. The reason is that $\mathbf{x}$ has a local self-intersection so that the projection to the xy-plane is non-injective. More precisely, the first two components of $\mathbf{x}(s,t)$ and $\mathbf{x}(-s,-t)$ coincide, while the third component has opposite sign. □

The next definition addresses this issue in the following way: first, $\mathbf{x}$ is moved by a Euclidean motion so that the central point is mapped to the origin, and the central normal is mapped to the third unit vector. In that way, the xy-plane becomes the tangent plane at the central point. Second, the xy-component of the resulting surface is checked for local injectivity.

Definition 2.11 (Single-sheetedness). Let $\mathbf{x} \in C_0^k(\boldsymbol{\Sigma}, \mathbb{R}^3)$ be normal continuous with central normal $\mathbf{n}^c$. For a pair $\mathbf{T}^c := [\mathbf{t}_1^c; \mathbf{t}_2^c]$ of orthonormal vectors in the tangent plane, $\mathbf{F}^c := [\mathbf{T}^c; \mathbf{n}^c]$ is an orthogonal (3×3)-matrix, called the *central frame*. The transformed surface

$$\mathbf{x}_* := (\mathbf{x} - \mathbf{x}^c) \cdot \mathbf{F}^c$$

has the *tangential component*

$$\boldsymbol{\xi}_* := (\mathbf{x} - \mathbf{x}^c) \cdot \mathbf{T}^c \in C_0^k(\boldsymbol{\Sigma}, \mathbb{R}^2)$$

and the *normal component*

$$z_* := (\mathbf{x} - \mathbf{x}^c) \cdot \mathbf{n}^c \in C_0^k(\boldsymbol{\Sigma}, \mathbb{R}).$$

If there exists an open connected neighborhood $\boldsymbol{\Sigma}_* \subset \boldsymbol{\Sigma}$ of the origin such that $\boldsymbol{\xi}_*$ restricted to $\boldsymbol{\Sigma}_*$ is injective, then $\mathbf{x}$ is called *single-sheeted* at the central point. If $\mathbf{x}$ is single-sheeted, then the *local height function*[4] h_* is defined by

$$h_* : \boldsymbol{\Xi}_* \ni \boldsymbol{\xi} \mapsto z_*\big(\boldsymbol{\sigma}_*(\boldsymbol{\xi})\big) \in \mathbb{R},$$

where the domain is $\boldsymbol{\Xi}_* := \boldsymbol{\xi}_*(\boldsymbol{\Sigma}_*) \subset \mathbb{R}^2$, and $\boldsymbol{\sigma}_* : \boldsymbol{\Xi}_* \to \boldsymbol{\Sigma}_*$ is the local inverse of $\boldsymbol{\xi}_*$.

In case of single-sheetedness, the surface can locally be represented with the help of the local height function h_*, see Fig. 2.4/27,

$$\mathbf{x}(\boldsymbol{\sigma}) = \tilde{\mathbf{x}}(\boldsymbol{\xi}) = \mathbf{x}^c + \boldsymbol{\xi}\mathbf{T}^c + h_*(\boldsymbol{\xi})\mathbf{n}^c, \quad \boldsymbol{\xi} \in \boldsymbol{\Xi}_*, \tag{2.16}$$

where $\boldsymbol{\sigma}$ and $\boldsymbol{\xi}$ are related by $\boldsymbol{\sigma} = \boldsymbol{\sigma}_*(\boldsymbol{\xi})$ and $\boldsymbol{\xi} = \boldsymbol{\xi}_*(\boldsymbol{\sigma})$. Accordingly,

$$\bar{\mathbf{x}}(\boldsymbol{\xi}) := \big(\tilde{\mathbf{x}}(\boldsymbol{\xi}) - \mathbf{x}^c\big) \cdot \mathbf{F}^c = [\boldsymbol{\xi}, h_*(\boldsymbol{\xi})] \tag{2.17}$$

is the *local Euler form* of $\mathbf{x}$, see (2.12/22). The surfaces $\mathbf{x}$ and $\bar{\mathbf{x}}$ are related by a regular affine map so that they share all shape properties. In particular, the Gauss maps and the embedded Weingarten maps are related by (2.9/21) and (2.10/21), respectively. We note that the parametrization $\bar{\mathbf{x}}$ is always regular so that, locally, we can identify smoothness properties of $\mathbf{x}$ and h_*.

Definition 2.12 (C_r^k-surface). Let $\mathbf{x} \in C_0^k(\boldsymbol{\Sigma}, \mathbb{R}^3)$ be normal continuous and single-sheeted. Then $\mathbf{x}$ is called a *C_r^k-surface* if the local height function is r-times continuously differentiable in a neighborhood of the origin.

In the sense of this definition, the two surfaces given in (2.15/23) are C_∞^0 and C_0^∞, respectively. In the forthcoming analysis of subdivision surfaces, the local height

[4] If we rotate the central frame about $\mathbf{n}^c$, we obtain a one parameter family of local height functions. However, any two members are equivalent in the sense that they differ only by a rotation of the variable $\boldsymbol{\xi}$ about the origin. In particular, all possible local height functions share the same smoothness properties. This justifies referring to a single representative.

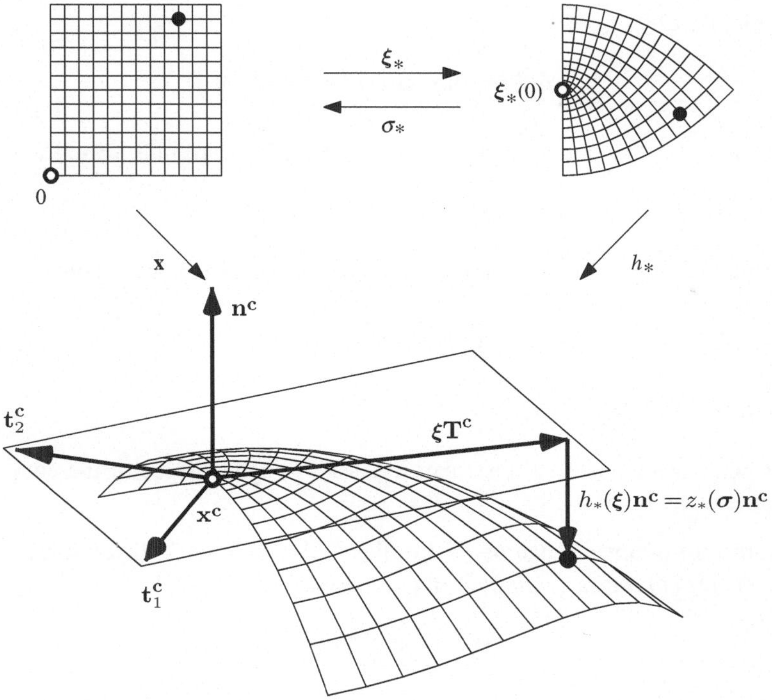

Fig. 2.4 Illustration of Definition 2.11/25: Central frame and local height function.

function is relatively hard to determine, while convenient formulas for the normal vector and the embedded Weingarten map are readily available. Therefore, we are now going to relate convergence properties of $\mathbf{n}$ and $\mathbf{W}$ to differentiability properties of h_*.

Theorem 2.13 (Normal continuity and single-sheetedness imply C_1^k). *If the surface $\mathbf{x} \in C_0^k(\Sigma, \mathbb{R}^3)$ is normal continuous and single-sheeted, then $\mathbf{x}$ is C_1^k. In particular, the local height function satisfies*

$$h_*(\mathbf{0}) = 0, \quad Dh_*(\mathbf{0}) = \mathbf{0}.$$

Proof. The equation

$$(\mathbf{x}(\boldsymbol{\sigma}) - \mathbf{x}^c) \cdot \mathbf{T}^c = \boldsymbol{\xi}$$

defines $\boldsymbol{\sigma} = \boldsymbol{\sigma}_*(\boldsymbol{\xi})$ as a function of $\boldsymbol{\xi}$. Since the domain Σ_* of $\boldsymbol{\xi}_*$ is assumed to be connected, $\boldsymbol{\sigma}_*$ is continuous with

$$\lim_{\boldsymbol{\xi} \to \mathbf{0}} \boldsymbol{\sigma}_*(\boldsymbol{\xi}) = \boldsymbol{\sigma}_*(\mathbf{0}) = \mathbf{0}.$$

Hence, $h_*(\mathbf{0}) = z_*(\mathbf{0}) = 0$. By the inverse function theorem, $\boldsymbol{\sigma}_*$ is C^k for $\boldsymbol{\xi} \neq 0$ because $D\mathbf{x}(\boldsymbol{\sigma}) \cdot \mathbf{T}^c$ has full rank for $\boldsymbol{\sigma} \neq \mathbf{0}$. Hence, $h_* = z_* \circ \boldsymbol{\sigma}_*$ is C^k away from the origin. By (2.9/21),

$$\lim_{\boldsymbol{\xi}\to\mathbf{0}} \bar{\mathbf{n}}(\boldsymbol{\xi}) = \lim_{\boldsymbol{\sigma}\to\mathbf{0}} \mathbf{n}(\boldsymbol{\sigma}) \cdot \mathbf{F}^{\mathrm{c}} = [0,\, 0,\, 1].$$

According to (2.13/23), the normal vector of the local Euler form is given by

$$\bar{\mathbf{n}} = w\,[-h_{*,u}, -h_{*,v}, 1], \quad w := (1 + (h_{*,u})^2 + (h_{*,v})^2)^{-(1/2)}$$

away from the origin. Hence, comparing the last two displays, we obtain

$$\lim_{\boldsymbol{\xi}\to\mathbf{0}} Dh_*(\boldsymbol{\xi}) = \mathbf{0}.$$

Because h_* is C^0 and Dh_* converges, h_* is C^1. □

Just as convergence of the Gauss map implies C^k_1, convergence of the embedded Weingarten map implies C^k_2.

Theorem 2.14 (Convergence of W implies C^k_2). Let $k \geq 2$. If the surface $\mathbf{x} \in C^k_1(\boldsymbol{\Sigma}, \mathbb{R}^3)$ is *curvature continuous* in the sense that the limit

$$\mathbf{W}^{\mathrm{c}} := \lim_{\boldsymbol{\sigma}\to\mathbf{0}} \mathbf{W}(\boldsymbol{\sigma})$$

exists, then $\mathbf{x}$ is C^k_2.

Proof. Let us consider the local Euler form $\bar{\mathbf{x}}(\boldsymbol{\xi}) = [\boldsymbol{\xi}, h_*(\boldsymbol{\xi})]$ of $\mathbf{x}$ according to (2.17/26), and its embedded Weingarten map $\bar{\mathbf{W}}$ according to (2.14/23). Below, we replace h_* by h to improve readability, and let $\boldsymbol{\xi} = (u, v)$. We extract three components from $\bar{\mathbf{W}}$ and write them in the form

$$\begin{bmatrix} \bar{\mathbf{W}}_{1,1} \\ \bar{\mathbf{W}}_{1,2} \\ \bar{\mathbf{W}}_{2,2} \end{bmatrix} = w^5 \begin{bmatrix} r_u^2 & -2h_u h_v r_u & h_u^2 h_v^2 \\ -h_u h_v r_u & h_u^2 h_v^2 + r_u r_v & -h_u h_v r_v \\ h_u^2 h_v^2 & -2h_u h_v r_v & r_v^2 \end{bmatrix} \begin{bmatrix} h_{uu} \\ h_{uv} \\ h_{vv} \end{bmatrix},$$

where w, r_u, r_v are defined as in Example 2.6/22. The matrix on the right hand side is always invertible, and we obtain

$$\begin{bmatrix} h_{uu} \\ h_{uv} \\ h_{vv} \end{bmatrix} = w^{-1} \begin{bmatrix} r_v^2 & 2h_u h_v r_v & h_u^2 h_v^2 \\ h_u h_v r_v & h_u^2 h_v^2 + r_u r_v & h_u h_v r_u \\ h_u^2 h_v^2 & 2h_u h_v r_u & r_u^2 \end{bmatrix} \begin{bmatrix} \bar{\mathbf{W}}_{1,1} \\ \bar{\mathbf{W}}_{1,2} \\ \bar{\mathbf{W}}_{2,2} \end{bmatrix}.$$

By assumption, h is C^1 everywhere, and C^2 away from the origin. Further, convergence of $\mathbf{W}$ implies convergence of $\bar{\mathbf{W}} = \mathbf{F}^{\mathrm{c}}\mathbf{W} \cdot \mathbf{F}^{\mathrm{c}}$, see (2.10/21). Hence, also h_{uu}, h_{uv}, and h_{vv} converge so that h is C^2 and $\mathbf{x}$ is C^k_2. □

It is important to notice that a similar result is *not* true for convergent principal curvatures alone. That is, there are C^k_1-surfaces with convergent principal curvatures that are not C^k_2. Let us illustrate the concepts developed so far by a simple example.

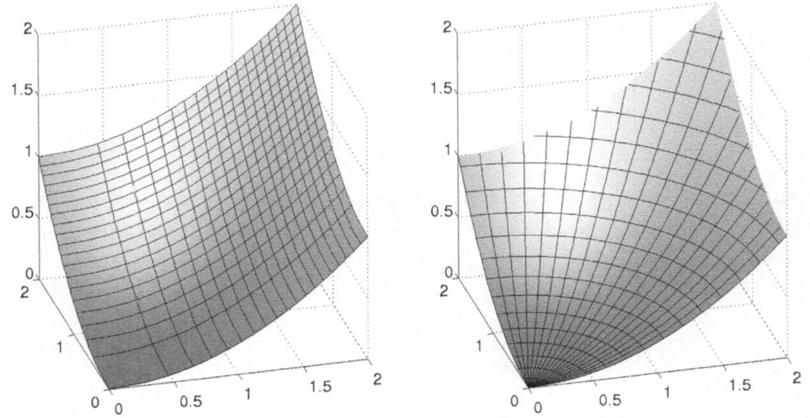

Fig. 2.5 Illustration of Example 2.15/29: Almost regular surface with (*left*) its parameter lines and (*right*) its confluent principal curvature lines.

Example 2.15 (Computing W). We consider the C_0^∞-surface

$$\mathbf{x}(s,t) := [2\sqrt{s},\ 2\sqrt{t},\ s+t], \quad (s,t) \in [0,1]^2,$$

see Fig. 2.5/29. The parametrization is not differentiable at the origin so that we do not know beforehand if the surface is normal continuous or curvature continuous. With $w := (1+s+t)^{-1/2}$, we obtain for $(s,t) \neq (0,0)$

$$D\mathbf{x} = \begin{bmatrix} 1/\sqrt{s} & 0 & 1 \\ 0 & 1/\sqrt{t} & 1 \end{bmatrix}, \quad \mathbf{n} = w\,[-\sqrt{s},\ -\sqrt{t},\ 1].$$

The normal vector converges according to

$$\lim_{(s,t)\to(0,0)} \mathbf{n}(s,t) = [0,\ 0,\ 1].$$

A suitable central frame $\mathbf{F}^c$ is given by the identity, and we obtain

$$\boldsymbol{\xi}_*(s,t) = [2\sqrt{s},\ 2\sqrt{t}], \quad z_*(s,t) = s+t.$$

The tangential component $\boldsymbol{\xi}_*$ is invertible so that $\mathbf{x}$ is single-sheeted. Hence, by Theorem 2.13/27, $\mathbf{x}$ is a C_1^∞-surface. The inverse of $\boldsymbol{\xi}_*$ and the local height function are

$$[s,t] = \boldsymbol{\sigma}_*(x,y) = [x^2/4,\ y^2/4], \quad h_*(x,y) = (x^2+y^2)/4,$$

showing that $\mathbf{x}$ is in fact C_∞^∞. However, in more complicated situations, such explicit knowledge on h_* is not readily available. Therefore, let us analyze curvature and re-establish curvature continuity that way. We obtain

$$I = \begin{bmatrix} 1+1/s & 1 \\ 1 & 1+1/t \end{bmatrix}, \quad II = \frac{w}{2}\begin{bmatrix} 1/s & 0 \\ 0 & 1/t \end{bmatrix}, \quad W = \frac{w^3}{2}\begin{bmatrix} 1+t & -t \\ -s & 1+s \end{bmatrix}.$$

Hence, the principal curvatures are $\kappa_1 = w/2$ and $\kappa_2 = w^3/2$. For $(s,t) \neq (0,0)$, the corresponding left eigenvectors are unique up to orientation, and we obtain

$$\mathbf{r}_1 = \frac{1}{\sqrt{s+t}}\,[\sqrt{t},\, -\sqrt{s},\, 0], \quad \mathbf{r}_2 = \frac{w}{\sqrt{s+t}}\,[\sqrt{s},\, \sqrt{t},\, s+t].$$

Obviously, these vectors do *not* converge as $(s,t) \to (0,0)$. Rather, the origin is an umbillic point, and any direction in the xy-plane is a principal direction. Now we compute the embedded Weingarten map. With $v := 2+s+t$, we find for $(s,t) \neq (0,0)$

$$D\mathbf{x}^+ = w^2 \begin{bmatrix} \sqrt{s}(1+t) & -\sqrt{s}t \\ -\sqrt{t}s & \sqrt{t}(1+s) \\ s & t \end{bmatrix}, \quad \mathbf{W} = \frac{w^5}{2} \begin{bmatrix} 1+vt & -v\sqrt{st} & \sqrt{s} \\ -v\sqrt{st} & 1+vs & \sqrt{t} \\ \sqrt{s} & \sqrt{t} & s+t \end{bmatrix}.$$

The eigenvalues of $\mathbf{W}$ are $\kappa_1, \kappa_2, 0$, and the corresponding eigenvectors are $\mathbf{r}_1, \mathbf{r}_2, \mathbf{n}$. At the origin, we obtain the limit

$$\mathbf{W}^c = \lim_{(s,t)\to(0,0)} \mathbf{W}(s,t) = \frac{1}{2} \begin{bmatrix} 1 & 0 & 0 \\ 0 & 1 & 0 \\ 0 & 0 & 0 \end{bmatrix}.$$

Hence, by Theorem 2.14/28, the surface is C_2^∞. One could argue that also the standard Weingarten map W converges as $(s,t) \to (0,0)$, and that curvature continuity follows equally from that. This is correct, but the situation changes if we consider the union of $\mathbf{x}$ and a second piece of surface,

$$\mathbf{x}'(s,t) := [-2\sqrt{t},\, 2\sqrt{s},\, s+t], \quad (s,t) \in [0,1]^2.$$

In the next chapter, we will identify such a construction as a spline surface. Here, $\mathbf{n}' = w\,[-\sqrt{t},\, \sqrt{s},\, s+t]$, and

$$W' = \frac{w^3}{2} \begin{bmatrix} 1+t & -t \\ -s & 1+s \end{bmatrix}, \quad \mathbf{W}' = \frac{w^5}{2} \begin{bmatrix} 1+vs & v\sqrt{st} & -\sqrt{t} \\ v\sqrt{st} & 1+vt & \sqrt{s} \\ -\sqrt{t} & \sqrt{s} & s+t \end{bmatrix}.$$

$\mathbf{x}$ and $\mathbf{x}'$ join normal continuous along the common boundary according to

$$\begin{aligned} \mathbf{x}(0,u) &= \mathbf{x}'(u,0) = [0,\, 2\sqrt{u},\, 0], \quad u \in [0,1] \\ \mathbf{n}(0,u) &= \mathbf{n}'(u,0) = (1+u)^{-1/2}\,[0,\, -\sqrt{u},\, 1]. \end{aligned}$$

The corresponding standard Weingarten maps differ at the common boundary,

$$W(0,u) = \frac{(1+u)^{-3/2}}{2} \begin{bmatrix} 1+u & -u \\ 0 & 1 \end{bmatrix}, \quad W'(u,0) = \frac{(1+u)^{-3/2}}{2} \begin{bmatrix} 1 & 0 \\ -u & 1+u \end{bmatrix},$$

while the embedded Weingarten maps coincide,

$$\mathbf{W}(0,u) = \mathbf{W}'(u,0) = \frac{(1+u)^{-5/2}}{2}\begin{bmatrix}(1+u)^2 & 0 & 0\\ 0 & 1 & \sqrt{u}\\ 0 & \sqrt{u} & u\end{bmatrix}.$$

This shows that $\mathbf{x}$ and $\mathbf{x}'$ join curvature continuously. Finally,

$$\lim_{(s,t)\to(0,0)} \mathbf{W}(s,t) = \lim_{(s,t)\to(0,0)} \mathbf{W}'(s,t) = \begin{bmatrix}1\,0\,0\\0\,1\,0\\0\,0\,0\end{bmatrix}$$

establishes the composed surface $\mathbf{x}([0,1]^2) \cup \mathbf{x}'([0,1]^2)$ as C_2^k in a generalized sense, which will be made precise in the next chapter. □

2.4 Criteria for Injectivity

The analysis of almost regular surfaces requires criteria for the injectivity of the tangential component $\boldsymbol{\xi}_* = (\mathbf{x}-\mathbf{x}^c)\cdot\mathbf{T}^c$ of $\mathbf{x}_*$ to establish single-sheetedness of the surface. At regular points, local injectivity of a function follows immediately from the inverse function theorem. At singularities, where the situation is much more complicated, an appropriate tool is provided by a concept from algebraic topology: the winding number. Below, we give a short introduction to the topic as far as it is required in this context.

If the surface $\mathbf{x}$ is almost regular then

$$^{\times}D\boldsymbol{\xi}_* = (^{\times}D\mathbf{x}) \cdot \mathbf{n}^c$$

shows that the tangential component $\boldsymbol{\xi}_*$ is almost regular in a vicinity of the origin. For simplicity, we assume $\boldsymbol{\Sigma} = \mathbb{R}^2$, i.e., $\boldsymbol{\xi} \in C_0^k(\mathbb{R}^2, \mathbb{R}^2)$. However, we will not take advantage of the unboundedness of the domain, but refer only to local properties. Further, we will omit the subscript star of $\boldsymbol{\xi}$ for the remainder of this chapter.

The map $\boldsymbol{\xi} = (\xi_1, \xi_2) : \mathbb{R}^2 \to \mathbb{R}^2$ can be identified in a natural way with a complex map $f : \mathbb{C} \to \mathbb{C}$ via

$$f(z) = \xi_1(\boldsymbol{\sigma}) + \mathbf{i}\xi_2(\boldsymbol{\sigma}), \quad z = \sigma_1 + \mathbf{i}\sigma_2. \tag{2.18}$$

Then the spaces of complex functions corresponding to $C^k(\mathbb{R}^2, \mathbb{R}^2)$ and $C_0^k(\mathbb{R}^2, \mathbb{R}^2)$ are denoted $C^k(\mathbb{C}, \mathbb{C})$ and $C_0^k(\mathbb{C}, \mathbb{C})$, respectively, and we can use notions for real functions $\boldsymbol{\xi}$ or for complex functions f accordingly, e.g.

$$^{\times}Df := {}^{\times}D\boldsymbol{\xi}.$$

Further, we assume

$$f(0) = 0$$

without loss of generality, throughout. We start with the fundamental definition of the winding number.

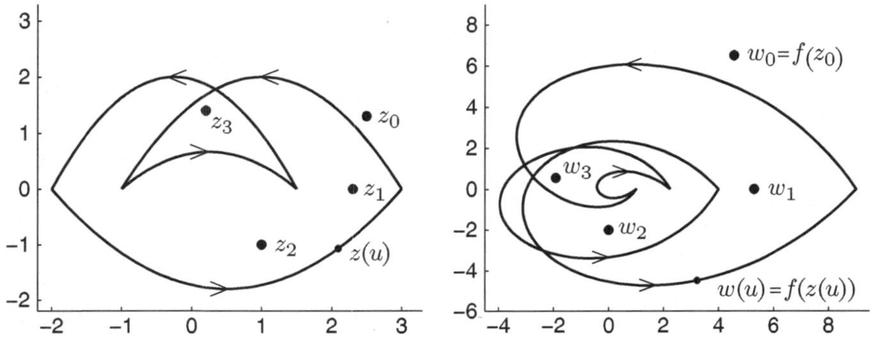

Fig. 2.6 Illustration of Definition 2.16/32: *(left)* Winding numbers $\nu(z, z_0) = 0$, $\nu(z, z_1) = \nu(z, z_2) = 1$, $\nu(z, z_3) = 2$ and *(right)* induced winding numbers $\nu(f, z, z_0) = 0$, $\nu(f, z, z_1) = 1$, $\nu(f, z, z_2) = 2$, $\nu(f, z, z_3) = 3$ for function $f(z) = z^2$.

Definition 2.16 (Winding number). For $U := [0, 1]$, let $z : U \to \mathbb{C}$ be a continuous, piecewise differentiable curve in the complex plane that is closed and does not contain the point z_0,

$$z(0) = z(1), \quad z_0 \notin z(U).$$

Then the *winding number* of z with respect to the point z_0 is defined by

$$\nu(z, z_0) := \frac{1}{2\pi\mathbf{i}} \int_0^1 \frac{z'(u)}{z(u) - z_0}\, du. \tag{2.19}$$

If the curve z passes through z_0, i.e., $z(u) = z_0$ for some $u \in U$, we formally set

$$\nu(z, z_0) := \infty$$

to indicate that, typically, the integral diverges. Further, for a function $f \in C^1(\mathbb{C}, \mathbb{C})$, we define the *induced winding number*

$$\nu(f, z, z_0) := \nu(f \circ z, f(z_0)) = \nu(w, w_0)$$

as the winding number of the curve $w := f \circ z$ with respect to the point $w_0 := f(z_0)$.

Later on, we will apply the above definitions also to functions, curves, and points in $\mathbb{R}^2$ using the standard identification with $\mathbb{C}$. As a consequence of the Cauchy integral formula, the winding number is always an integer. Roughly speaking, the winding number counts how many times the point $z(u)$ is orbiting counter-clockwise around the point z_0 for u from 0 to 1. Figure 2.6/32 shows a few examples.

The winding number $\nu(f, z, z_0)$ depends continuously on its arguments, wherever it is finite. Hence, because it can attain only integer values, it is piecewise constant, where the pieces are separated by infinite values. More precisely, we state without formal proof:

Lemma 2.17 (Persistence of winding number). *If $f^\alpha, z^\alpha, z_0^\alpha$ are families of functions, curves, and points, respectively that depend, with respect to the sup-norm, continuously on the parameter $\alpha \in [0,1]$ and satisfy*

$$f^\alpha(z_0^\alpha) \notin f^\alpha(z^\alpha(U))$$

for all α, then the induced winding number is independent of α,

$$\nu(f^\alpha, z^\alpha, z_0^\alpha) = \nu(f^0, z^0, z_0^0).$$

In particular, if f^α is consistently the identity, then $\nu(z^\alpha, z_0^\alpha) = \nu(z^0, z_0^0)$.

The assumption that f be C^1 is made only to ensure that $w = f \circ z$ is piecewise differentiable. Hence, we can relax this assumption, and require differentiability of f only in a neighborhood of $z(U)$. In particular, the induced winding number $\nu(f, z, z_0)$ is well-defined for $f \in C_0^1(\mathbb{C}, \mathbb{C})$ as given by (2.18/31), provided that $0 \notin z(U)$. However, for the theory to be derived now, we have to extend the definition of the induced winding number to the case where $z_0 \neq 0$ and the curve z passes through the origin. Here, the integral defining $\nu(w, w_0)$ does not necessarily exist, but we can resort to the following smoothing process: Let f^α be a continuous family of C^1-functions converging uniformly according to

$$\lim_{\alpha \to 0} \|f^\alpha - f\|_\infty = 0.$$

Then, for f^α sufficiently close to f, the number of orbits of $f^\alpha(z)$ around the point $f(z_0)$ depends neither on α nor on the particular choice of f^α. This observation justifies the definition

$$\nu(f, z, z_0) := \lim_{\alpha \to 0} \nu(f^\alpha, z, z_0)$$

for functions $f \in C_0^1(\mathbb{C}, \mathbb{C})$. Thus, Lemma 2.17/33 remains valid also in this more general setting.

Now, we consider an almost regular function $f \in C_0^1(\mathbb{C}, \mathbb{C})$. Because ${}^{\times}\mathcal{D}f$ has no zeros on the connected set $\mathbb{C}\backslash\{0\}$, the sign of ${}^{\times}\mathcal{D}f$ is constant,

$$s_f := \operatorname{sign} {}^{\times}\mathcal{D}f(z_0) \equiv \pm 1, \quad z_0 \in \mathbb{C}\backslash\{0\}.$$

For a value $w_0 \neq 0$, the preimage does not contain the origin, but only regular points,

$$f(z_0) = w_0 \quad \Rightarrow \quad {}^{\times}\mathcal{D}f(z_0) \neq 0.$$

By the inverse function theorem, there exists a neighborhood $\Gamma(z_0)$ of z_0 such that f restricted to $\Gamma(z_0)$ is a diffeomorphism. Hence, for any curve z in $\Gamma(z_0)$ encircling the point z_0 once, i.e., $\nu(z, z_0) = 1$, also the image $w = f \circ z$ is encircling the point $w_0 := f(z_0)$ once. The orientation of the image curve depends on the sign of ${}^{\times}\mathcal{D}f$ so that

$$\nu(f, z, z_0) = \nu(f \circ z, w_0) = s_f.$$

This value is independent of the particular choice of the curve z as long as z is sufficiently close to z_0. It is called the *index* of f at z_0. Indices and winding numbers are related by a fundamental formula. In our setting, it takes on the following simple form.

Lemma 2.18 (Number of preimages). *Let $f \in C^1_0(\mathbb{C},\mathbb{C})$ be almost regular, $f(z_0) = w_0 \neq 0$ be the image of a regular point, and z be a piecewise differentiable Jordan curve with winding number $\nu(z, z_0) = 1$. Then the set*

$$\{z_0, \dots, z_{m-1}\} := \{z_* \in f^{-1}(w_0) : \nu(z, z_*) = 1\}$$

of preimages in the interior of the set bounded by z consists of

$$m = s_f\, \nu(f, z, z_0)$$

elements.

Proof. We assume $s_f = 1$; the other case is analogous. In the following, z^α is always a family of curves that depend continuously on $\alpha \in [0,1]$, where

- $z^0 = z$ coincides with the given curve,
- $f(z^\alpha(U))$ does not contain w_0. By Lemma 2.17/33,

$$\nu(z^1, z_0) = \nu(z, z_0) = 1 \quad \text{and} \quad \nu(f, z^1, z_0) = \nu(f, z, z_0).$$

Hence, we have to show that $\nu(f, z^1, z_0) = m$. The proof is by induction on m, starting from $m = 1$. In this case, z^α is chosen such that z^1 is shrunk to a circle lying entirely in the neighborhood $\Gamma(z_0)$. Hence, $\nu(f, z^1, z_0) = 1$. Now, assume that the assertion is true for $\leq m$ preimages, and consider the case of $m+1$ preimages. Then, as shown in Fig. 2.7/35, z^α is chosen such that z^1 consists of two closed curves z^1_0 and z^1_1. The curve z^1_0 encloses the point z_0, and z^1_1 encloses the remaining points $z_1, \dots, z_m$. We split the integral defining $\nu(f, z^1, z_0)$ into two parts. Using the induction hypothesis, we find

$$\nu(f, z^1, z_0) = \nu(f, z^1_0, z_0) + \nu(f, z^1_1, z_0) = 1 + m,$$

and the proof is complete. □

This lemma has an important consequence. It reduces the task of establishing injectivity of an almost regular function f in the vicinity of the origin to checking the winding number of a single curve.

Theorem 2.19 (Injectivity of an almost regular function). *Let $f \in C^1_0(\mathbb{C},\mathbb{C})$ be an almost regular function, and let $z : U \to \mathbb{C}$ be a piecewise differentiable Jordan curve with $\nu(z, 0) = 1$.*

- *If $|\nu(f, z, 0)| = 1$ then the restriction of f to a sufficiently small neighborhood $\Gamma(0)$ of the origin is injective.*
- *If $|\nu(f, z, 0)| \neq 1$ then the restriction of f to any neighborhood of the origin is not injective.*

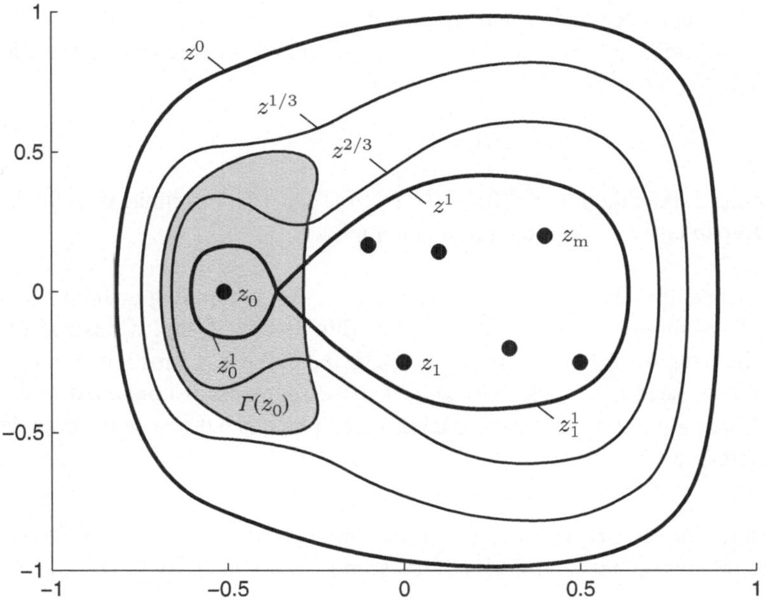

Fig. 2.7 Homotopy used in the proof of Lemma 2.18/34: The given curve $z = z^0$ is transformed into the target curve z^1, which separates z_0 from $z_1, \ldots, z_m$.

Proof. We consider the first statement, and assume $|\nu(f, z, 0)| = 1$. Since the compact set $f(z(U))$ does not contain the point $f(0) = 0$, there exists $\varepsilon_0 > 0$ such that the neighborhood

$$\Gamma(0) := \{z \in \mathbb{C} : |z| < \varepsilon_0\}$$

of the origin and $f(z(U))$ are disjoint. Let $w_0 = f(z_0)$ be an arbitrary point in the image of $\Gamma(0)$.

First, we consider the case $w_0 \neq 0$ of a regular value. The family $z_0^\alpha := \alpha z_0$ of points satisfies the assumptions of Lemma 2.17/33. Hence,

$$\nu(z, z_0) = \nu(z, 0) = 1, \quad |\nu(f, z, z_0)| = |\nu(f, z, 0)|.$$

By Lemma 2.18/34, w_0 has exactly one preimage in $\Gamma(0)$.

Second, consider the irregular point $w_0 = f(0) = 0$. Suppose there exists another point $z_1 \neq 0$ in $\Gamma(0)$ with $f(z_1) = 0$. By the inverse function theorem, there exists $\varepsilon_1 \in (0, |z_1|/2)$ such that the neighborhood

$$\Gamma(z_1) := \{z \in \Gamma(0) : |z - z_1| < \varepsilon_1\}$$

of z_1 is mapped to a neighborhood of the origin. Now, consider the sequence of values $f(1/r), r \in \mathbb{N}$, converging to the origin. For r sufficiently large, $1/r \notin \Gamma(z_1)$, while $f(1/r) \in f(\Gamma(z_1))$. Hence, the regular value $f(1/r)$ has at least two preimages, contradicting the result of the first part of the proof.

Now, we consider the second statement, and assume $|\nu(f,z,0)| \neq 1$. For any $\varepsilon > 0$, the family $z^\alpha := \big((1-\alpha)\varepsilon + \alpha\big)z$ of curves satisfies the assumptions of Lemma 2.17/33. Hence,

$$\nu(\varepsilon z, 0) = \nu(z,0) = 1, \quad |\nu(f,\varepsilon z,0)| = |\nu(f,z,0)|.$$

By Lemma 2.18/34, the point $f(0) = 0$ has $|\nu(f,z,0)| > 1$ preimages in the interior of the Jordan curve εz, and the proof is complete. □

We conclude this chapter by providing a tool for computing winding numbers $\nu(z,0)$ by summing up a finite number of differences of angles, instead of evaluating the integral (2.19/32). This is possible if the curve z is partitioned into a finite number n of segments such that each of these segments is contained in a sliced plane. Given a point $h_j \neq 0$ in $\mathbb{C}$, such a sliced plane is defined as the complement of the half-line

$$H_j := \{rh_j : r \geq 0\}.$$

Ambiguities, as they typically arise for the complex logarithm can be avoided when considering only points $z \in \mathbb{C}\backslash H_j$ in the sector complementary to H_j. Using

$$\arg \frac{z}{h_j} := \varphi \in [0, 2\pi), \quad \frac{z}{h_j} = \left|\frac{z}{h_j}\right| e^{\mathrm{i}\varphi},$$

the complex logarithm

$$\ln : \mathbb{C}\backslash H_j \ni z \mapsto \ln|z| + \mathbf{i} \arg \frac{z}{h_j} \tag{2.20}$$

is a uniquely defined smooth function on $\mathbb{C}\backslash H_j$ with $\ln'(z) = 1/z$.

Lemma 2.20 (Winding number via arguments). *Let $z : U \to \mathbb{C}$ be a closed piecewise differentiable curve. If there exist points $0 = u_0 < u_1 < \cdots < u_{n-1} < u_n = 1$ in U and half-lines $H_1, \ldots, H_n$ such that*

$$z([u_{j-1}, u_j]) \cap H_j = \emptyset$$

for all $j = 1, \ldots, n$, then

$$\nu(z,0) = \frac{1}{2\pi} \sum_{j=1}^{n} \left(\arg \frac{z_j}{h_j} - \arg \frac{z_{j-1}}{h_j} \right),$$

where $z_j := z(u_j)$ (cf. Fig. 2.8/37).

Proof. For fixed j, the curve segment $z([u_{j-1}, u_j])$, is contained in $\mathbb{C}\backslash H_j$. Hence, we can use the complex logarithm (2.20/36) to evaluate the contour integral

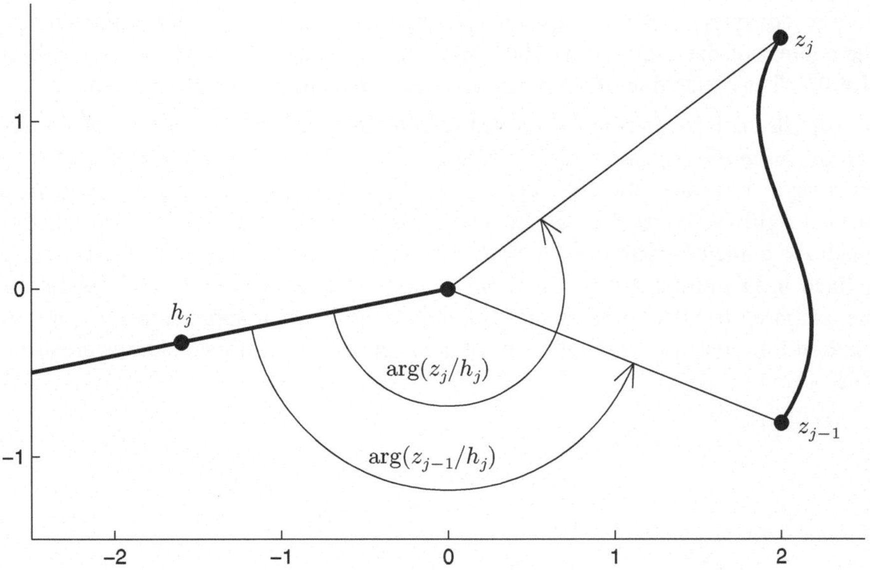

Fig. 2.8 Illustration of Lemma 2.20/36: Half-line H_j through h_j and disjoint curve segment $z([u_{j-1}, u_j])$ with endpoints z_{j-1}, z_j and angles $\arg(z_{j-1}/h_j)$, $\arg(z_j/h_j)$.

$$\begin{aligned} r_j &:= \int_{u_{j-1}}^{u_j} \frac{z'(u)}{z(u)}\, du = \ln(z(u))\Big|_{u_{j-1}}^{u_j} = \ln z_j - \ln z_{j-1} \\ &= \ln|z_j| - \ln|z_{j-1}| + \mathbf{i}\Big(\arg\frac{z_j}{h_j} - \arg\frac{z_{j-1}}{h_j}\Big). \end{aligned}$$

Summing over j, the real part vanishes since $\ln|z_0| = \ln|z_n|$, and we obtain

$$2\pi\mathbf{i}\,\nu(z,0) = \sum_{j=1}^{n} r_j = \mathbf{i}\sum_{j=1}^{n}\Big(\arg\frac{z_j}{h_j} - \arg\frac{z_{j-1}}{h_j}\Big),$$

as stated. □

Bibliographic Notes

1. For a detailed introduction to the differential geometry of surfaces, we refer, e.g., to the classical work [dC76]. However, as in most other text books on the topic, the focus in [dC76] is on regularly parametrized surfaces.

2. The winding number is a concept from algebraic topology; see [Ful95] for a good introduction.

3. The advantages of the embedded Weingarten map over earlier approaches in the context of the analysis of subdivision surfaces were observed only recently in [Rei07]. The exposition of the material given here is inspired by that paper.

4. An alternative criterion for curvature continuity is given by means of the *anchored osculating quadratic*, as introduced in [KP08]. For a single-sheeted, almost regular surface, the anchored osculating quadratic is defined in the vicinity of the origin. Using the formalism derived in this chapter, the anchored osculating quadratic coincides with the quadratic Taylor jet of the local height function h_* as defined in Definition 2.11/25. Curvature continuity is equivalent to convergence of the anchored osculating quadratic to a unique limit when approaching the origin. Explicit formulas for the coefficients of the anchored osculating quadratic are available.

Chapter 3
Generalized Splines

In this chapter, we define bivariate splines. The term spline is often used synonymous with linear combinations of B-splines and hence piecewise polynomials. We will define splines in a less restrictive fashion to include, for example, trigonometric splines and functions generated by interpolating refinement algorithms. This will allow us to cover the shared underlying fundamentals once and for all. Specifically, splines are defined as continuous functions on a domain that is a topological space. This domain is the result of gluing together indexed copies of the unit square, and is locally homeomorphic to the domain of standard bivariate tensor product spline spaces – except at extraordinary knots where more or fewer than four unit squares join up. Consequently, we can focus on characterizing analytical and differential-geometric properties of such splines at and near these isolated singularities.

To lead up to this generalized definition of splines, Sect. 3.1/40 re-interprets the familiar uniform univariate spline as a spline over a piecewise domain. Section 3.2/41, devoted to the bivariate setting, formalizes the familiar view of spline continuity as the joining of pairs of patches along common boundary curves. This is made precise by a relation that gives the domain the structure of a topological space. Once that topology is defined, we can introduce splines as continuous functions on this space. In Sect. 3.3/44, C^k-splines are defined in terms of standard differentiability properties of joint patches, which are obtained by embedding neighboring cells of the domain into $\mathbb{R}^2$ in a specific way. This approach turns out to be equivalent to the familiar characterization of C^k-splines via the agreement of transversal derivatives along patch boundaries. Of course, in the case of spline surfaces, the analytical definition fails to yield geometric information at points where the parametrization is singular.

With the machinery of the preceding chapter, Sect. 3.4/47 characterizes properties of the spline surface at an extraordinary knot where the parametrization is necessarily singular. Note that this is not yet a characterization of smoothness and shape of recursive subdivision surfaces, the main topic of this book. However, it captures essential concepts in a simpler setting. Finally, in Sect. 3.5/53, all new concepts are illustrated by a singularly parametrized cubic splines in Bernstein–Bézier form.

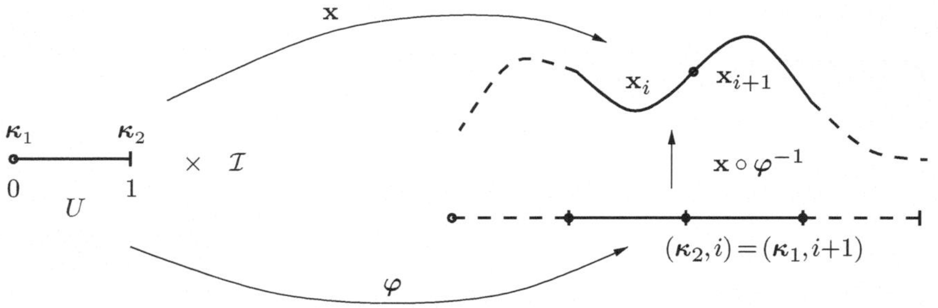

Fig. 3.1 Illustration of Sect. 3.1/40: The spline domain $\mathbf{S} := U \times \mathcal{I}$ is embedded by φ into $\mathbb{R}$, and mapped by $\mathbf{x}$ to $\mathbb{R}^d$. The map $\mathbf{x}$ is a C^k-spline curve, if $\mathbf{x} \circ \varphi^{-1}$ is k-times continuously differentiable.

3.1 An Alternative View of Spline Curves

Usually, a spline curve $\mathbf{x}$ is viewed as an $\mathbb{R}^d$-valued function defined piecewise over the real line. The abscissae separating its segments are called *knots*. Between two consecutive knots, the spline curve coincides with a function of a fixed type, say a polynomial of a certain degree. At knots, the spline has to satisfy smoothness conditions.

To prepare our analysis of bivariate splines, and to motivate the setting to be developed then, we take an alternative view. Let $U := [0, 1]$ denote the unit interval, with $\boldsymbol{\kappa}_1 = 0, \boldsymbol{\kappa}_2 = 1$ its left and right endpoint.[1] Further, $\mathcal{I} := \mathbb{Z}$ is an index set enumerating *segments* $\mathbf{x}_i : U \to \mathbb{R}^d, i \in \mathcal{I}$, of the spline curve $\mathbf{x}$. The spline domain

$$\mathbf{S} := U \times \mathcal{I}$$

consists of indexed copies $(U, i) := U \times \{i\}$ of the unit interval (Fig. 3.1/40). We call these copies *cells*. Then $\mathbf{x}$ is a function of both the index $i \in \mathcal{I}$ indicating the segment and $u \in U$ parametrizing the segment,

$$\mathbf{x} : \mathbf{S} \ni (u, i) \mapsto \mathbf{x}_i(u) \in \mathbb{R}^d.$$

Let us assume that the different segments $\mathbf{x}_i$ are linked so that the right endpoint of the ith segment coincides with the left endpoint of the $(i+1)$st segment,

$$\mathbf{x}_i(\boldsymbol{\kappa}_2) = \mathbf{x}_{i+1}(\boldsymbol{\kappa}_1), \quad i \in \mathcal{I}. \tag{3.1}$$

Intuitively, matching values lead to a continuous spline curve. How can this be made precise? Instead of identifying function values, we identify domain endpoints,

[1] According to our conventions, bold greek characters are reserved for objects in $\mathbb{R}^2$. This and other deviations from our notational conventions are restricted to this section, and justified by the fact that we are preparing for the bivariate setting.

$$(\boldsymbol{\kappa}_2, i) = (\boldsymbol{\kappa}_1, i+1), \quad i \in \mathcal{I}. \tag{3.2}$$

The pairs of identified endpoints play the role of knots, i.e., a knot is an equivalence class of identified interval endpoints. By this definition, a knot $\{(\boldsymbol{\kappa}_2, i), (\boldsymbol{\kappa}_1, i+1)\}$ is a single argument, which must be assigned a single function value. Hence, (3.1/40) is not an imposed condition, but an elementary property, which we refer to as *consistency*.

To give $\mathbf{S}$ the structure of a topological space, we consider a subset $\mathbf{S}' \subset \mathbf{S}$. $\mathbf{S}'$ is defined to be *open* if and only if U_i', defined by $(U_i', i) := \mathbf{S}' \cap (U, i)$, is open for all $i \in \mathcal{I}$ in the natural topology of U. This process is a standard technique in topology, where it is called *gluing*. For instance, the set $\mathbf{S}' := \big((1/2, 1], 5\big) \cup \big([0, 1/2), 6\big)$ is open since $U_5' := (1/2, 1]$ and $U_6' := [0, 1/2)$ are open in $[0, 1]$, and all other U_i' are empty, hence open, too. But $\mathbf{S}' := \big((1/2, 1], 7\big)$ is not open since, by identification of end points, $U_8' := \{0\}$, which is not open but closed.

By defining open sets, $\mathbf{S}$ obtains the structure of a topological space. Continuity of $\mathbf{x}$ with respect to this topology is equivalent to continuity of all segments $\mathbf{x}_i$ together with consistency expressed by (3.1/40). To declare higher order smoothness, we define an *embedding* $\varphi : \mathbf{S} \to \mathbb{R}$ of the domain $\mathbf{S}$ into $\mathbb{R}$ as an injective, continuous, and real-valued spline. The simplest choice is

$$\varphi : \mathbf{S} \ni (u, i) \mapsto u + i \in \mathbb{R}.$$

Now, $\mathbf{x}$ is called a C^k-spline curve, if the composed map

$$\mathbf{x}_\varphi := \mathbf{x} \circ \varphi^{-1} : \mathbb{R} \to \mathbb{R}^d$$

is k-times continuously differentiable. It is easily shown that this is the case if and only if all segments are C^k, and if the values of the derivatives of two neighboring segments always agree at the common knot. The above choice of φ corresponds to the familiar class of uniform splines. Non-uniform splines correspond to linear segments of φ with arbitrary slopes, while non-linear segments of φ lead to the concept of geometric smoothness.

Obviously, the complexity of the setup described so far is excessive for uniform spline curves, but anyway, it is merely meant to be a motivation for the less straightforward bivariate setting. The requirements of an analysis of general spline surfaces necessarily lead to non-trivial topological concepts, any much of the early literature on the topic suffers from a lack of clean foundations.

3.2 Continuous Bivariate Splines

Analogous to the univariate case, we describe continuous bivariate splines in topological terms. Let $\mathcal{I} \subset \mathbb{Z}$ be a set of indices, and denote by

$$U := [0, 1] \quad \text{and} \quad \Sigma := U \times U$$

the unit interval in $\mathbb{R}$ and the unit square in $\mathbb{R}^2$, respectively. Then $\boldsymbol{\Sigma}$ has corners $\boldsymbol{\kappa}_\ell$ and *counterclockwise oriented edges* $\boldsymbol{\varepsilon}_\ell$, where the index $\ell \in \{1,2,3,4\}$ is always understood modulo 4,

$$\begin{aligned}
\boldsymbol{\kappa}_1 &:= (0,0)\,, & \boldsymbol{\varepsilon}_1(u) &:= (u,0) \\
\boldsymbol{\kappa}_2 &:= (1,0)\,, & \boldsymbol{\varepsilon}_2(u) &:= (1,u) \\
\boldsymbol{\kappa}_3 &:= (1,1)\,, & \boldsymbol{\varepsilon}_3(u) &:= (1-u,1) \\
\boldsymbol{\kappa}_4 &:= (0,1)\,, & \boldsymbol{\varepsilon}_4(u) &:= (0,1-u), \qquad u \in U,
\end{aligned} \tag{3.3}$$

see Fig. 3.2/43. The pair $(\boldsymbol{\Sigma}, i) := \boldsymbol{\Sigma} \times \{i\}, i \in \mathcal{I}$, is a *cell* and the union of all cells forms the *spline domain*

$$\mathbf{S} := \boldsymbol{\Sigma} \times \mathcal{I}.$$

A 'patch layout' is stamped on $\mathbf{S}$ by defining a *neighbor relation*

$$(\boldsymbol{\varepsilon}_\ell, i) \sim (\boldsymbol{\varepsilon}_{\ell'}, i')$$

on the set of edges $\{\boldsymbol{\varepsilon}_1, \ldots, \boldsymbol{\varepsilon}_4\} \times \mathcal{I}$, and identifying points on these edges according to[2]

$$(\boldsymbol{\varepsilon}_\ell(u), i) = (\boldsymbol{\varepsilon}_{\ell'}(1-u), i'), \quad u \in U. \tag{3.4}$$

The pointwise identification of edges induces an equivalence relation

$$(\boldsymbol{\varepsilon}_\ell, i) \sim (\boldsymbol{\varepsilon}_{\ell'}, i') \;\Rightarrow\; (\boldsymbol{\kappa}_\ell, i) = (\boldsymbol{\kappa}_{\ell'+1}, i')$$

on the set $\mathbf{K} := \{\boldsymbol{\kappa}_1, \ldots, \boldsymbol{\kappa}_4\} \times \mathcal{I}$ of corners. Each such equivalence class is called a *knot*. The number of elements in the equivalence class is the *valence* of a knot and denoted by n throughout the book. Pairs of related edges are called *knot lines*.

To avoid meaningless structures, we make the following assumptions on the neighbor relation.

- No pair of related edges belongs to the same cell, i.e., $i \neq i'$.
- For a spline domain *without boundary*, the valence n of each knot is finite and at least 3. Each knot line corresponds to two edges. That is, each edge is related to exactly one other edge.
- A spline domain $\mathbf{S}' = \boldsymbol{\Sigma} \times \mathcal{I}'$ *with boundary* is a proper subset of a spline domain $\mathbf{S}$ without boundary, endowed with the same rules for identification.

The pointwise identification (3.4/42) of the edges of the cells gives $\mathbf{S}$ the structure of a *topological space* via gluing as explained in the univariate setting. We always refer to this topology in the following. The elements of $\boldsymbol{\Sigma}$ and $\mathbf{S}$,

$$\boldsymbol{\sigma} = (s,t) \in \boldsymbol{\Sigma}, \quad \mathrm{s} = (\boldsymbol{\sigma}, i) = (s,t,i) \in \mathbf{S},$$

[2] The special choice of identification leads to orientable surfaces. A generalization of the concept would complicate the formal setup without yielding additional insight.

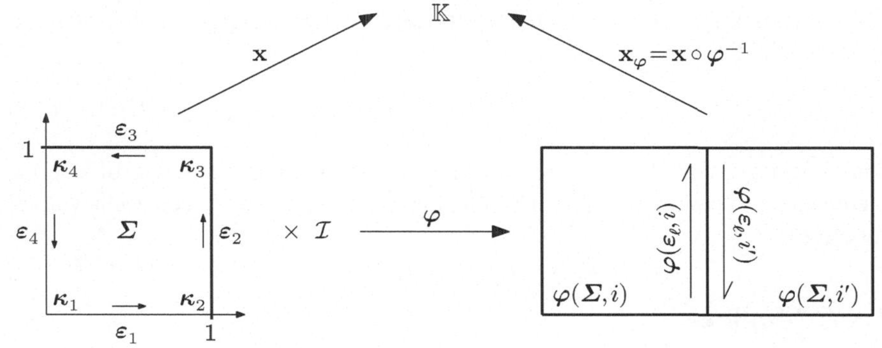

Fig. 3.2 Illustration of Definition 3.3/44: The spline domain $\mathbf{S} := \boldsymbol{\Sigma} \times \mathcal{I}$ is rigidly embedded by φ into $\mathbb{R}^2$ and mapped by $\mathbf{x}$ to $\mathbb{K}$. Gluing identifies knots $(\boldsymbol{\kappa}_\ell, i) = (\boldsymbol{\kappa}_{\ell'+1}, i')$ and $(\boldsymbol{\kappa}_{\ell+1}, i) = (\boldsymbol{\kappa}_{\ell'}, i')$, and edges $(\boldsymbol{\varepsilon}_\ell, i) \sim (\boldsymbol{\varepsilon}_{\ell'}, i')$ (see (3.4/42)).

respectively, are called *parameters*. To simplify notation later on, we use the convention that multiplication of s by a scalar applies only to its first component,

$$c\mathbf{s} = c(\boldsymbol{\sigma}, i) = (c\boldsymbol{\sigma}, i), \quad c \in \mathbb{R}. \tag{3.5}$$

Now we define splines as continuous functions over the domain $\mathbf{S}$.

Definition 3.1 (Spline). Let $\mathbf{S}$ be a spline domain and $\mathbb{K} = \mathbb{R}^d$ or $\mathbb{K} = \mathbb{C}^d$. A $\mathbb{K}$-valued *spline* is a continuous map[3]

$$\mathbf{x} : \mathbf{S} \to \mathbb{K}.$$

The restriction

$$\mathbf{x}_i : \boldsymbol{\Sigma} \ni \boldsymbol{\sigma} \mapsto \mathbf{x}(\boldsymbol{\sigma}, i) \in \mathbb{K}$$

of $\mathbf{x}$ to the cell $(\boldsymbol{\Sigma}, i)$ is called a *patch*.[4] If $\mathbb{K} = \mathbb{R}^3$, then $\mathbf{x}(\mathbf{S})$ is called a real *spline surface*. The set of all $\mathbb{K}$-valued splines defined on $\mathbf{S}$ is denoted by $C^0(\mathbf{S}, \mathbb{K})$.

The case $\mathbb{K} = \mathbb{R}$ corresponds to coordinate functions of higher dimensional splines. In particular, it will serve to construct systems of generating splines. The case $\mathbb{K} = \mathbb{R}^3$ corresponds to spline surfaces, as does $\mathbb{K} = \mathbb{R}^4$ for the homogeneous representation of rational surfaces. The case $\mathbb{K} = \mathbb{R}^2$, or equivalently $\mathbb{K} = \mathbb{C}$, occurs as a change of parameter and to represent the local structure of the spline domain.

---

[3] In the classical theory of spline functions, continuity is not required. However, in the context of spline *surfaces*, discontinuous parametrizations make little sense and are therefore excluded a priori.

[4] For the sake of generality, we do not assume that the patches are polynomials or belong to some other finite-dimensional function space. It is arguable if the object defined here should still be called 'spline'. However, we find it appropriate since the common characteristics of the many residents in the 'zoo of splines' is segmentation, which is at the core of our definition.

Corresponding to the pointwise identification of related edges, any two abutting patches $\mathbf{x}_i$ and $\mathbf{x}_{i'}$ of a spline $\mathbf{x}$ satisfy

$$(\varepsilon_\ell, i) \sim (\varepsilon_{\ell'}, i') \;\Rightarrow\; \mathbf{x}_i(\varepsilon_\ell(u)) = \mathbf{x}_{i'}(\varepsilon_{\ell'}(1-u)), \quad u \in U. \tag{3.6}$$

This property is called *consistency*, and it is easy to see that continuity of a spline is equivalent to continuity of its individual patches together with consistency at all knot lines.

3.3 C^k-Splines

As in the univariate case, differentiability properties are studied by embedding. That is, parts of the spline domain are mapped to $\mathbb{R}^2$ in order to construct a suitable reparametrization of the spline that can be analyzed by standard tools.

Definition 3.2 (Embedding). Let $\mathbf{S}$ be a spline domain and $\mathbf{S}' = \boldsymbol{\Sigma} \times \mathcal{I}'$ a subset inheriting the topology. Then $\mathbf{S}'$ is called a *sub-domain.* A spline $\varphi \in C^0(\mathbf{S}', \mathbb{R}^2)$ is an *embedding* of $\mathbf{S}'$ if φ is injective. If all patches of φ are rigid motions of the unit square $\boldsymbol{\Sigma}$ in $\mathbb{R}^2$, then φ is called a *rigid embedding.*

In other words, an embedding φ identifies a subset of $\mathbb{R}^2$ with a subset of $\mathbf{S}$. In the following, we will *use only rigid embeddings* to define smoothness of splines.[5]

Definition 3.3 (C^k-spline). Let $(\varepsilon_\ell, i) \sim (\varepsilon_{\ell'}, i')$ be a knot line, $\mathbf{S}' := (\boldsymbol{\Sigma}, i) \cup (\boldsymbol{\Sigma}, i')$ the corresponding pair of the abutting cells, and φ a rigid embedding of $\mathbf{S}'$. A spline $\mathbf{x} \in C^0(\mathbf{S}, \mathbb{K})$, $\mathbb{K} \in \{\mathbb{R}^d, \mathbb{C}^d\}$, is called C^k *on* $\mathbf{S}'$, if the composed map

$$\mathbf{x}_\varphi := \mathbf{x} \circ \varphi^{-1} : \mathbb{R}^2 \supset \varphi(\mathbf{S}') \to \mathbb{K}$$

is k-times continuously differentiable, see Fig. 3.2/43. $\mathbf{x}$ is a C^k-*spline*, if it is C^k on all pairs of abutting cells. The space of all C^k-splines on $\mathbf{S}$ with values in $\mathbb{K}$ is denoted by $C^k(\mathbf{S}, \mathbb{K})$.

Since $\mathbf{x}_\varphi$ is k-times continuously differentiable either for all or for no rigid embedding of two abutting cells, the definition of a C^k-spline is independent of the particular choice of a rigid φ. The following characterization of C^k-splines shows that the notion of smoothness introduced above is equivalent to the notion of parametric smoothness as it appears in the literature. That is, for a C^k-spline, appropriately chosen directional derivatives up to order k of abutting patches agree along the common edge up to sign. Below, differentiation in the direction of the edge ε_ℓ is expressed by means of the operator $\vec{D}_\ell$, defined by

$$\vec{D}_\ell \mathbf{x} := \frac{d}{du}\mathbf{x}(\cdot + \varepsilon_\ell(u))\big|_{u=0}.$$

---

[5] General embeddings can be used to formalize the concept of 'geometric smoothness' but are not needed here.

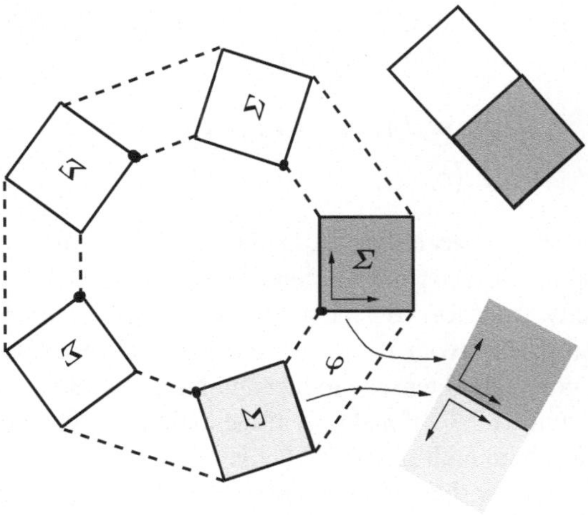

Fig. 3.3 Illustration of (3.9/45): Piecewise rigid embedding of the spline domain $\mathbf{S} := U \times \mathcal{I}$.

Lemma 3.4 (Derivatives at knot lines). *A spline* $\mathbf{x} \in C^0(\mathbf{S}, \mathbb{K})$, $\mathbb{K} \in \{\mathbb{R}^d, \mathbb{C}^d\}$, *is a* C^k*-spline if and only if the following holds:*

- *Each patch* $\mathbf{x}_i$ *is a* C^k*-function on* $\boldsymbol{\Sigma}$.
- *For all knot lines* $(\varepsilon_\ell, i) \sim (\varepsilon_{\ell'}, i')$ *and all* μ, ν *with* $0 \le \mu + \nu \le k$,

$$\vec{D}_\ell^\mu \vec{D}_{\ell+1}^\nu \mathbf{x}_i\big(\varepsilon_\ell(u)\big) = (-1)^{\mu+\nu} \vec{D}_{\ell'}^\mu \vec{D}_{\ell'+1}^\nu \mathbf{x}_{i'}\big(\varepsilon_{\ell'}(1-u)\big), \quad u \in U. \tag{3.7}$$

This system is referred to as the C^k-smoothness conditions.

Proof. Let $(\varepsilon_\ell, i) \sim (\varepsilon_{\ell'}, i')$ and $\mathbf{S}' := (\boldsymbol{\Sigma}, i) \cup (\boldsymbol{\Sigma}, i')$. Further, T denotes the translation by $(1,0)$ and

$$R = \begin{bmatrix} 0 & -1 \\ 1 & 0 \end{bmatrix} \tag{3.8}$$

the planar rotation by the angle $\pi/2$ about the origin. Then we define a rigid embedding of $\mathbf{S}'$ by

$$\varphi(\boldsymbol{\sigma}, \iota) = \begin{cases} \boldsymbol{\sigma}\,(RT)^{-\ell} & \text{if } \iota = i, \\ \boldsymbol{\sigma}\,(RT)^{1-\ell'} R & \text{if } \iota = i'. \end{cases} \tag{3.9}$$

In particular, $\varphi(\mathbf{S}') = [-1, 1] \times U$ and

$$\varphi\big(\varepsilon_\ell(u)\big) = \varphi\big(\varepsilon_{\ell'}(1-u)\big) = (0, 1-u), \quad u \in U,$$

as illustrated in Fig. 3.3/45. The equivalence of k-fold continuous differentiability of $\mathbf{x} \circ \varphi^{-1}$ on $\{0\} \times [0,1]$ and condition (3.7/45) follows from the chain rule. □

As an important special case of (3.7/45), the first order derivatives along and transversal to a common edge satisfy the C^1-conditions

$$\begin{aligned}\vec{D}_\ell \mathbf{x}_i\big(\varepsilon_\ell(u)\big) &= -\vec{D}_{\ell'} \mathbf{x}_{i'}\big(\varepsilon_{\ell'}(1-u)\big)\\ \vec{D}_{\ell+1} \mathbf{x}_i\big(\varepsilon_\ell(u)\big) &= -\vec{D}_{\ell'+1} \mathbf{x}_{i'}\big(\varepsilon_{\ell'}(1-u)\big).\end{aligned} \tag{3.10}$$

Parametric smoothness according to Definition 3.3/44 is not necessary for the smoothness of a spline surface from a differential geometric point of view; however, despite its simplicity, it suffices to generate free-form splines of arbitrary topology, i.e., splines whose patch layout is not restricted to a checkerboard pattern. As a generic example, polynomial tensor-product splines of coordinate degree d are C^k in the sense of Definition 3.3/44 if and only if the underlying univariate splines have equally spaced knots with multiplicity $\leq d-k$.

Definition 3.3/44 implies that the elementary algebraic properties of the space of k-times continuously differentiable functions in $\mathbb{K}$ are inherited by the spline space $C^k(\mathbf{S},\mathbb{K})$. First, $C^k(\mathbf{S},\mathbb{K})$ is a *linear space* and *affine invariant*, i.e., closed under affine transformations of the image space $\mathbb{K}$. Second, $C^k(\mathbf{S},\mathbb{K})$ with coordinate-wise multiplication is a *commutative ring with unit*; the latter is the spline that is constant equal 1 in all coordinates.

Since the construction of continuous splines on the spline domain is straightforward, we now consider at least once continuously differentiable splines. In other words, unless explicitly specified otherwise, in the following

$$\text{we assume an order of differentiability } k \geq 1. \tag{3.11}$$

It should be noted that it does not make sense to apply the operators $\vec{D}_\ell$ to a spline because, typically, inconsistencies would arise at knot lines. By contrast, the *partial cross product operator*

$${}^{\times}\!D = D_1 \times D_2,$$

as introduced in Sect. 2.1/16, is well defined on the space of C^1-splines. More precisely, if $\mathbf{x}$ is a C^k-spline, then the patches ${}^{\times}\!D\mathbf{x}_i$ satisfy the smoothness conditions (3.7/45) up to order $k-1$.

Lemma 3.5 (Smoothness of ${}^{\times}\!D\mathbf{x}$). For $d \in \{2,3\}$, let $\mathbf{x} \in C^k(\mathbf{S},\mathbb{R}^d)$. Then the map

$${}^{\times}\!D\mathbf{x} : \mathbf{S} \ni (\boldsymbol{\sigma}, i) \mapsto {}^{\times}\!D\mathbf{x}_i(\boldsymbol{\sigma}) \in \begin{cases}\mathbb{R} & \text{if } d=2\\ \mathbb{R}^3 & \text{if } d=3\end{cases}$$

is a C^{k-1}-spline, i.e., ${}^{\times}\!D\mathbf{x} \in C^{k-1}(\mathbf{S},\mathbb{R}^{2d-3})$.

Proof. Obviously, ${}^{\times}\!D = \vec{D}_\ell \times \vec{D}_{\ell+1}$ for all $\ell = \{1,\ldots,4\}$. Let $(\varepsilon_\ell, i) \sim (\varepsilon_{\ell'}, i')$ be a knot line. For $\mu + \nu < k$, the product rule yields

$$\vec{D}_\ell^{\mu} \vec{D}_{\ell+1}^{\nu}\, {}^{\times}\!D\mathbf{x}_i = \sum_{m=0}^{\mu} \sum_{n=0}^{\nu} \binom{\mu}{m}\binom{\nu}{n} \vec{D}_\ell^{m+1} \vec{D}_{\ell+1}^{\nu-n} \mathbf{x}_i \times \vec{D}_\ell^{\mu-m} \vec{D}_{\ell+1}^{n+1} \mathbf{x}_i$$

and equally

$$\vec{D}^{\mu}_{\ell'}\vec{D}^{\nu}_{\ell'+1}{}^{\times}\!D\mathbf{x}_{i'} = \sum_{m=0}^{\mu}\sum_{n=0}^{\nu}\binom{\mu}{m}\binom{\nu}{n}\vec{D}^{m+1}_{\ell'}\vec{D}^{\nu-n}_{\ell'+1}\mathbf{x}_{i'} \times \vec{D}^{\mu-m}_{\ell'}\vec{D}^{n+1}_{\ell'+1}\mathbf{x}_{i'}.$$

If we evaluate these expressions at $\varepsilon_\ell(u)$ and $\varepsilon_{\ell'}(1-u)$, respectively, then Lemma 3.4/45 shows that all corresponding cross products agree since the factors are equal up to the necessary multiplier $(-1)^{\mu+\nu}$. □

Having cross products of partial derivatives at our disposal, regularity of a spline surface can be defined just as in Definition 2.1/18 for ordinary surfaces.

Definition 3.6 (Regular surface). A C^1-spline surface $\mathbf{x}$ is called *regular*, if ${}^{\times}\!D\mathbf{x}(\mathbf{s}) \neq \mathbf{0}$ for all $\mathbf{s} \in \mathbf{S}$.

For a regular spline surface $\mathbf{x}$, the *Gauss map*

$$\mathbf{n} : \mathbf{S} \ni \mathbf{s} \mapsto \frac{{}^{\times}\!D\mathbf{x}(\mathbf{s})}{\|{}^{\times}\!D\mathbf{x}(\mathbf{s})\|} \in \mathbb{R}^3$$

is well defined. This follows immediately from Lemma 3.5/46. An equally valid argument would be to employ a rigid embedding φ of abutting cells. The resulting surface $\mathbf{x}\circ\varphi^{-1}$ has a well defined normal vector. By invariance of the normal vector according to Theorem 2.2/18, the patches equally have a well defined normal vector which satisfies the consistency condition. Just in the same way, one can show using Theorem 2.5/22 that for a regular C^2-spline surface $\mathbf{x}$ the *embedded Weingarten map*

$$\mathbf{W} : \mathbf{S} \ni (\sigma, i) \mapsto \mathbf{W}_i(\sigma) \in \mathbb{R}^{3\times 3}$$

is well defined as a spline in $C^0(\mathbf{S}, \mathbb{R}^{3\times 3})$, where $\mathbf{W}_i$ denotes the embedded Weingarten map of the patch $\mathbf{x}_i$.

3.4 C^k_r-Splines

As stated earlier, C^1-continuity of splines is not necessary for geometric smoothness, but is assumed for convenience. Conversely, C^1-continuity is not sufficient to imply geometric smoothness due to possible linear dependencies of the partial derivatives. Such dependencies can occur in any affine invariant space of surface parametrizations; for example, if all coordinate functions are equal. In applications this is typically considered a curiosity rather than a problem, but for C^k-splines with extraordinary knots, singularities in the parametrization are a necessity, as Lemma 3.7/48 below will show. Therefore, following Sect. 2.3/23, we now derive a concept of *geometric smoothness* that applies to spline surfaces with an isolated singularity at a knot.

Let us consider an interior knot of valence n. Since, for the forthcoming analysis of subdivision surfaces, the continuity properties of interest are local, we can restrict the analysis to the n cells that share the knot. With $\mathbb{Z}_n$ the integers modulo n, we define the *local domain* of n cells by

$$\mathbf{S}_n := \boldsymbol{\Sigma} \times \mathbb{Z}_n. \tag{3.12}$$

A spline $\mathbf{x}$ on $\mathbf{S}_n$ is composed of the patches

$$\mathbf{x}_0, \ldots, \mathbf{x}_{n-1}.$$

Without loss of generality, we assume that the neighbor relation has the cyclic structure

$$(\varepsilon_4, j) \sim (\varepsilon_1, j+1), \quad j \in \mathbb{Z}_n. \tag{3.13}$$

For brevity, we denote the *central knot* by

$$\mathbf{0} := (\mathbf{0}, 0) = \cdots = (\mathbf{0}, n-1),$$

and its image by

$$\mathbf{x}^{\mathrm{c}} := \mathbf{x}(\mathbf{0}) = \mathbf{x}_0(\mathbf{0}) = \cdots = \mathbf{x}_{n-1}(\mathbf{0}).$$

The following lemma points to the trade-off between geometric and parametric singularities.

Lemma 3.7 (Forced singularities). *If* $\mathbf{x} \in C^1(\mathbf{S}_n, \mathbb{R}^3)$ *and* $n \neq 4$ *then either*

$$^{\times}D\mathbf{x}_0(0,0) = \cdots = {}^{\times}D\mathbf{x}_{n-1}(0,0) = \mathbf{0},$$

or $\mathbf{x}$ *has no injective projection into the tangent plane at* $\mathbf{x}^{\mathrm{c}}$.

Proof. Since directional and partial derivatives are related by

$$\vec{D}_1 = D_1, \quad \vec{D}_4 = -D_2,$$

the smoothness conditions (3.7/45) on $\mathbf{S}_n$ now read

$$D_1^{\mu} D_2^{\nu} \mathbf{x}_j(0,u) = (-1)^{\mu} D_1^{\nu} D_2^{\mu} \mathbf{x}_{j+1}(u,0), \tag{3.14}$$

where

$$\nu + \mu \leq k, \quad u \in U := [0,1], \quad j \in \mathbb{Z}_n.$$

For $\mu = \nu = 0$, this yields the consistency conditions

$$\mathbf{x}_j(0,u) = \mathbf{x}_{j+1}(u,0).$$

For the first order derivatives, with $R = [0,-1;1,0]$ as in (3.8/45),

$$D\mathbf{x}_j(0,u) = R\, D\mathbf{x}_{j+1}(u,0),$$

so that at $u = 0$

$$D\mathbf{x}_j(0,0) = R\,D\mathbf{x}_{j+1}(0,0) = \cdots = R^4\,D\mathbf{x}_{j+4}(0,0) = D\mathbf{x}_{j+4}(0,0)$$

since $R^4 = \mathbb{1}$ is the identity. The sequence $D\mathbf{x}_j(0,0)$ is therefore 4-periodic with respect to the index j. However, by construction, it is n-periodic as well. For $n \neq 4$, this implies that for any $j \in \mathbb{Z}_n$ there exists an index $j' \neq j$ such that $D\mathbf{x}_j(0,0) = D\mathbf{x}_{j'}(0,0)$. If the matrix $D\mathbf{x}_j(0,0) = D\mathbf{x}_{j'}(0,0)$ has full rank, then the implicit function theorem shows that the projections of the patches $\mathbf{x}_j$ and $\mathbf{x}_{j'}$ to the tangent plane at $\mathbf{x}^c$ overlap. □

Lemma 3.7/48 motivates the following definition.

Definition 3.8 (Ordinary and extraordinary knot). A knot with valence $n = 4$ is called *ordinary*; a knot with valence $n \neq 4$ is called *extraordinary*.

Spline surfaces with all ordinary knots are well understood and need not be discussed in the following. In the extraordinary case, $n \neq 4$, Lemma 3.7/48 shows that we either have an undesirable multi-sheeted surface or a C^k-spline surface with a singular parametrization at the central knot. In the latter case, we cannot deduce geometric smoothness of the surface from the analytic smoothness of the parametrization. *Therefore, there is no point in requiring the spline to be differentiable at an extraordinary knot.*

Parts of the material presented now are almost verbatim transcriptions from Sect. 2.3/23. The difference is that parameters are now not points in a subset $\boldsymbol{\Sigma}$ of $\mathbb{R}^2$, but in the spline domain $\mathbf{S}_n = \boldsymbol{\Sigma} \times \mathbb{Z}_n$. Since the proofs need not be carried out again, arrows in the headings of definitions and theorems point to their counterparts in the preceding chapter.

Definition 3.9 (C^k_0-spline → Definition 2.8/24). A spline $\mathbf{x} \in C^0(\mathbf{S}_n, \mathbb{K})$ is called C^k_0 if it is C^k everywhere except for the central knot. That is, using the notation of Definition 3.3/44, the composed map

$$\mathbf{x}_\varphi := \mathbf{x} \circ \varphi^{-1}$$

is k-times continuously differentiable on $\varphi(\mathbf{S}')\backslash\varphi(\mathbf{0})$. The space of all such functions is denoted by $C^k_0(\mathbf{S}_n, \mathbb{K})$. The image

$$\mathbf{x}^c := \mathbf{x}(\mathbf{0})$$

is called the *central point* of $\mathbf{x}$. For $\mathbb{K} \in \{\mathbb{R}^2, \mathbb{R}^3, \mathbb{C}\}$, the spline $\mathbf{x} \in C^k_0(\mathbf{S}_n, \mathbb{K})$ is called *almost regular* if ${}^{\times}\!D\mathbf{x}(\mathbf{s}) \neq 0$ for all $\mathbf{s} \in \mathbf{S}_n\backslash\{\mathbf{0}\}$.

Obviously, the relaxation of smoothness requirements at the central knot implies slight modifications of the results given in Lemmas 3.4/45 and 3.5/46. In particular,

- The smoothness conditions (3.14/48) do no longer apply at $u = 0$ if $\nu + \mu > 0$
- If $\mathbf{x} \in C^k_0(\mathbf{S}_n, \mathbb{R}^d)$, then ${}^{\times}\!D\mathbf{x} \in C^{k-1}_0(\mathbf{S}_n, \mathbb{R}^3)$

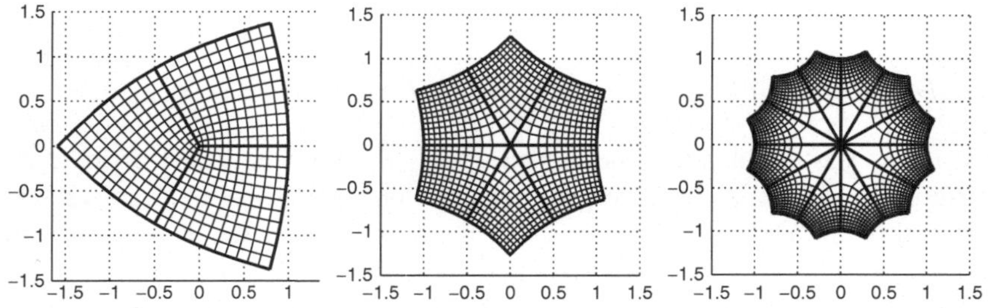

Fig. 3.4 Illustration of Example 3.10/50: Fractional power embedding for *(left)* $n = 3$, *(middle)* $n = 6$, and *(right)* $n = 12$.

The following example of an almost regular function on $\mathbf{S}_n$ does not only illustrate the definition above, but will later on also serve as a canonical embedding of $\mathbf{S}_n$.

Example 3.10 (Fractional power embedding). Let $w_n := \exp(2\pi\mathbf{i}/n)$ denote the primitive nth root of unity. We define the complex-valued spline

$$p : \mathbf{S}_n \ni (s,t,j) \mapsto w_n^j (s + \mathbf{i}t)^{4/n} \in \mathbb{C} \tag{3.15}$$

and its real-valued equivalent by

$$\boldsymbol{\pi} := [\mathrm{Re}\, p, \mathrm{Im}\, p], \tag{3.16}$$

see Fig. 3.4/50. Obviously, p is continuous, and C^∞ away from the origin, i.e., $p \in C_0^\infty(\mathbf{S}_n, \mathbb{C})$. Its Jacobian determinant is given by

$$^\times\!Dp(s,t,j) := {}^\times\!D\boldsymbol{\pi}(s,t,j) = \frac{16}{n^2}\,(s^2+t^2)^{4/n-1},$$

showing that p is almost regular. Further, p is injective so that it is an embedding of the whole domain $\mathbf{S}_n$ in the sense of Definition 3.2/44. Due to its definition via the complex power function, we call p and also $\boldsymbol{\pi}$ the *fractional power embedding* of $\mathbf{S}_n$.

Now, we consider a closed piecewise differentiable curve $\mathbf{c} : U := [0,1] \to \mathbf{S}_n$ which does not contain the origin, $\mathbf{0} \notin \mathbf{c}(U)$, and the curve $z := p \circ \mathbf{c} : U \to \mathbb{C}$ in the complex plane. If there exists a sequence of break-points $0 = u_0 < u_1 < \cdots < u_n = 1$ such that

$$\mathbf{c}(u) \in (\boldsymbol{\Sigma}, j), \quad u \in [u_j, u_{j+1}],$$

for all $j \in \mathbb{Z}_n$, then the segments $z([u_{j-1}, u_j])$ lie in the sectors $p((\boldsymbol{\Sigma}, j))$. Since p is injective, $z(U)$ does not contain the origin, and Lemma 2.20/36 easily yields the winding number

$$\nu(z, 0) = 1.$$

□

Typically, it is impossible to define a normal vector of a C_0^k-spline surface at the central point in a meaningful way. However, in special situations it may still be possible. The following definition addresses such a case.

Definition 3.11 (Normal continuity and single-sheetedness → Definitions 2.9/24, 2.11/25). An almost regular spline surface $\mathbf{x} \in C^1_0(\mathbf{S}_n, \mathbb{R}^3)$ is *normal continuous* if the limit

$$\mathbf{n}^c := \lim_{\mathbf{s}\to\mathbf{0}} \mathbf{n}(\mathbf{s}),$$

called the *central normal*, exists. For a pair $\mathbf{T}^c := [\mathbf{t}^c_1; \mathbf{t}^c_2]$ of orthonormal vectors in the tangent plane, $\mathbf{F}^c := [\mathbf{T}^c; \mathbf{n}^c]$ is an orthogonal (3×3)-matrix, called the *central frame*. The transformed spline

$$\mathbf{x}_* := (\mathbf{x} - \mathbf{x}^c) \cdot \mathbf{F}^c$$

has the *tangential component*

$$\boldsymbol{\xi}_* := (\mathbf{x} - \mathbf{x}^c) \cdot \mathbf{T}^c \in C^k_0(\mathbf{S}_n, \mathbb{R}^2)$$

and the *normal component*

$$z_* := (\mathbf{x} - \mathbf{x}^c) \cdot \mathbf{n}^c \in C^k_0(\mathbf{S}_n, \mathbb{R}).$$

If there exists an open connected neighborhood $\mathbf{S}_* \subset \mathbf{S}_n$ of the central knot such that $\boldsymbol{\xi}_*$ restricted to $\mathbf{S}_*$ is injective, then $\mathbf{x}$ is called *single-sheeted* at the central point. If $\mathbf{x}$ is single-sheeted, then the *local height function* h_* is defined by

$$h_* : \Xi_* \ni \boldsymbol{\xi} \mapsto z_*\big(\boldsymbol{\sigma}_*(\boldsymbol{\xi})\big) \in \mathbb{R},$$

where the domain is $\Xi_* := \boldsymbol{\xi}_*(\mathbf{S}_*) \subset \mathbb{R}^2$, and $\boldsymbol{\sigma}_* : \Xi_* \to \mathbf{S}_*$ is the local inverse of $\boldsymbol{\xi}_*$.

In other words, $\mathbf{x}$ is single-sheeted, if the tangential component $\boldsymbol{\xi}_*$ is an embedding. In this case, the spline surface can locally be represented with the help of the local height function h_*,

$$\mathbf{x}(\mathbf{s}) = \tilde{\mathbf{x}}(\boldsymbol{\xi}) = \mathbf{x}^c + \boldsymbol{\xi}\mathbf{T}^c + h_*(\boldsymbol{\xi})\mathbf{n}^c, \quad \boldsymbol{\xi} \in \Xi_*, \tag{3.17}$$

where $\mathbf{s}$ and $\boldsymbol{\xi}$ are related by $\mathbf{s} = \boldsymbol{\sigma}_*(\boldsymbol{\xi})$ and $\boldsymbol{\xi} = \boldsymbol{\xi}_*(\mathbf{s})$. The *local Euler form* is defined by

$$\bar{\mathbf{x}}(\boldsymbol{\xi}) := \big(\tilde{\mathbf{x}}(\boldsymbol{\xi}) - \mathbf{x}^c\big) \cdot \mathbf{F}^c = [\boldsymbol{\xi}, h^c(\boldsymbol{\xi})],$$

which is no longer a spline, but a standard surface, as considered in the preceding chapter. We note that both $\tilde{\mathbf{x}}$ and $\bar{\mathbf{x}}$ are always regular so that, locally, we can identify smoothness properties of $\mathbf{x}$ and h_*.

Definition 3.12 (C^k_r-spline surface → Definition 2.12/26). Let $\mathbf{x} \in C^k_0(\mathbf{S}_n, \mathbb{R}^3)$ be normal continuous and single-sheeted. Then $\mathbf{x}$ is called a *C^k_r-spline surface* if the local height function h_* is r-times continuously differentiable in a neighborhood of the origin.

Following exactly the same arguments as in Sect. 2.3/23, it is possible to relate convergence properties of the Gauss map and the embedded Weingarten map to the regularity of a spline surface at the central point.

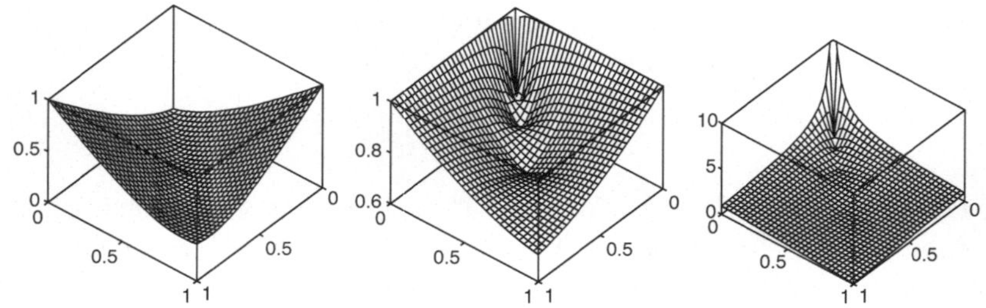

Fig. 3.5 Illustration of Example 3.14/52: The gradient near the central knot is (*left*) continuous with limit $\mathbf{0}$; (*middle*) not continuous but bounded, and (*right*) divergent.

Theorem 3.13 (Conditions for C_1^k- and C_2^k-spline surfaces → Theorems 2.13/27, 2.14/28). *If a spline surface* $\mathbf{x} \in C_0^k(\mathbf{S}_n, \mathbb{R}^3)$ *is normal continuous and single-sheeted, then it is* C_1^k*. Further, if a spline surface* $\mathbf{x} \in C_1^k(\mathbf{S}_n, \mathbb{R}^3)$, $k \geq 2$, *is* curvature continuous *in the sense that the limit*

$$\mathbf{W}^{\mathrm{c}} := \lim_{\mathbf{s}\to\mathbf{0}} \mathbf{W}(\mathbf{s})$$

exists, then $\mathbf{x}$ *is* C_2^k.

As in the proof of Theorem 2.13/27, the inverse $\boldsymbol{\sigma}_*$ of $\boldsymbol{\xi}_*$ is continuous with

$$\lim_{\boldsymbol{\xi}\to\mathbf{0}} \boldsymbol{\sigma}_*(\boldsymbol{\xi}) = \boldsymbol{\sigma}_*(\mathbf{0}) = \mathbf{0},$$

and the local height function satisfies

$$h_*(\mathbf{0}) = 0, \quad Dh_*(\mathbf{0}) = \mathbf{0}.$$

The notion of C_r^k-smoothness is particularly well suited for analyzing subdivision surfaces. We illustrate this claim by the following example:

Example 3.14 (Gradients near the central knot). Drawing on the discussion of generalized biquadratic subdivision in Sect. 6.2/116, we consider three variants that all yield C_1^1-surfaces. Figure 3.5/52 shows the norm of the gradient of scalar patches generated by these algorithms. On the left, the subdominant eigenvalue is $\lambda = 0.3$ and the gradient vanishes at the origin. In the middle, $\lambda = 0.5$ and the gradient is bounded but discontinuous. On the right, $\lambda = 0.8$ and the gradient diverges. □

Checking projections of almost regular spline surfaces for injectivity is facilitated by the concept of winding numbers, as introduced in Sect. 2.4/31. Since the definitions in the preceding chapter do not immediately apply to closed curves in $\mathbf{S}_n$, we employ an embedding $\boldsymbol{\pi}$ of the domain $\mathbf{S}_n$, for instance the fractional power embedding $\boldsymbol{\pi}$ according to Example 3.10/50. On one hand, the reparametrization $\boldsymbol{\xi}_* \circ \boldsymbol{\pi}^{-1}$ is an almost regular function defined in a neighborhood of the origin in $\mathbb{R}^2$. On the other hand, if $\mathbf{c} : U = [0, 1] \to \mathbf{S}_n$ is a piecewise differentiable closed curve in the

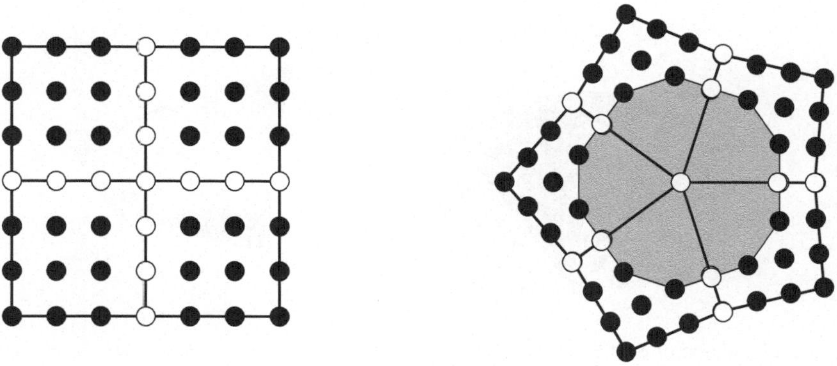

Fig. 3.6 Illustration of Sect. 3.5/53: Free control points • and dependent control points ◦ for (*left*) the ordinary case and (*right*) the extraordinary case. The control points bounding the *shaded region* must form a convex planar polygon.

spline domain, then $\boldsymbol{\pi} \circ \mathbf{c}$ is a piecewise differentiable closed curve in $\mathbb{R}^2$. Recalling that Definition 2.16/32 can be equally applied to functions, curves, and points in $\mathbb{R}^2$ using the standard identification with $\mathbb{C}$, we note that Theorem 2.19/34 can be applied to the function $\boldsymbol{\xi}_* \circ \boldsymbol{\pi}^{-1}$ when considering the winding numbers $\nu(\boldsymbol{\pi} \circ \mathbf{c}, 0)$ and $\nu(\boldsymbol{\xi}_* \circ \boldsymbol{\pi}^{-1}, \boldsymbol{\pi} \circ \mathbf{c}, 0)$. For the first one, we write briefly

$$\nu(\mathbf{c}, \mathbf{0}) := \nu(\boldsymbol{\pi} \circ \mathbf{c}, 0), \tag{3.18}$$

while, by definition, the second one is just

$$\nu(\boldsymbol{\xi}_* \circ \boldsymbol{\pi}^{-1}, \boldsymbol{\pi} \circ \mathbf{c}, 0) = \nu(\boldsymbol{\xi}_* \circ \mathbf{c}, 0).$$

Theorem 3.15 (Single-sheetedness via winding number). *Let* $\mathbf{c} : U = [0,1] \to \mathbf{S}_n$ *be a piecewise differentiable Jordan curve with* $\nu(\mathbf{c}, \mathbf{0}) = 1$. *Then the normal continuous spline surface* $\mathbf{x} \in C^1_0(\mathbf{S}_n, \mathbb{R}^3)$ *is single-sheeted if and only if*

$$|\nu(\boldsymbol{\xi}_* \circ \mathbf{c}, 0)| = 1.$$

Proof. Since $\boldsymbol{\pi}$ is injective, $\boldsymbol{\xi}_*$ is injective if and only if $\boldsymbol{\xi}_* \circ \boldsymbol{\pi}$ is injective. By Theorem 2.19/34, the latter function is injective if and only if $|\nu(\boldsymbol{\xi}_* \circ \mathbf{c}, 0)| = 1$. □

3.5 A Bicubic Illustration

We illustrate the framework developed so far with a concrete construction taken from [Rei97]. The example presumes some basic knowledge of Bézier techniques, see [Far97, PBP02] for an introduction (Figs. 3.6/53 and 3.7/54).

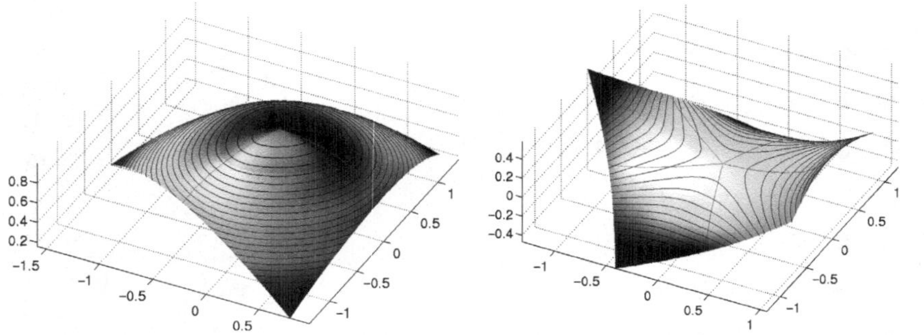

Fig. 3.7 Illustration of Sect. 3.5/53: Bicubic C^1_1-surfaces for (*left*) $n = 3$ and (*right*) $n = 5$ with patch boundaries (*red*) and level lines (*black*).

For $n \neq 4$, we consider the domain $\mathbf{S}_n$ provided with the cyclic neighbor relation (3.13/48),

$$(\varepsilon_4, j) \sim (\varepsilon_1, j+1), \quad j \in \mathbb{Z}_n.$$

The patches

$$\mathbf{x}_j(s,t) := \sum_{\nu=0}^{3} \sum_{\mu=0}^{3} b^3_\nu(s) b^3_\mu(t) \mathbf{q}^j_{\nu,\mu}$$

of the spline surface $\mathbf{x}$ are bicubic polynomials with *control points* $\mathbf{q}^j_{\nu,\mu} \in \mathbb{R}^3$. The corresponding basis functions are given as products of cubic Bernstein polynomials,

$$b^3_\nu(u) := \binom{3}{\nu}(1-u)^{3-\nu}u^\nu, \quad u \in U,\ \nu = 0,\dots,3.$$

Consistency of $\mathbf{x}$ requires that control points corresponding to related edges coincide,

$$\mathbf{q}^j_{0,\nu} = \mathbf{q}^{j+1}_{\nu,0}.$$

Further, $\mathbf{x}$ is a C^1*-spline surface* if and only if the control points on edges are the midpoints of neighboring interior control points,

$$2\mathbf{q}^j_{0,\nu} = \mathbf{q}^j_{1,\nu} + \mathbf{q}^{j+1}_{\nu,1}. \tag{3.19}$$

This implies that the full set of control points can be determined from the subset

$$\{\mathbf{q}^j_{\nu,\mu} : j \in \mathbb{Z}_n, (\nu,\mu) \in \{1,2,3\}^2\}$$

by a simple averaging process. However, since the conditions for $\nu, \mu \leq 1$ are coupled, the points $\mathbf{q}^j_{1,1}$ cannot be chosen freely. A simple argument shows that the sequence of these points is both n-periodic and 4-periodic. Single-sheetedness is possible only if we resort to the trivial solution

$$\mathbf{q}^0_{1,1} = \dots = \mathbf{q}^{n-1}_{1,1},$$

yielding coalescing control points according to

$$\mathbf{x}^{\mathrm{c}} = \mathbf{q}^j_{0,0} = \mathbf{q}^j_{1,0} = \mathbf{q}^j_{0,1} = \mathbf{q}^j_{1,1}, \quad j \in \mathbb{Z}_n.$$

As predicted, the collapse of control points at the extraordinary knot implies a singular parametrization of all patches at $\mathbf{x}^{\mathrm{c}}$. Typically, the so-constructed spline surfaces are continuous and almost regular, but not normal continuous. However, C^1_1*-spline surfaces* can still be generated if some additional constraints on the control points $\mathbf{q}^j_{1,2}, \mathbf{q}^j_{2,1}$ are satisfied. For instance, it suffices to require that the points

$$\mathbf{q}^0_{2,1}, \mathbf{q}^0_{1,2}, \mathbf{q}^1_{2,1}, \mathbf{q}^2_{1,2}, \ldots, \mathbf{q}^{n-1}_{2,1}, \mathbf{q}^{n-1}_{1,2}$$

are the corners of a convex planar polygon which also contains the central point $\mathbf{x}^{\mathrm{c}}$. The proof, which is non-trivial, can be found in [BR97].

Bibliographic Notes

1. The very general definition of a spline as a continuous function on a domain consisting of cells is new. The importance of endowing the domain with a topological structure was, however, already noted explicitly in [Rei93], as well as in [Zor97, Zor00b].

2. Reif's thesis [Rei93] and later [Rei95c] contain the observation that normal continuity alone is not sufficient for smoothness of a spline in the sense of differential geometry.

3. Singular parameterizations have been used to build smooth piecewise polynomial surfaces with arbitrary patch layout [Pet91, BR97, Rei95a, Rei97, Pra97, NP94]. [BR97] even features singularly parametrized Bézier patches forming a C^2_2-spline surface. Singularity was also explicitly leveraged in the construction of TURBS [Rei96b, Rei98].

4. We learned about the elegant fractional power embedding (3.10/50) from David Levin in 1993.

5. As mentioned in the footnote on page 44, non-rigid embeddings can be used to formalize the concept of geometric continuity. An overview on that topic is given in [Pet02b].

Chapter 4
Subdivision Surfaces

Subdivision derives its name from a splitting of the domain. A spline $\mathbf{x}$ on the initial domain $\mathbf{S}$ is mapped to a finer domain $\tilde{\mathbf{S}}$ where it is represented by more, smaller pieces. This chapter focuses on such refinement, in particular near extraordinary knots. We will not yet discuss specific algorithms.

Section 4.1/58 motivates the framework of subdivision by formalizing the refinement of spline domains: the basic step is to replaced each cell of the given domain by four new ones. In Sect. 4.2/59, we study a special reparametrization of splines, which is facilitated by iterated domain refinement. If exactly one of the corners of the initial square is an extraordinary knot, one of the four new cells inherits this knot while the other three, which have only ordinary knots, combine to an L-shape. Accordingly, the initial surface patch is split into a smaller patch with an extraordinary point, and an L-shaped *segment*. Repeating the refinement for the new extraordinary patch yields another patch and another segment of even smaller size. If this process is iterated ad infinitum, the initial patch is eventually replaced by a sequence of smaller and smaller segments, and the extraordinary point itself. If we consider a spline surface $\mathbf{x}$ consisting of n patches $\mathbf{x}_1, \ldots, \mathbf{x}_n$ sharing a common central point $\mathbf{x}^c$, always n segments at refinement level m form an annular piece of surface $\mathbf{x}^m$, called a *ring*. As illustrated by Fig. 4.3/61 (top), the sequence of rings is nested, and contracts towards the central point $\mathbf{x}^c$. The representation of a spline as the union of rings and a central point is called a *spline in subdivision form*. Thus, spline surfaces in subdivision form, as they are generated by many popular algorithms, can be understood by analyzing this sequence. In particular, the conditions for continuity, smoothness and single-sheetedness can all be reduced to conditions on rings.

In Sect. 4.3/65, we represent a ring $\mathbf{x}^m = G\mathbf{Q}^m$ in terms of a vector $\mathbf{Q}^m$ of *coefficients* $\mathbf{q}_\ell^m \in \mathbb{R}^d$ and a vector G of *generating rings* g_ℓ. Typically, we think of $\mathbf{q}_\ell^m$ as points in 3-space. But $\mathbf{q}_\ell^m$ can just as well represent derivative data, or color and texture information so that the setup conveniently covers a very general setting. In many practical algorithms, the generating rings are built from box-splines and form a basis. We emphasize, however, that we assume neither that the generating rings are piecewise polynomial nor that they are linearly independent. Joining the rings $\mathbf{x}^m = G\mathbf{Q}^m$, we obtain the representation of the spline $\mathbf{x} := B\mathbf{Q}$ as a linear combination of *generating splines* b_ℓ.

In Sect. 4.4/67, subdivision algorithms are characterized as recursions for rings. The recursion is governed by a *subdivision matrix*. Since the subdivision matrix is applied over and over again, it is natural to introduce at this point notational and algebraic tools: the asymptotic equivalence of expansions in Sect. 4.5/71 and the Jordan decomposition of matrices in Sect. 4.6/72. In particular, the subdivision matrix is decomposed into $A = VJV^{-1}$, where V is a matrix of eigenvectors and generalized eigenvectors. Correspondingly, we introduce *eigenrings* $F = GV$ and *eigensplines* $E = BV$. In Sect. 4.7/75, we can then relate properties of the subdivision matrix to properties of the limit surface. In the process, we see examples of the insufficiency of an analysis based solely on the control points. For example, so-called ineffective eigenvectors have to be removed before we can claim the leading eigenvalue of the subdivision matrix needs to be 1.

4.1 Refinability

A major feature of C_0^k-splines is their refinability, i.e., the fact that a C_0^k-spline surface $\mathbf{x}(\mathbf{S})$ can be reproduced by a spline surface $\tilde{\mathbf{x}}(\tilde{\mathbf{S}})$ with a finer patch structure. The splitting of patches corresponds to a refinement of the cells of the given spline domain. Every cell $(\boldsymbol{\Sigma}, i)$ is split into four new cells $(\boldsymbol{\Sigma}, 4i+1), \ldots, (\boldsymbol{\Sigma}, 4i+4)$, as illustrated in Fig. 4.1/59.

Definition 4.1 (Refined domain). Let $\mathbf{S} := \boldsymbol{\Sigma} \times \mathcal{I}$ be a spline domain. The *refined domain* $\tilde{\mathbf{S}} = \Sigma \times \tilde{\mathcal{I}}$ is characterized as follows:

- The new index set is $\tilde{\mathcal{I}} := \{4i + \ell : i \in I,\ \ell = 1, \ldots, 4\}$.
- The original and the new domain are linked by the isomorphism

$$\mathbf{r} : \tilde{\mathbf{S}} \ni (s, t, 4i + \ell) \mapsto \begin{cases} (s/2, t/2, i) & \text{for } \ell = 1 \\ (\bar{s}/2, t/2, i) & \text{for } \ell = 2 \\ (\bar{s}/2, \bar{t}/2, i) & \text{for } \ell = 3 \\ (s/2, \bar{t}/2, i) & \text{for } \ell = 4 \end{cases} \in \mathbf{S}, \tag{4.1}$$

  where $\bar{s} := 1 + s, \bar{t} := 1 + t$, see Fig. 4.1/59.
- The neighbor relation on $\tilde{\mathbf{S}}$ is defined by

$$(\varepsilon_\ell, i) \sim (\varepsilon_{\ell'}, i') \iff \mathbf{r}(\varepsilon_\ell(u), i) = \mathbf{r}(\varepsilon_{\ell'}(1-u), i'),\ u \in U.$$

We note that all four new cells are of standard size. They are not shrunk to quarters or the like. Rather, the process represents a *topological split*. This split is expressed by means of the *inverse* of $\mathbf{r}$ because, according to the above definition, $\mathbf{r}$ describes the merging of always four refined cells. To motivate the framework of the following sections, we briefly discuss some further properties of domain refinement. Every

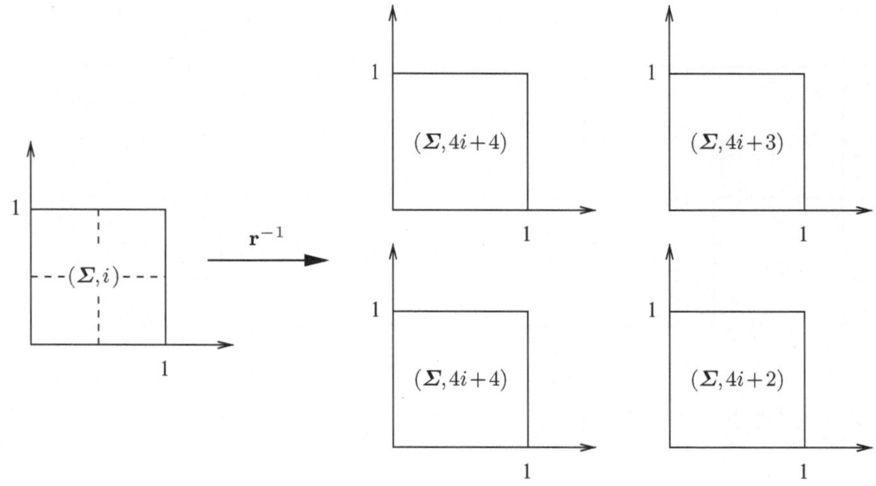

Fig. 4.1 Illustration of (4.1/58): Refinement of a cell.

knot of valence n in $\mathbf{S}$ is mapped to a knot of valence n in $\tilde{\mathbf{S}}$. All new knots in $\tilde{\mathbf{S}}$ have valence 4. Hence, the number and valences of extraordinary knots are fixed under refinement, while the number of ordinary knots is increasing. When refinement is iterated, the extraordinary knots become more and more separated in the sense that every extraordinary knot is surrounded by more and more layers of ordinary cells that contain exclusively ordinary knots. The algorithms that we are going to consider in the following are based on a fixed number of neighboring cells. The non-trivial part of the analysis concerns the long term behavior of iterated refinement. In this respect, any finite number of initial refinement steps is irrelevant (see, however, Example 8.10/173). As a consequence, the limit behavior of the algorithms will not depend on the interplay between various extraordinary knots, but only on the characteristics of the situation near an extraordinary knot of valence n. For this reason, we will confine our analysis to splines with one isolated extraordinary knotof valence n. That is, in the following, $\mathbf{x}$ is always assumed to be defined on the local domain $\mathbf{S}_n = \Sigma \times \mathbb{Z}_n$, as defined in (3.12/48).

4.2 Segments and Rings

In this section, we define a *spline in subdivision form* as a special representation of a standard spline using segments and rings as building blocks. Further, we relate properties of splines, such as normal continuity or single-sheetedness, with properties of these objects.

To fix ideas, let us consider a spline surface $\mathbf{x} : \mathbf{S}_n \to \mathbb{R}^3$. If we apply one step of refinement, as introduced in the preceding section, to the local domain $\mathbf{S}_n$, we

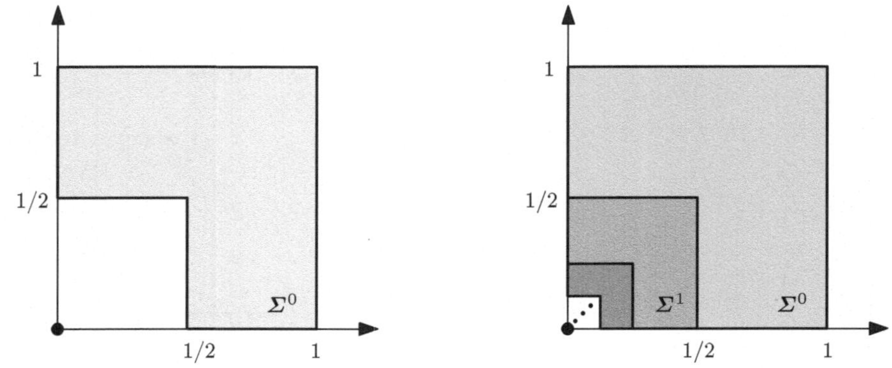

Fig. 4.2 Illustration of (4.2/60): Partitioning of Σ into scaled copies of Σ^0.

obtain a new domain with $4n$ cells: $3n$ outer cells that are ordinary, and n inner cells sharing the central knot. Due to its annular shape, the restriction $\mathbf{x}^0$ of $\mathbf{x}$ to the outer cells is called a *ring*. The n L-shaped parts of the ring corresponding to the initial patches are called *segments*. We keep the ring $\mathbf{x}$ and its domain and repeat the refinement process only for the inner part to obtain a second ring $\mathbf{x}^1$ and an even smaller inner part. In this way, iterated refinement generates a sequence of rings, which eventually covers all of the surface with the exception of the central point $\mathbf{x}^c$ itself. The alternative representation of $\mathbf{x}$ in terms of rings and segments is called its subdivision form. More precisely, we subdivide Σ and $\mathbf{S}_n$ as follows (cf. Fig. 4.2/60). Let

$$\Sigma^0 := [0,1]^2 \backslash [0,1/2)^2, \quad \Sigma^m := 2^{-m}\Sigma^0, \quad \mathbf{S}_n^m := \Sigma^m \times \mathbb{Z}_n, \quad m \in \mathbb{N}_0. \tag{4.2}$$

Then

$$\Sigma = \bigcup_{m\in\mathbb{N}_0} \Sigma^m \cup \{\mathbf{0}\}, \quad \mathbf{S}_n = \bigcup_{m\in\mathbb{N}_0} \mathbf{S}_n^m \cup \{\mathbf{0}\}. \tag{4.3}$$

We note that all $\mathbf{S}_n^m$, $m \in \mathbb{N}_0$, are just scaled copies of the prototype $\mathbf{S}_n^0$, which we call the *ring domain*. Points in $\mathbf{S}_n^0$ corresponding to knot lines in $\mathbf{S}_n$ are identified in the natural way,

$$(0,u,j) = (u,0,j+1), \quad u \in [1/2,1],\ j \in \mathbb{Z}_n.$$

Definition 4.2 (Segment and ring). Let $\mathbf{x} \in C_0^k(\mathbf{S}_n,\mathbb{K})$, $\mathbb{K} \in \{\mathbb{R}^d,\mathbb{C}^d\}$, be a spline with patches $\mathbf{x}_j$. For $m \in \mathbb{N}_0$ and $j \in \mathbb{Z}_n$, the *segment* $\mathbf{x}_j^m$ is defined by

$$\mathbf{x}_j^m : \Sigma^0 \ni \sigma \mapsto \mathbf{x}_j(2^{-m}\sigma), \tag{4.4}$$

and the *ring* $\mathbf{x}^m$ is defined by

$$\mathbf{x}^m : \mathbf{S}_n^0 \ni \mathbf{s} \mapsto \mathbf{x}(2^{-m}\mathbf{s}), \tag{4.5}$$

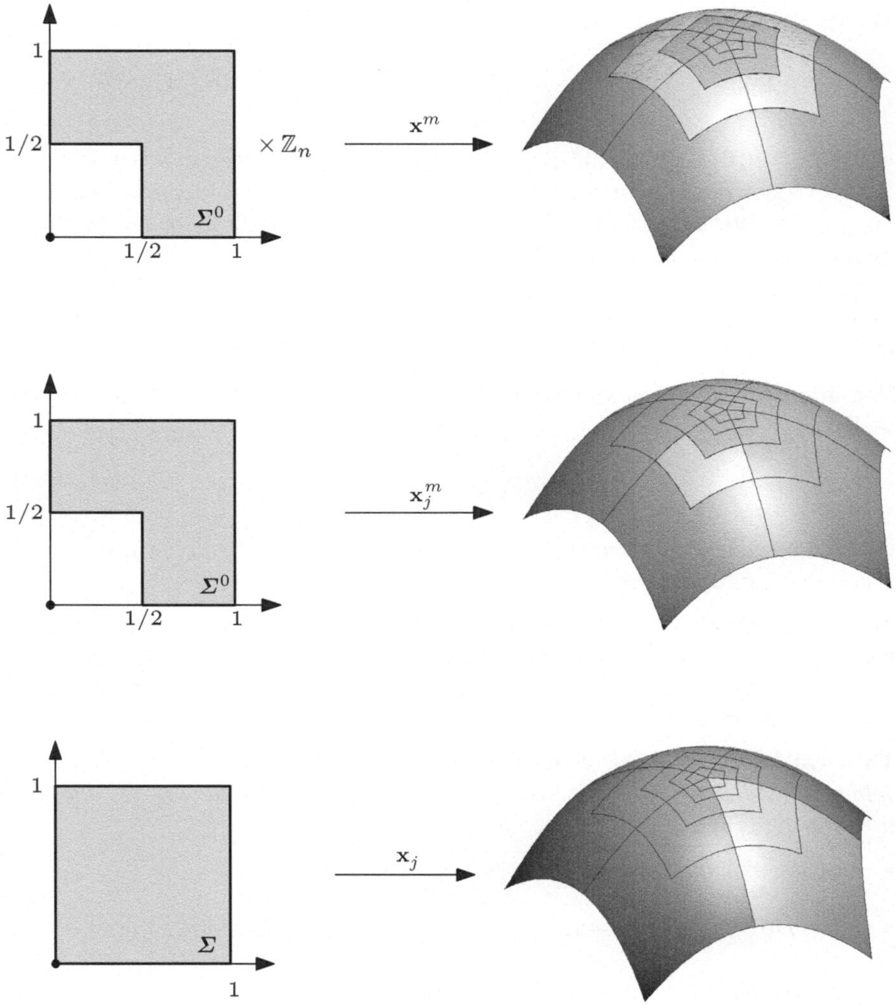

Fig. 4.3 Illustration of Definition 4.2/60: (*top*) Ring $\mathbf{x}^m$, (*middle*) segment $\mathbf{x}_j^m$, and (*bottom*) patch $\mathbf{x}_j$ of a subdivision surface $\mathbf{x}$. The rings $\mathbf{x}^0, \mathbf{x}^1, \ldots$ form the spline $\mathbf{x}$; the patches $\mathbf{x}_0, \ldots, \mathbf{x}_{n-1}$ form the spline $\mathbf{x}$; the segments $\mathbf{x}_j^0, \mathbf{x}_j^1, \ldots$ form the patch $\mathbf{x}_j$; the segments $\mathbf{x}_0^m, \ldots, \mathbf{x}_{n-1}^m$ form the ring $\mathbf{x}^m$.

where, recalling (3.5/43), $2^{-m}\mathbf{s} = 2^{-m}(\boldsymbol{\sigma}, j) = (2^{-m}\boldsymbol{\sigma}, j)$. The space of all C^k-rings is denoted by $C^k(\mathbf{S}_n^0, \mathbb{K})$.

The segment $\mathbf{x}_j^m$ corresponds to the restriction of the patch $\mathbf{x}_j$ to the set $\boldsymbol{\Sigma}^m$, i.e., $\mathbf{x}_j^m(\boldsymbol{\Sigma}^0) = \mathbf{x}_j(\boldsymbol{\Sigma}^m)$, see Fig. 4.3/61. The re-scaling of arguments facilitates the use of a common domain for all m. Analogously, the ring $\mathbf{x}^m$ corresponds to the restriction of the spline $\mathbf{x}$ to the set $\mathbf{S}_n^m$, i.e., $\mathbf{x}^m(\mathbf{S}_n^0) = \mathbf{x}(\mathbf{S}_n^m)$. With (4.3/60), this implies

$$\mathbf{x}_j(\boldsymbol{\Sigma}) = \bigcup_{m\in\mathbb{N}_0} \mathbf{x}_j^m(\boldsymbol{\Sigma}^0)\cup\{\mathbf{x}^c\},\quad \mathbf{x}(\mathbf{S}_n) = \bigcup_{m\in\mathbb{N}_0} \mathbf{x}^m(\mathbf{S}_n^0)\cup\{\mathbf{x}^c\},$$

where $\mathbf{x}^c = \mathbf{x}(\mathbf{0})$ is the central point. An alternate way of looking at it is as follows. With $\tilde{\mathbf{x}} = \mathbf{x}\circ\mathbf{r}$ the refinement of $\mathbf{x}$, the first ring $\mathbf{x}^1$ corresponds to $\tilde{\mathbf{x}}$ restricted to its $3n$ ordinary patches. Accordingly, the mth ring $\mathbf{x}^m$ corresponds to the $3n$ innermost ordinary patches of the m-fold refined spline $\mathbf{x}\circ\mathbf{r}^m$.

The partition of a spline into rings and segments leads to the notion of subdivision. It refers to a special way of representing splines rather than to a new class of objects.

Definition 4.3 (Spline in subdivision form). The spline $\mathbf{x} \in C_0^k(\mathbf{S}_n, \mathbb{R}^d)$ represented as

$$\mathbf{x} : \mathbf{S}_n \ni \mathbf{s} \mapsto \begin{cases} \mathbf{x}^m(2^m\mathbf{s}) & \text{if} \quad \mathbf{s}\in\mathbf{S}_n^m \\ \mathbf{x}^c & \text{if} \quad \mathbf{s} = \mathbf{0} \end{cases} \tag{4.6}$$

is called a *spline in subdivision form with ring sequence* $\{\mathbf{x}^m\}_m$. For $d = 3$, $\mathbf{x}$ is also called a *subdivision surface.*

The principal task of analyzing splines in subdivision form is to deduce their geometric properties from analytic properties of the ring sequence. Being parts of a C_0^k-spline $\mathbf{x}$, the segments $\mathbf{x}_j^m$ fulfill certain consistency and differentiability conditions, both inside a ring and between consecutive rings.

Theorem 4.4 (Smoothness conditions for segments). *Let* $\mathbf{x} \in C_0^k(\mathbf{S}_n, \mathbb{R}^d)$ *be a spline in subdivision form with segments* $\mathbf{x}_j^m$*. Then all segments are* C^k *and, for* $u \in U = [0,1]$ *and* $\mu+\nu\le k$*, we have:*

- *All pairs of neighboring segments* $\mathbf{x}_j^m, \mathbf{x}_{j+1}^m$ *satisfy*

$$D_1^\mu D_2^\nu \mathbf{x}_j^m(0, 1/2+u/2) = (-1)^\mu D_1^\nu D_2^\mu \mathbf{x}_{j+1}^m(1/2+u/2, 0), \tag{4.7}$$

- *All pairs of consecutive segments* $\mathbf{x}_j^m, \mathbf{x}_j^{m+1}$ *satisfy*

$$\begin{aligned} D_1^\mu D_2^\nu \mathbf{x}_j^m(1/2, u/2) &= 2^{\mu+\nu} D_1^\mu D_2^\nu \mathbf{x}_j^{m+1}(1, u) \\ D_1^\mu D_2^\nu \mathbf{x}_j^m(u/2, 1/2) &= 2^{\mu+\nu} D_1^\mu D_2^\nu \mathbf{x}_j^{m+1}(u, 1). \end{aligned} \tag{4.8}$$

Proof. The first equality follows from (3.14/48), while the other one is a straightforward application of the chain rule. □

In particular, for $\mu = \nu = 0$, we obtain the consistency conditions

$$\begin{aligned} \mathbf{x}_j^m(0, 1/2+u/2) &= \mathbf{x}_{j+1}^m(1/2+u/2, 0) \\ \mathbf{x}_j^m(1/2, u/2) &= \mathbf{x}_j^{m+1}(1, u) \\ \mathbf{x}_j^m(u/2, 1/2) &= \mathbf{x}_j^{m+1}(u, 1), \end{aligned} \tag{4.9}$$

and $\mu + \nu = 1$ yields with $R = [0, -1; 1, 0]$

$$\begin{aligned} D\mathbf{x}_j^m(0, 1/2 + u/2) &= RD\mathbf{x}_{j+1}^m(1/2 + u/2, 0) \\ D\mathbf{x}_j^m(1/2, u/2) &= 2D\mathbf{x}_j^{m+1}(1, u) \\ D\mathbf{x}_j^m(u/2, 1/2) &= 2D\mathbf{x}_j^{m+1}(u, 1). \end{aligned} \tag{4.10}$$

The question arises whether the conditions given in the last theorem are also sufficient for a set of segments to form a C_0^k-spline $\mathbf{x}$. Obviously, these conditions guarantee $\mathbf{x}$ to be C^k on all of $\mathbf{S}_n$, except for the extraordinary knot $\mathbf{0}$. According to Definition 3.9/49, no differentiability conditions apply at that point, but consistency still has to be ensured.

Theorem 4.5 (From C^k-rings to C_0^k-splines). *Let $\{\mathbf{x}^m\}_m$ be a sequence of rings in $C^k(\mathbf{S}_n^0, \mathbb{R}^d)$ with segments $\mathbf{x}_j^m$ satisfying* (4.7/62) *and* (4.8/62). *If, in addition, there exists a point $\mathbf{x}^c \in \mathbb{R}^d$ such that, for any $j \in \mathbb{Z}_n$ and any sequence $\{\boldsymbol{\sigma}^m\}_m$ in $\boldsymbol{\Sigma}^0$,*

$$\mathbf{x}^c = \lim_{m \to \infty} \mathbf{x}_j^m(\boldsymbol{\sigma}^m),$$

then these rings form a C_0^k-spline in subdivision form (4.6/62).

Proof. The only non-trivial part of the proof concerns continuity at the extraordinary knot. Let $\{\mathbf{s}^\ell = (\boldsymbol{\sigma}^\ell, j^\ell)\}_\ell$ be any sequence in $\mathbf{S}_n$ converging to $\mathbf{0}$. Then this sequence can be partitioned into subsequences, indexed by $\ell_j^m(\mu)$, with $\boldsymbol{\sigma}^{\ell_j^m(\mu)} \in \boldsymbol{\Sigma}^m$ and $j^{\ell_j^m(\mu)} = j$. By assumption, for all such subsequences,

$$\lim_{\mu \to \infty} \mathbf{x}(\mathbf{s}^{\ell_j^m(\mu)}) = \mathbf{x}^c$$

so that

$$\lim_{\ell \to \infty} \mathbf{x}(\mathbf{s}^\ell) = \mathbf{x}^c.$$

□

Now, we are going to transfer the results on normal continuity and single-sheetedness of splines, as derived in the preceding chapter, to the subdivision setting. Because the properties in question are local, we can slightly relax the assumptions on almost regularity. It is no longer required on all of $\mathbf{S}_n$, but only in a vicinity of the central knot.

Definition 4.6 (Local almost regularity). A spline $\mathbf{x} \in C_0^1(\mathbf{S}_n, \mathbb{R}^d), d \in \{2, 3\}$, is *locally almost regular* if there exists $m_0 \in \mathbb{N}_0$ such that

$$^{\times}\!D\mathbf{x}(\mathbf{s}) \neq \mathbf{0} \quad \text{for all} \quad \mathbf{s} \in \bigcup_{m \geq m_0} \mathbf{S}_n^m.$$

In this case, we also say that $\mathbf{x}$ is m_0-*almost regular*.

In the following, we denote by $\mathbf{n}^m$ and $\mathbf{n}_j^m$ the rings and the segments of the Gauss map $\mathbf{n}$ of the spline surface $\mathbf{x}$, respectively. Of course, in case of local almost regularity, existence is ensured only for $m \geq m_0$.

Theorem 4.7 (Normal continuity of subdivision surfaces). *A subdivision surface* $\mathbf{x} \in C_0^k(\mathbf{S}_n, \mathbb{R}^3)$ *is*

- m_0*-almost regular, if the rings* $\mathbf{x}^m$ *are regular for* $m \geq m_0$*, i.e.,*

$$^{\times}\!D\mathbf{x}^m(\mathbf{s}) \neq \mathbf{0} \quad \textit{for all} \quad \mathbf{s} \in \mathbf{S}_n^0, \quad m \geq m_0;$$

- *normal continuous, if it is* m_0*-almost regular, and there exists a vector* $\mathbf{n}^c \in \mathbb{R}^3$*, called the* central normal, *such that for all* $j \in \mathbb{Z}_n$ *and any sequence* $\{\boldsymbol{\sigma}^m\}_{m \geq m_0}$ *in* $\boldsymbol{\Sigma}^0$

$$\mathbf{n}^c = \lim_{m \to \infty} \mathbf{n}_j^m(\boldsymbol{\sigma}^m).$$

Proof. Using the same arguments concerning subsequences as in the previous theorem, the results follow from Definitions 3.9/49 and 3.11/51. □

Following Definition 3.11/51, if $\mathbf{x}$ is normal continuous, we denote by $\boldsymbol{\xi}_*^m$ and z_*^m the rings of the tangential and the normal component of $\mathbf{x}_* = [\boldsymbol{\xi}_*, z_*]$, respectively,

$$\begin{aligned} \boldsymbol{\xi}_*^m &:= (\mathbf{x}^m - \mathbf{x}^c) \cdot \mathbf{T}^c \\ z_*^m &:= (\mathbf{x}^m - \mathbf{x}^c) \cdot \mathbf{n}^c. \end{aligned} \tag{4.11}$$

We can use the sequence $\boldsymbol{\xi}_*^m$ to formulate a criterion for single-sheetedness by specializing the curve $\mathbf{c}$ used in Theorem 3.15/53 to a curve $\mathbf{c} : U \to \mathbf{S}_n^0$ in the domain of rings. With $\boldsymbol{\pi}^0$ denoting the initial ring of the fractional power embedding (3.16/50), we define the winding number

$$\ell(\mathbf{c}, \mathbf{0}) := \ell(\boldsymbol{\pi}^0 \circ \mathbf{c}, 0), \quad \mathbf{0} \notin \mathbf{c}(U),$$

just as in (3.18/53).

Theorem 4.8 (Single-sheetedness of subdivision surfaces). *Let the subdivision surface* $\mathbf{x} \in C_0^1(\mathbf{S}_n, \mathbb{R}^3)$ *be* m_0*-almost regular and normal continuous. Further, let* $\mathbf{c} : U = [0,1] \to \mathbf{S}_n^0$ *be a piecewise differentiable Jordan curve with* $\ell(\mathbf{c}, \mathbf{0}) = 1$*. Then* $\mathbf{x}$ *is single-sheeted if and only if*

$$|\ell(\boldsymbol{\xi}_*^{m_0} \circ \mathbf{c}, 0)| = 1.$$

Proof. The rings

$$\tilde{\mathbf{x}}^m := \mathbf{x}^{m+m_0}, \quad m \in \mathbb{N}_0$$

define a spline surface $\tilde{\mathbf{x}} \in C_0^1(\mathbf{S}_n, \mathbb{R}^3)$ which is almost regular and normal continuous. In fact, it is nothing but a reparametrized version of the regular part of $\mathbf{x}$. Then $\boldsymbol{\xi}_*^{m_0} = \tilde{\boldsymbol{\xi}}_*^0$ implies

$$|\ell(\boldsymbol{\xi}_*^{m_0} \circ \mathbf{c}, 0)| = |\ell(\tilde{\boldsymbol{\xi}}_*^0 \circ \mathbf{c}, 0)| = |\ell(\tilde{\boldsymbol{\xi}}_* \circ \mathbf{c}, 0)|.$$

From Theorem 3.15/53, we conclude that $\mathbf{x}_*$, and hence also $\mathbf{x}$, is single-sheeted if and only if this number equals 1. □

4.3 Splines in Finite-Dimensional Subspaces

Due to the refinement property, any C_0^k-spline on $\mathbf{S}_n^0$ can be represented as the union of rings. Subdivision algorithms in the literature, as well as a large class of generalizations, are distinguished by a further property: all rings belong to a common linear function space that is spanned by a finite set of real-valued functions. In many cases of practical relevance, these functions are obtained as an appropriate arrangement of B-splines, and in particular, they are linearly independent. In the forthcoming analysis, however, we explicitly drop such assumptions for the sake of generality: the *generating rings* spanning the space of rings can be linearly dependent, and they need not be piecewise polynomial or the like. Thus, subdivision algorithms generalizing box spline subdivision, interpolatory subdivision, and many other generalizations are covered.

Definition 4.9 (Generating rings). Let

$$G := [g_0, \ldots, g_{\bar{\ell}}], \quad g_\ell \in C^k(\mathbf{S}_n^0, \mathbb{R}),\ \ell = 0, \ldots, \bar{\ell},$$

be a row-vector of scalar-valued rings that form a partition of unity,

$$\sum_{\ell=0}^{\bar{\ell}} g_\ell(\mathbf{s}) = 1, \quad \mathbf{s} \in \mathbf{S}_n^0. \tag{4.12}$$

Then G is a C^k*-system of generating rings* g_ℓ, spanning the function space

$$C^k(\mathbf{S}_n^0, \mathbb{K}, G) := \left\{ \mathbf{x}^m = \sum_{\ell=0}^{\bar{\ell}} g_\ell \mathbf{q}_\ell : \mathbf{q}_\ell \in \mathbb{K} \right\} \subset C^k(\mathbf{S}_n^0, \mathbb{K}), \tag{4.13}$$

see Fig. 4.4/66.

Of course, adding linearly dependent functions to a given system of generating rings does not change the corresponding space of rings. However, the class of subdivision surfaces which can be generated by linear stationary algorithms, as defined below, becomes substantially richer. Example 4.14/70 will illustrate this fact. Moreover, many box splines that give rise to subdivision algorithms lead to linearly dependent generating rings in a natural way.

We endow the finite-dimensional linear function space $C^k(\mathbf{S}_n^0, \mathbb{R}^d, G)$ with the max-norm

$$\|\mathbf{x}^m\|_\infty := \max_{\mathbf{s} \in \mathbf{S}_n^0} \|\mathbf{x}^m(\mathbf{s})\|,$$

where $\|\cdot\|$ denotes the Euclidean norm in $\mathbb{R}^d$. Limits of sequences of rings are always understood with respect to this norm. For instance, the condition of Theorem 4.5/63 for consistency and continuity of a sequence of rings is equivalent to

$$\lim_{m \to \infty} \mathbf{x}^m = \mathbf{x}^c. \tag{4.14}$$

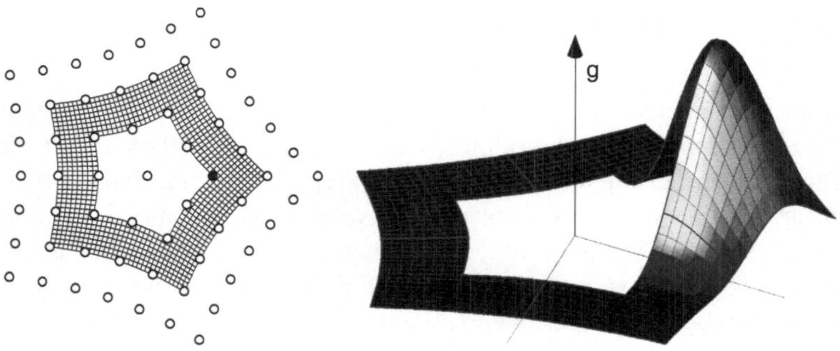

Fig. 4.4 Illustration of Definition 4.9/65: (*left*) Embedding of $\mathbf{S}_n^0$ and natural arrangement of B-spline coefficients for bicubic rings. (*right*) Generating ring g_ℓ corresponding to the coefficient •, plotted above the embedding.

Here, the point $\mathbf{x}^c \in \mathbb{R}^d$ on the right hand side is identified with the constant ring

$$\mathbf{x}^c = \sum_{\ell=0}^{\bar\ell} g_\ell \mathbf{x}^c \in C^k(\mathbf{S}_n^0, \mathbb{R}^d, G) \tag{4.15}$$

via (4.12/65). A ring $\mathbf{x}^m \in C^k(\mathbf{S}_n^0, \mathbb{R}^d, G)$ is written as a sum or in matrix notation as

$$\mathbf{x}^m = \sum_{\ell=0}^{\bar\ell} g_\ell \mathbf{q}_\ell^m = G\mathbf{Q}^m.$$

The points $\mathbf{q}_\ell^m \in \mathbb{R}^d$ are called the *coefficients*[1] of $\mathbf{x}^m$. These coefficients are collected in an $((\bar\ell+1)\times d)$-matrix[2] $\mathbf{Q}^m := [\mathbf{q}_0^m; \ldots; \mathbf{q}_{\bar\ell}^m]$ to obtain the short expression $\mathbf{x}^m = G\mathbf{Q}^m$. Specifying arguments, we have

$$\mathbf{x}^m(\mathbf{s}) = \mathbf{x}_j^m(\boldsymbol{\sigma}) = G(\mathbf{s})\mathbf{Q}^m, \quad \mathbf{s} = (\boldsymbol{\sigma}, j) \in \mathbf{S}_n^0.$$

As usual, the fact that the generating rings form a partition of unity implies *affine invariance* of the given representation. The following observation is as simple as it is important since it ensures a free choice of coordinates when investigating sequences of rings.

Theorem 4.10 (Affine invariance of rings). *With a $(d \times d')$-matrix $\mathbf{H}$ and a vector $\mathbf{h} \in \mathbb{R}^{d'}$, define the affine map $\mathbf{L} : \mathbb{R}^d \to \mathbb{R}^{d'}$ by $\mathbf{L}(\mathbf{p}) := \mathbf{pH} + \mathbf{h}$. Accordingly, $\mathbf{L}(\mathbf{Q}^m) := [\mathbf{L}(\mathbf{q}_0^m); \ldots; \mathbf{L}(\mathbf{q}_{\bar\ell}^m)]$. Then*

$$\mathbf{L}(G\mathbf{Q}^m) = G\mathbf{L}(\mathbf{Q}^m). \tag{4.16}$$

---

[1] In applications, the coefficients $\mathbf{q}_\ell^m$ are also called *control points*. However, this word is misleading here since the coefficients do not necessarily have a geometric meaning as they do have, for instance, in the B-spline setup.

[2] Instead of regarding $\mathbf{Q}^m$ as a matrix of reals, it is also convenient to think of it as a column vector of points in $\mathbb{R}^d$.

Proof. Using (4.15/66), we obtain

$$\mathbf{L}(G\mathbf{Q}^m) = \left(\sum_{\ell=0}^{\bar{\ell}} g_\ell \mathbf{q}_q^m\right)\mathbf{H} + \mathbf{h} = \sum_{\ell=0}^{\bar{\ell}} g_\ell(\mathbf{q}_\ell^m \mathbf{H} + \mathbf{h}) = G\mathbf{L}(\mathbf{Q}^m).$$

□

4.4 Subdivision Algorithms

A subdivision algorithm is an iterative prescription for generating a sequence of rings starting from some initial data (cf. Fig. 4.5/68). This definition is in sync with the earlier definition of subdivision surfaces as the union of a sequence of rings and their limit, but is in contrast with the commonly held view of subdivision algorithms as rules for refining spatial nets.

Definition 4.11 (Subdivision algorithm, preliminary). Let A be a square matrix with all rows summing up to 1, and G a system of generating rings of according size. Given a column vector $\mathbf{Q} = [\mathbf{q}_0; \dots; \mathbf{q}_{\bar{\ell}}]$ of coefficients $\mathbf{q}_\ell \in \mathbb{R}^d$, we define the corresponding sequence of rings

$$\mathbf{x}^m := G\mathbf{Q}^m, \quad \mathbf{Q}^m := A^m\mathbf{Q}, \quad m \in \mathbb{N}_0,$$

in $C^k(\mathbf{S}_n^0, \mathbb{R}^d, G)$. If, for any choice of $\mathbf{Q}$, this sequence satisfies the assumptions of Theorem 4.5/63, i.e., if the sequence constitutes a spline $\mathbf{x} \in C_0^k(\mathbf{S}_n, \mathbb{R}^d)$ in subdivision form, then (A, G) defines a (linear stationary C^k-)*subdivision algorithm*. Because $\mathbf{x}^0 = G\mathbf{Q}$, the coefficients $\mathbf{Q}$ are also called the *initial data of* $\mathbf{x}$.

This definition is preliminary because the conditions on A are not specific enough to exclude unwanted situations, such as non-smooth contact between consecutive rings or spurious components in the spectrum. At the end of this chapter, the final version will be presented in Definition 4.27/80.

Of course, generalizations are conceivable. For instance, one could consider non-stationary algorithms, where the sequence of coefficients is defined recursively by $\mathbf{Q}^{m+1} = A(m)\mathbf{Q}^m$ with a sequence $\{A(m)\}_m$ of matrices. If $A(m)$ has a repeating pattern then all results apply to the finite product of matrices that generates the pattern. Even algorithms of the form $\mathbf{Q}^{m+1} = f^m(\mathbf{Q}^m)$, using non-linear functions f^m, are conceivable. However, our understanding of such algorithms is still very limited, and most algorithms currently in use fit Definition 4.11/67.

Definition 4.11/67 imposes conditions on the matrix A in conjunction with G.

- Consecutive rings $\mathbf{x}^m$ and $\mathbf{x}^{m+1}$ must join smoothly according to (4.8/62). Typically, this is easy to achieve by using the regular subdivision rules of the underlying spline space.

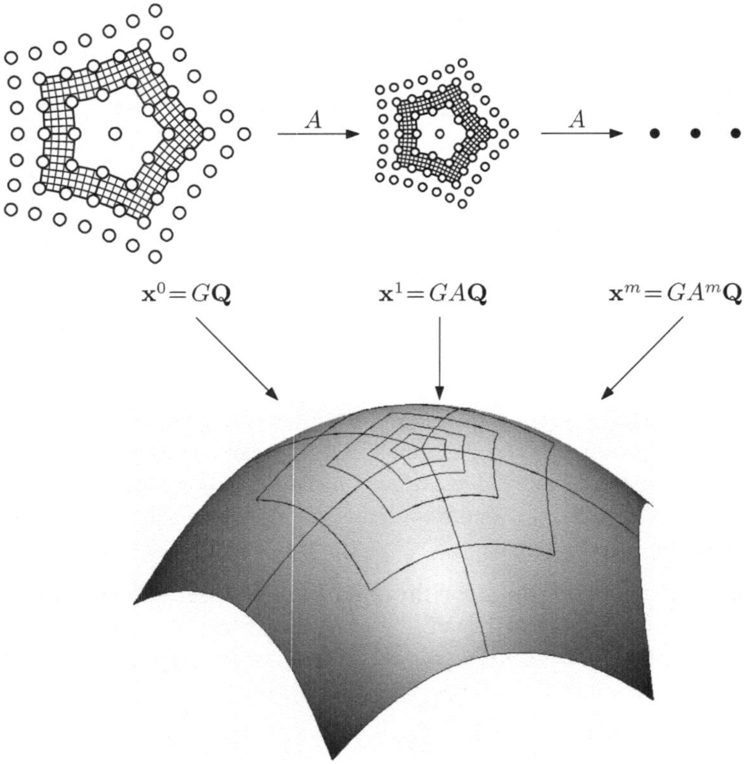

Fig. 4.5 Illustration of Definition 4.11/67: Subdivision algorithm (A, G) generating a sequence $\mathbf{x}^m$ of rings.

- By Theorem 4.5/63, the rings $\mathbf{x}^m$ form a C_0^k-spline if and only if there exists a point $\mathbf{x}^c \in \mathbb{R}^d$ such that
$$\lim_{m\to\infty} \mathbf{x}^m = \mathbf{x}^c.$$
In Sect. 4.7/75, we show that this condition is met if the dominant eigenvalue of A is $\lambda_0 = 1$.
- In Theorem 4.13/69, the condition that the rows of A sum to 1 is crucial for showing that the generating splines form a partition of unity. Just as in Theorem 4.10/66, the condition also implies that splines in subdivision form are affine invariant.

The correspondence between the initial data $\mathbf{Q}$ and the resulting spline $\mathbf{x}$ can be formalized as follows (Fig. 4.6/69).

Definition 4.12 (Generating spline). Let δ_ℓ denote the ℓth unit vector in $\mathbb{R}^{\bar{\ell}+1}$. The *generating spline* $b_\ell \in C_0^k(\mathbf{S}_n, \mathbb{R})$ of the subdivision algorithm (A, G) is defined as the spline corresponding to the initial data δ_ℓ,

$$b_\ell(\mathbf{s}) := G(2^m\mathbf{s})A^m\delta_\ell \quad \text{for} \quad \mathbf{s} \in \mathbf{S}_n^m.$$

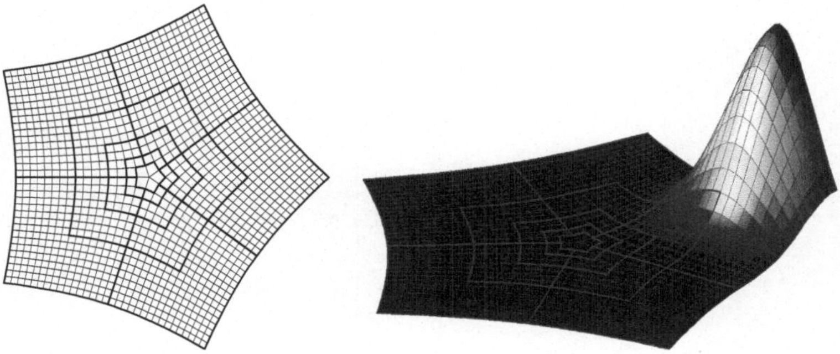

Fig. 4.6 Illustration of Definition 4.12/68: (*left*) Embedding of $\mathbf{S}_n$ and (*right*) generating spline b_ℓ plotted above the embedding.

The row-vector of all generating splines is denoted by

$$B := [b_0, \ldots, b_{\bar{\ell}}].$$

In the definition, the value of $b_\ell^c = b_\ell(\mathbf{0})$ at the central knot is not specified explicitly. However, its existence and uniqueness is ensured because, by Definition 4.11/67, the consistency condition (4.14/65) is satisfied. A formula will be derived later. The following theorem summarizes some basic properties of generating splines.

Theorem 4.13 (Properties of generating splines). *The generating splines B of a subdivision algorithm (A, G)*

(1) *span the space of splines generated by the subdivision algorithm via the representation*

$$\mathbf{x} = B\mathbf{Q};$$

(2) *satisfy the scaling relation*

$$B(2^{-m}\mathbf{s}) = B(\mathbf{s})A^m, \quad \mathbf{s} \in \mathbf{S}_n,\ m \in \mathbb{N}_0; \tag{4.17}$$

(3) *form a partition of unity,* $\sum_\ell b_\ell = 1$;
(4) *are affine invariant, i.e.*

$$\mathbf{L}(B\mathbf{Q}) = B\mathbf{L}(\mathbf{Q})$$

for any affine map $\mathbf{L} : \mathbb{R}^d \to \mathbb{R}^{d'}$.

Proof. Part (1) follows immediately from linearity of the whole process. To prove (2), let $\boldsymbol{\sigma} \in \boldsymbol{\Sigma}^{m'}$. Then (4.6/62) yields $B(\boldsymbol{\sigma}, \cdot) = G(2^{m'}\boldsymbol{\sigma}, \cdot)A^{m'}$. For $m \in \mathbb{N}_0$, it is $2^{-m}\boldsymbol{\sigma} \in \boldsymbol{\Sigma}^{m+m'}$, and accordingly $B(2^{-m}\boldsymbol{\sigma}, \cdot) = G(2^{m'}\boldsymbol{\sigma}, \cdot)A^{m+m'}$. Comparison of the two equations, which hold for any $m' \in \mathbb{N}$, establishes (4.17/69). To prove part (3) we denote by $\delta = [1; \ldots; 1]$ the one-vector in

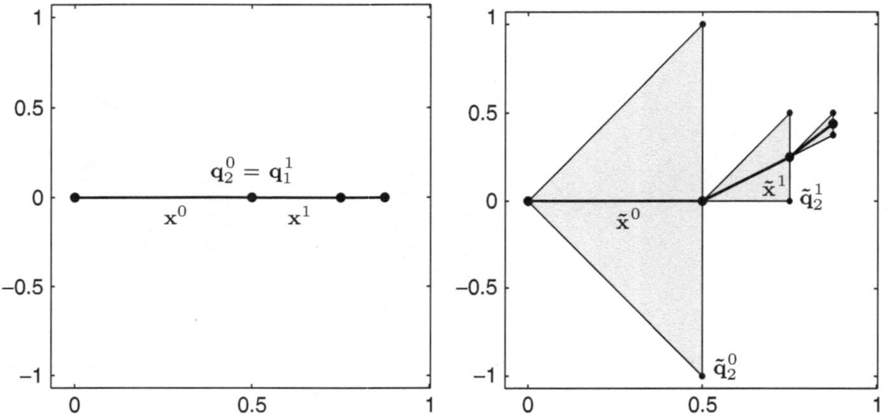

Fig. 4.7 Illustration of Example 4.14/70: *(left)* The minimal generating system necessarily yields collinear line segments, while *(right)* the overcomplete system generates a *broken line*. The *shaded triangles* are spanned by the coefficients $\tilde{\mathbf{q}}_1^m$, $\tilde{\mathbf{q}}_2^m$, $\tilde{\mathbf{q}}_3^m$ of $\tilde{\mathbf{Q}}^m$ at level m.

$\mathbb{R}^{\bar{\ell}+1}$. Because the rows in A sum to one, we have $A^m\delta = \delta$. For $\boldsymbol{\sigma} \in \boldsymbol{\Sigma}^m$,

$$\sum_\ell b_\ell(\boldsymbol{\sigma}, j) = B(\boldsymbol{\sigma}, j)\delta = B(2^{-m}\boldsymbol{\sigma}, j)\delta = G(2^{-m}\boldsymbol{\sigma}, j)\delta = 1$$

follows from (4.17/69) and (4.12/65). The same arguments as in Theorem 4.10/66 imply (4). □

We do not elaborate on generating splines, but continue our analysis of subdivision surfaces based on generating rings. However, in Sects. 8.2/169 and 8.1/157, generating splines will play an important role for investigating approximation and interpolation properties.

Before we continue our analysis of subdivision algorithms, we want to emphasize that Definition 4.9/65 does *not* require G to be linearly independent. Apparently, if G and $\tilde{G}$ are linearly dependent in the sense that each function of one system can be written as linear combination of the other, then the corresponding spaces of rings coincide: $C^k(\mathbf{S}_n^0, \mathbb{R}^d, G) = C^k(\mathbf{S}_n^0, \mathbb{R}^d, \tilde{G})$. So it is tempting to simplify the analysis considerably by confining it to linearly independent systems. However, this would miss a number of potentially useful algorithms. Let us assume that $\tilde{G}$ has more entries than G. Then, given a matrix $\tilde{A}$, one cannot guarantee that there exists a matrix A such that the sequences of rings generated by (A, G) always equal those generated by $(\tilde{A}, \tilde{G})$. In other words, there need not exist A and $\mathbf{Q}$ such that $GA^m\mathbf{Q} = \tilde{G}\tilde{A}^m\tilde{\mathbf{Q}}$ for all m. Instead of detailing the linear algebra of this statement, we illustrate the difference by Example 4.14/70 and Fig. 4.7/70.

Example 4.14 (Usefulness of overcomplete systems of generating rings). To keep the example simple, we consider a univariate case with the system $G := [1 - \mathrm{u}, u]$, $u \in$

$[0, 1]$, of generating rings spanning the space of linear functions. The matrix A is to be chosen such that the sequence $\mathbf{x}^m = GA^m\mathbf{Q} = G[\mathbf{q}_1^m; \mathbf{q}_2^m]$ forms a continuous, piecewise linear function in the sense of Sect. 3.1/40. That is, $\mathbf{x}^m(1) = \mathbf{x}^{m+1}(0)$, and the most general form of A is

$$A := \begin{bmatrix} 0 & 1 \\ 1-a & a \end{bmatrix}$$

with $a \in \mathbb{R}$. As an overcomplete alternative, we extend G artificially to $\tilde{G} := [1-u, u/2, u/2]$. Obviously, the additional third component of $\tilde{G}$ does not influence the shape of a single piece $\tilde{\mathbf{x}}^m$. However, it substantially enriches the possible behavior of the sequence $\{\tilde{\mathbf{x}}^m\}_m$. To show this, we consider

$$\tilde{A} := \frac{1}{4}\begin{bmatrix} 0 & 2 & 2 \\ -2 & 3 & 3 \\ -2 & 2 & 4 \end{bmatrix}, \quad \tilde{\mathbf{Q}} := \begin{bmatrix} 0 & 0 \\ 1 & -1 \\ 1 & 1 \end{bmatrix}.$$

We check that $\tilde{\mathbf{x}}_0$ connects $[0, 0]$ to $[1, 0]$, while $\tilde{\mathbf{x}}_1$ connects $[1, 0]$ to $[3/2, 1/4]$. Let us try to find A and $\mathbf{Q}$ such that $\mathbf{x}^m = \tilde{\mathbf{x}}^m$. In order to match $\mathbf{x}^0$, we have to choose $\mathbf{Q} = [0, 0;\ 1, 0]$, and the endpoint of $\mathbf{x}^1$ becomes $[a, 0] \neq [3/2, 1/4]$. So, no matter how we choose a, all points generated by A will lie on the x-axis, while those generated by $\tilde{A}$ do not. Figure 4.7/70 illustrates this observation. □

We note that linear dependent systems of generating rings are especially useful when representing vector-valued subdivision algorithms, or the like. There might exist components of these vectors which are used for the recursion, but not for the parametrization of the rings. Accordingly, the generating system can be padded with zeros to facilitate the representation in the form given here. This aspect is discussed also in Sect. 9.1/176.

4.5 Asymptotic Expansion of Sequences

In the following analysis, we are rarely interested in the full expression of a sequence of rings, but only in the asymptotic behavior of the dominant terms as m tends to infinity. To efficiently deal with such *asymptotic expansions*, we introduce a relation for sequences of functions with coinciding leading terms.

Definition 4.15 (Equivalence of sequences). Let $\{a^m\}_m$, $\{b^m\}_m$, and $\{c^m\}_m$ be sequences in m. We write

$$a^m \overset{c^m}{=} b^m \quad \text{iff} \quad a^m - b^m = o(c^m),$$

where, by definition, $o(c^m)/c^m$ converges uniformly to zero as $m \to \infty$. In particular, $a^m \overset{1}{=} a$ means that $\{a^m\}_m$ converges to a. For simplicity, $\overset{c^m}{=}$ is mostly replaced by the symbol $\overset{*}{=}$ with the understanding that the little star refers to the lowest order term specified explicitly on the right hand side of a relation.

For example,

$$\frac{1}{2^m+u} \stackrel{*}{=} \frac{1}{2^m} - \frac{u}{4^m}, \quad u \in U, \tag{4.18}$$

implies equivalence up to order $o(4^{-m})$. For vector-valued expressions, the equivalence relation is understood component-wise.

To compare the decay of sequences of numbers or functions, we write

$$\begin{aligned}
a^m \succ b^m \quad &\text{iff} \quad \lim_{m\to\infty} |b^m/a^m| = 0 \\
a^m \succcurlyeq b^m \quad &\text{iff} \quad \limsup_{m\to\infty} |b^m/a^m| < \infty \\
a^m \sim b^m \quad &\text{iff} \quad (a^m \succcurlyeq b^m \text{ and } b^m \succcurlyeq a^m) \text{ or } (a^m = b^m = 0).
\end{aligned}$$

In other words, for sequences converging to 0, $a^m \succ b^m$ means that b^m decays faster than a^m, $a^m \succcurlyeq b^m$ means that a^m does not decay faster than b^m, and $a^m \sim b^m$ means that both sequences decay at equal rates. For example, anticipating (4.21/73),

$$\lambda^m m^\ell \sim \lambda^{m,\ell} \succ \lambda^{m,\ell-1} \tag{4.19}$$

for $\lambda \neq 0$ and $\ell \geq 0$ or, recalling (4.18/72),

$$\frac{1}{2^m} \succ \frac{1}{4^m} \sim \frac{1}{2^m+u} - \frac{1}{2^m}, \quad u \in U.$$

4.6 Jordan Decomposition

In order to compute and analyze powers of A, it is convenient to use its *Jordan decomposition*. The *Jordan block* $J(\lambda,\ell)$ is defined as the $(\ell+1)\times(\ell+1)$-matrix

$$J(\lambda,\ell) := \begin{bmatrix} \lambda & 1 & 0 & \cdots & 0 \\ 0 & \lambda & 1 & \cdots & 0 \\ \vdots & \ddots & \ddots & \ddots & \vdots \\ 0 & \cdots & 0 & \lambda & 1 \\ 0 & \cdots & 0 & 0 & \lambda \end{bmatrix}. \tag{4.20}$$

As is easily shown by induction, its powers are

$$J^m(\lambda,\ell) = \begin{bmatrix} \lambda^{m,0} & \lambda^{m,1} & \lambda^{m,2} & \dots & \lambda^{m,\ell} \\ 0 & \lambda^{m,0} & \lambda^{m,1} & \dots & \lambda^{m,\ell-1} \\ \vdots & \ddots & \ddots & \ddots & \vdots \\ 0 & \cdots & 0 & \lambda^{m,0} & \lambda^{m,1} \\ 0 & \cdots & 0 & 0 & \lambda^{m,0} \end{bmatrix},$$

where

$$\lambda^{m,\ell} := \begin{cases} \binom{m}{\ell}\lambda^{m-\ell} & \text{if } \lambda \neq 0 \text{ and } 0 \leq \ell \leq m \\ 1 & \text{if } \lambda = 0 \text{ and } \ell = m \\ 0 & \text{else.} \end{cases} \tag{4.21}$$

Defining the relations

$$\begin{array}{lll} (\lambda,\ell) \sim (\tilde\lambda,\tilde\ell) & \text{iff} & \lambda^{m,\ell} \sim \tilde\lambda^{m,\tilde\ell}, \text{ i.e., } |\lambda| = |\tilde\lambda|,\ \ell = \tilde\ell \text{ or } \lambda = \tilde\lambda = 0 \\ (\lambda,\ell) \succ (\tilde\lambda,\tilde\ell) & \text{iff} & \lambda^{m,\ell} \succ \tilde\lambda^{m,\tilde\ell}, \text{ i.e., } |\lambda| = |\tilde\lambda| > 0,\ \ell > \tilde\ell \text{ or } |\lambda| > |\tilde\lambda| \\ (\lambda,\ell) \succeq (\tilde\lambda,\tilde\ell) & \text{iff} & (\lambda,\ell) \sim (\tilde\lambda,\tilde\ell) \text{ or } (\lambda,\ell) \succ (\tilde\lambda,\tilde\ell) \end{array}$$

for pairs $(\lambda, \ell) \in \mathbb{C} \times \mathbb{N}_0$, we can conveniently compare sequences of type $\lambda^{m,\ell}$,

$$\begin{array}{lll} (\lambda,\ell) \sim (\tilde\lambda,\tilde\ell) & \Leftrightarrow & \lambda^{m,\ell} \sim \lambda^{m,\tilde\ell} \\ (\lambda,\ell) \succ (\tilde\lambda,\tilde\ell) & \Leftrightarrow & \lambda^{m,\ell} \succ \lambda^{m,\tilde\ell} \\ (\lambda,\ell) \succeq (\tilde\lambda,\tilde\ell) & \Leftrightarrow & \lambda^{m,\ell} \succeq \lambda^{m,\tilde\ell} \end{array}$$

The Jordan decomposition of A is

$$A = VJV^{-1}, \quad J = \operatorname{diag}(J_0, J_1, \ldots, J_{\bar r}), \quad J_r = J(\lambda_r, \ell_r), \tag{4.22}$$

that is, the rth Jordan block J_r has dimension $(\ell_r + 1)$ and corresponds to the eigenvalue λ_r. Without loss of generality, we may assume that the Jordan blocks are ordered such that

$$(\lambda_0, \ell_0) \succeq (\lambda_1, \ell_1) \succeq \cdots \succeq (\lambda_{\bar r}, \ell_{\bar r}).$$

According to the block structure of J, the matrix V is partitioned into

$$V = [V_0, V_1, \ldots, V_{\bar r}], \quad V_r = [v_r^0, \ldots, v_r^{\ell_r}].$$

The eigenvector v_r^0 to the eigenvalue λ_r satisfies

$$Av_r^0 = \lambda_p v_r^0,$$

and for $\ell_r>0$, the other vectors $v_r^1, \ldots, v_r^{\ell_r}$ form a chain of generalized eigenvectors,

$$Av_r^\ell = \lambda_r v_r^\ell + v_r^{\ell-1}, \quad \ell = 1, \ldots, \ell_r.$$

As usual, we assume that V_r is chosen real if λ is real. Powers of A are given by

$$A^m = VJ^mV^{-1}, \quad J^m = \operatorname{diag}(J_0^m, J_1^m, \ldots, J_{\bar r}^m).$$

Thus, the mth ring becomes $\mathbf{x}^m = GVJ^mV^{-1}\mathbf{Q}$. This expression is most convenient for analytical purposes since J^m is known explicitly. The factors to the left and to the right of J^m are referred to as follows:

Definition 4.16 (Eigenrings and eigencoefficients). Consider a subdivision algorithm (A, G) and the Jordan decomposition $A = VJV^{-1}$. The vector F of *eigenrings* and the vector $\mathbf{P}$ of *eigencoefficients* corresponding to the initial data $\mathbf{Q}$ are defined by

$$F := GV, \quad \mathbf{P} := V^{-1}\mathbf{Q}, \tag{4.23}$$

respectively.

With these settings, the sequence of rings reads

$$\mathbf{x}^m = FJ^m\mathbf{P}. \tag{4.24}$$

The vectors F and $\mathbf{P}$ are partitioned into blocks according to the structure of V,

$$F =: [F_0, F_1, \ldots, F_{\bar{r}}], \quad F_r =: [f_r^0, f_r^1, \ldots, f_r^{\ell_r}]$$
$$\mathbf{P} =: \begin{bmatrix} \mathbf{P}_0 \\ \mathbf{P}_1 \\ \vdots \\ \mathbf{P}_{\bar{r}} \end{bmatrix}, \quad \mathbf{P}_r =: \begin{bmatrix} \mathbf{p}_r^0 \\ \mathbf{p}_r^1 \\ \vdots \\ \mathbf{p}_r^{\ell_r} \end{bmatrix}, \tag{4.25}$$

where the individual eigenrings are $f_r^\ell = Gv_r^\ell$. For brevity, we further define

$$v_r := v_r^0, \quad f_r := f_r^0, \quad \mathbf{p}_r := \mathbf{p}_r^{\ell_r}.$$

In general, V is complex, so that also the eigenrings are complex,

$$f_r^\ell \in C^k(\mathbf{S}_n^0, \mathbb{C}, G).$$

Now, (4.24/74) becomes

$$\mathbf{x}^m = \sum_{r=0}^{\bar{r}} F_r J_r^m \mathbf{P}_r. \tag{4.26}$$

Collecting factors of type $\lambda_r^{m,\ell}$, we find for the summands

$$F_r J_r^m \mathbf{P}_r = \sum_{\ell=0}^{\ell_r} \lambda_r^{m,\ell} \sum_{i=\ell}^{\ell_r} f_r^{i-\ell} \mathbf{p}_r^i. \tag{4.27}$$

Equation (4.19/72) implies that the dominant term in this sum corresponds to $\ell = \ell_r$. Hence, we obtain

$$F_r J_r^m \mathbf{P}_r \stackrel{*}{=} \lambda_r^{m,\ell_r} f_r \mathbf{p}_r. \tag{4.28}$$

An even simpler expression is obtained for $\ell_r = 0$. In this case,

$$F_r J_r^m \mathbf{P}_r = \lambda_r^m f_r \mathbf{p}_r. \tag{4.29}$$

Since A is real, the eigenvalues and eigenvectors come in complex conjugate pairs, and we assume without loss of generality that such pairs have consecutive indices.

If $\overline{\lambda_r} = \lambda_{r+1} \notin \mathbb{R}$ is such a pair, then

$$\overline{J_r} = J_{r+1}, \quad \overline{V_r} = V_{r+1}, \quad \overline{F_r} = F_{r+1}.$$

Moreover, if $\mathbf{Q}$ is real, we have also $\overline{\mathbf{P}_r} = \mathbf{P}_{r+1}$ so that the summands $F_r J_r^m \mathbf{P}_r$ and $F_{r+1} J_{r+1}^m \mathbf{P}_{r+1}$ combine to

$$F_r J_r^m \mathbf{P}_r + F_{r+1} J_{r+1}^m \mathbf{P}_{r+1} = 2\,\mathrm{Re}(F_r J_r^m \mathbf{P}_r) \stackrel{*}{=} 2|\lambda_r^{m,\ell_r}|\,\mathrm{Re}(d_r^{m-\ell_r} f_r \mathbf{p}_r), \tag{4.30}$$

where the *direction*

$$d_r := \frac{\lambda_r}{|\lambda_r|} \tag{4.31}$$

of λ_r is a number on the complex unit circle.

4.7 The Subdivision Matrix

It is our goal to relate properties of the matrix A as introduced in Definition 4.11/67 to properties of the generated spline $\mathbf{x} = B\mathbf{Q}$. The first issue is consistency at the central knot $\mathbf{0}$. The matrix $\mathbf{A}$ is called *consistent* if, for any set of coefficients $\mathbf{Q}$, there exists a unique limit

$$\mathbf{x}(\mathbf{0}) := \mathbf{x}^{\mathrm{c}} := \lim_{m\to\infty} GA^m\mathbf{Q}.$$

Since the rows of A sum to 1, we have

$$A\delta = \delta, \quad \delta := [1;\ldots;1].$$

Hence, 1 is always an eigenvalue of A. If this eigenvalue is *dominant* in the sense that all other eigenvalues of A are smaller than 1 in modulus, then it is easily shown that A is consistent.

Lemma 4.17 (Dominant eigenvalue and consistency). *If the dominant eigenvalue if A is*

$$(1,0) = (\lambda_0, \ell_0) \succ (\lambda_1, \ell_1),$$

then A is consistent. Further, if $w_0^{\mathrm{t}} = w_0^{\mathrm{t}} A$ is the left eigenvector of A corresponding to $\lambda_0 = 1$, then the unique limit of rings is given by the first eigencoefficient,

$$\mathbf{x}^{\mathrm{c}} = \mathbf{p}_0 = \frac{w_0^{\mathrm{t}}\mathbf{Q}}{w_0^{\mathrm{t}}\delta}. \tag{4.32}$$

Proof. The first column of V is $v_0^0 = \delta$, while the first row of V^{-1} is $w_0^{\mathrm{t}}/(w_0^{\mathrm{t}}\delta)$. This implies $f_0 = G\delta = 1$ and $\mathbf{p}_0 = (w_0^{\mathrm{t}}\mathbf{Q})/(w_0^{\mathrm{t}}\delta)$. By (4.19/72), $J_r^m \stackrel{1}{=} 0$ for $r \geq 1$. Hence, (4.26/74) yields $\mathbf{x}^m \stackrel{1}{=} f_0\mathbf{p}_0 = \mathbf{p}_0$. □

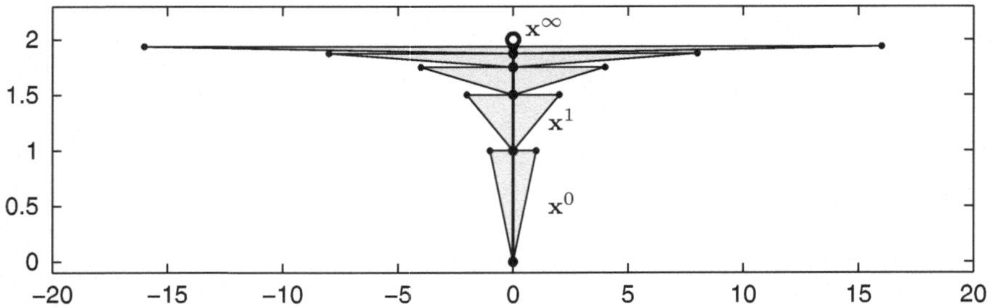

Fig. 4.8 Illustration of Example 4.18/76: Consistent univariate subdivision algorithm with dominant eigenvalue 2. The *shaded triangles* are spanned by the coefficients $\mathbf{Q}^m$ at level m.

Intuitively, one might expect that a dominant eigenvalue 1 is also necessary for consistency – but this is by no means true.

Example 4.18 (Large dominant eigenvalue permits consistency). We consider the generating rings $G = [1-u,\ u/2,\ u/2]$ introduced in Example 4.14/70. Let

$$A := \frac{1}{4}\begin{bmatrix} 0 & 2 & 2 \\ -2 & 7 & -1 \\ -2 & -1 & 7 \end{bmatrix}. \tag{4.33}$$

The eigenvalues are $\lambda_0 = 2, \lambda_1 = 1, \lambda_2 = 1/2$ so that, in general, $A^m\mathbf{Q}$ diverges (see the increasing width of the triangles in Fig. 4.8/76). However,

$$GA^m = [-1,\ 1,\ 1] + 2^{-m}[2-u,\ u/2-1,\ u/2-1].$$

Hence, despite the dominant eigenvalue 2, the matrix A is consistent with

$$\mathbf{x}^c = \lim_{m\to\infty} GA^m\mathbf{Q} = [-1,\ 1,\ 1]\,\mathbf{Q}.$$

This phenomenon is possible since the eigenvector $v_0^0 = [0; 1; -1]$ corresponding to the dominant eigenvalue $\lambda_0 = 2$ is in the kernel of G, i.e., $Gv_0^0 = 0$. In other words, the divergent component of $A^m\mathbf{Q}$ is annihilated by G. The remaining components are well behaved. □

More generally, if G is linearly dependent, it is conceivable that there exist possibly different matrices A and $\tilde{A}$ that generate equal sequences of rings. To address this issue, we say that two subdivision algorithms (A, G) and $(\tilde{A}, G)$ are *equivalent*, if

$$GA^m = G\tilde{A}^m \quad \text{for all} \quad m \in \mathbb{N}_0.$$

This concept will now be used for removing eigenvectors to non-zero eigenvalues of A that are annihilated by G.

Definition 4.19 (Ineffective eigenvector). The vector $v \neq 0$ is an *ineffective eigenvector* of the subdivision algorithm (A, G) if

$$Av = \lambda v, \quad \lambda \neq 0, \quad \text{and} \quad Gv = 0. \tag{4.34}$$

The notion of ineffective eigenvectors is definitely essential for the further development of a spectral theory of subdivision algorithms. So far, however, they have not been observed in applications. But even if a given algorithm happens to reveal ineffective eigenvectors, they can be removed without further consequences. The following theorem addresses this issue.

Theorem 4.20 (Removal of ineffective eigenvectors). *For any subdivision algorithm (A,G) there exists an equivalent algorithm $(\tilde{A},G)$ without ineffective eigenvectors.*

Proof. The proof is constructive. We recursively construct a finite sequence $\{(A_\ell,G)\}_\ell$ of algorithms starting from $A_0 := A$ as follows. If A_ℓ has no ineffective eigenvectors, set $\tilde{A} := A_\ell$ and stop. Otherwise, let v be an ineffective eigenvector of (A_ℓ,G) corresponding to the eigenvalue $\lambda \neq 0$. Then there exists a vector w such that $w^{\mathrm{t}}v = \lambda$, and $w^{\mathrm{t}}v' = 0$ for any other eigenvector v' of A_ℓ. Now, set $A_{\ell+1} := A_\ell - vw^{\mathrm{t}}$. The matrix $A_{\ell+1}$ has the following properties:

- v remains an eigenvector of $A_{\ell+1}$, but is no longer ineffective since $A_{\ell+1}v = (A_\ell - vw^{\mathrm{t}})v = \lambda v - \lambda v = 0$.
- All other eigenvalues and eigenvectors are retained. That is, for any eigenvector $v' \neq v$ of A_ℓ, we have $A_{\ell+1}v' = (A_\ell - vw^{\mathrm{t}})v' = A_\ell v' = \lambda' v'$.
- If the rows of A_ℓ sum up to 1, so do the rows of $A_{\ell+1}$. This follows immediately from specializing the preceding statement to $A_{\ell+1}\delta = A_\ell\delta = \delta$.
- $GA_{\ell+1}^m = GA_\ell^m$ for all $m \in \mathbb{N}_0$. To show this, we observe that for any m there exists a vector w_m such that $(A_\ell - vw^{\mathrm{t}})^m = A_\ell^m - vw_m^{\mathrm{t}}$. Hence, $GA_{\ell+1}^m = GA_\ell^m - Gvw_m^{\mathrm{t}} = GA_\ell^m$.

By induction, all algorithms $(A_0,G),(A_1,G),\ldots$ are equivalent. A_ℓ has the eigenvalue 1 and at least an ℓ-fold eigenvalue 0. Hence, the iteration terminates after at most $\bar{\ell}$ steps. □

Example 4.21 (Removal of ineffective eigenvectors). Let us illustrate the algorithm given in the proof by Example 4.18/76. The matrix $A_0 = A$ has the ineffective eigenvector $v = v_0^0 = [0;1;-1]$ corresponding to $\lambda_0 = 2$. The other two eigenvectors corresponding to $\lambda_1 = 1$ and $\lambda_2 = 1/2$ are $v_1^0 = [1;1;1]$ and $v_2^0 = [2;1;1]$. The conditions $w^{\mathrm{t}}v = 2$ and $w^{\mathrm{t}}v_1^0 = w^{\mathrm{t}}v_2^0 = 0$ yield $w := [0;1;-1]$. We define

$$A_1 := A_0 - vw^{\mathrm{t}} = \frac{1}{4}\begin{bmatrix} 0 & 2 & 2 \\ -2 & 3 & 3 \\ -2 & 3 & 3 \end{bmatrix}$$

and obtain $A_1 v_0^0 = 0$. The other two eigenvectors v_1^0 and v_2^0 satisfy $Gv_1^0 = 1$ and $Gv_2^0 = 2 - u$. Hence, $(\tilde{A},G) := (A_1,G)$ is equivalent to (A,G) and has no ineffective eigenvectors. □

To be able to link the spectrum of A to properties of the resulting subdivision surfaces, we will assume that (A,G) has no ineffective eigenvectors. Before we proceed in that direction, we establish an important result on the linear independence of eigenrings, which, in particular, implies that eigenrings f_r corresponding to non-zero eigenvalues cannot vanish.

Lemma 4.22 (Linear independence of eigenrings). *Let (A, G) be a subdivision algorithm without ineffective eigenvectors. If $v_{r_1}^0, \dots, v_{r_k}^0$ are linearly independent eigenvectors of A to the same eigenvalue $\lambda \neq 0$, then the corresponding eigenrings $f_{r_1}, \dots, f_{r_k}$ are linearly independent. Moreover, if the generating rings G are linearly independent, so are the eigenrings F.*

Proof. Let $f := c_{r_1} f_{r_1} + \cdots + c_{r_k} f_{r_k}$ and $v := c_{r_1} v_{r_1}^0 + \cdots + c_{r_k} v_{r_k}^0$. Then $f = Gv$ and $Av = \lambda v$. Since $\lambda \neq 0$, $f = 0$ is possible only if $v = 0$ and $c_{r_1} = \cdots = c_{r_k} = 0$.

The second statement of the lemma follows immediately from $F = GV$ and the linear independence of the generalized eigenvectors forming V. □

The following example shows that eigenrings corresponding to *different* eigenvalues can be linearly dependent even if there are no ineffective eigenvectors.

Example 4.23 (Linear dependence of eigenrings). For

$$G := [1 - u, u/2, u/2], \quad A := \frac{1}{2} \begin{bmatrix} 0 & 1 & 1 \\ 2 & 0 & 0 \\ -1 & 1 & 2 \end{bmatrix},$$

we have

$$\begin{aligned} \lambda_0 &= 1, & v_0 &= [1, 1, 1]^{\mathrm{t}}, & f_0 &= 1 \\ \lambda_1 &= 1/2, & v_1 &= [1, 2, -1]^{\mathrm{t}}, & f_1 &= 1 - u/2 \\ \lambda_2 &= -1/2, & v_2 &= [1, -2, 1]^{\mathrm{t}}, & f_2 &= 1 - 3u/2. \end{aligned}$$

Clearly, there is no ineffective eigenvector, but $2f_0 - 3f_1 + f_2 = 0$. □

Just as the eigenrings are derived from the generating rings, we derive eigensplines from generating splines.

Definition 4.24 (Eigensplines). The generating splines B of a subdivision algorithm (A, G) define the *eigensplines* $E = [e_0, \dots, e_{\bar{\ell}}]$ via

$$E := BV.$$

Unlike eigenrings, the eigensplines e_ℓ corresponding to non-zero eigenvalues are linearly independent.

Lemma 4.25 (Linear independence of eigensplines). *Let E be partitioned into blocks E_r corresponding to the Jordan blocks of A. The eigensplines $E' := [E_0, \dots, E_{r'}]$ corresponding to nonzero eigenvalues are linearly independent if (A, G) has no ineffective eigenvectors.*

Proof. Since eigensplines to the eigenvalue 0 are not involved, we assume $E' = E$ to simplify notation. Assume that E' is dependent, and let $w \neq 0$ be a vector such that $EV^{-1}w = Bw = 0$. Since $A^m w \neq 0$, the sequence

$$w^m := \frac{A^m w}{\|A^m w\|}$$

is well defined and bounded, hence has a convergent subsequence with limit $w^\infty \neq 0$. Because all rings of Bw vanish, we have

$$0 = \frac{(Bw)^m}{\|A^m w\|} = \frac{GA^m w}{\|A^m w\|} = Gw^m, \quad m \in \mathbb{N}_0.$$

Passing to the limit of the subsequence, we find $Gw^\infty = 0$. This implies that w^∞ is an ineffective eigenvector of (A, G), contradicting the assumptions of the lemma. □

Now, we are prepared to prove the fundamental theorem on the dominant eigenvalue 1 of a consistent matrix A. Although this result seems to suggest itself, it should be emphasized that it is essentially based on the absence of ineffective eigenvectors. Even then, a concise verification is not completely straightforward.

Theorem 4.26 (Unique dominant eigenvalue). *Let (A, G) be a subdivision algorithm according to Definition* 4.11/67 *without ineffective eigenvectors. Then A is consistent if and only if*

$$(1, 0) = (\lambda_0, \ell_0) \succ (\lambda_1, \ell_1).$$

Proof. Sufficiency of the given condition has already been established in Lemma 4.17/75. To prove necessity, it suffices to consider the case $d = 1$ of real-valued coefficients. We eliminate all cases contradicting the given condition. First, we exclude the case $(\lambda_0, \ell_0) \succ (1, 0)$.

- Let $(\lambda_0, \ell_0) \succ (1, 0)$, and $\lambda \in \mathbb{R}$. We choose $\mathbf{p}_0 = 1$ and set all other eigencoefficients to 0. The corresponding sequence $\mathbf{x}^m \stackrel{*}{=} \lambda_0^{m,\ell_0} f_0$ of rings, see (4.28/74), is divergent since $|\lambda_0^{m,\ell_0}| \to \infty$ by (4.19/72) and $f_0 \neq 0$ by Lemma 4.22/78.
- Let $(\lambda_0, \ell_0) \succ (1, 0)$, and $\lambda \notin \mathbb{R}$. Since $f_0 \neq 0$, there exists $\mathbf{s}_0 \in \mathbf{S}_n$ with $f_0(\mathbf{s}_0) \neq 0$. We choose $\overline{\mathbf{p}_0} = \mathbf{p}_1 = f_0(\mathbf{s}_0)$ and set all other eigencoefficients to 0. The sequence $\mathbf{x}^m \stackrel{*}{=} 2|\lambda_0^{m,\ell_0}| \operatorname{Re}(d_0^{m-\ell_0} f_0 \mathbf{p}_0)$ of rings, see (4.30/75), is divergent for the following reasons: first, $|\lambda_0^{m,\ell_0}| \to \infty$; second, there exists a subsequence of $\{d_0^{m-\ell_0}\}_m$ converging to 1; third, $\operatorname{Re}(f_0 \mathbf{p}_0) \neq 0$ since $f_0(\mathbf{s}_0)\mathbf{p}_0 = |f_0(\mathbf{s}_0)|^2 \neq 0$.

Since δ is always an eigenvector of A to the eigenvalue 1, we may assume $(\lambda_0, \ell_0) = (1, 0)$ and $v_0^0 = \delta$ for the remaining cases, where we have to exclude $(\lambda_1, \ell_1) \sim (1, 0)$.

- Let $(\lambda_1, \ell_1) = (1, 0)$. We choose $\mathbf{p}_1 = 1$ and set all other eigencoefficients to 0. $\mathbf{x}^m = f_1$ cannot be constant since f_1 is linearly independent of $f_0 = G\delta = 1$ by Lemma 4.22/78.
- Let $(\lambda_1, \ell_1) = (-1, 0)$. We choose $\mathbf{p}_1 = 1$ and set all other eigencoefficients to 0. $\mathbf{x}^m = (-1)^m f_1$ does not converge since $f_1 \neq 0$.
- Let $(\lambda_1, \ell_1) = (\overline{\lambda_2}, \ell_2) \sim (1, 0)$ and $\lambda_1 \notin \mathbb{R}$. We choose $\mathbf{p}_0 = \mathbf{p}_1 = 1$ and set all other eigencoefficients to 0. $\mathbf{x}^m = 2\,\mathrm{Re}(d_1^m f_1)$ does not converge since $f_1 \neq 0$. □

In summary, the necessary and sufficient conditions for smoothness and consistency derived in this chapter lead to the following final definition of a subdivision algorithm, which from now on replaces the tentative draft of Sect. 4.4/67.

Definition 4.27 (Subdivision algorithm, final). For $k \geq 1$, let $G = [g_0, \ldots, g_{\bar{\ell}}]$ be a system of generating rings $g_\ell \in C^k(\mathbf{S}_n^0, \mathbb{R})$, and let $A \in \mathbb{R}^{\bar{\ell} \times \bar{\ell}}$ be a square matrix with all rows summing up to 1, i.e., $A\delta = \delta$. If

- the matrix A has no ineffective eigenvectors, and
- the eigenvalues of A satisfy $(1, 0) = (\lambda_0, \ell_0) \succ (\lambda_1, \ell_1)$, and
- for any choice of initial data $\mathbf{Q}$, the generated sequence $\mathbf{x}^m = GA^m\mathbf{Q}$ of rings satisfies the C^k-conditions (4.7/62) and (4.8/62).

then (A, G) is a C_0^k-*subdivision algorithm*, and A is its *subdivision matrix*.

We recall from Theorem 4.20/77 that the absence of ineffective eigenvectors can be required without loss of generality. In view of Theorem 4.26/79, the assumption concerning the unique dominant eigenvalue $\lambda_0 = 1$ is necessary and sufficient for consistency at the central knot. Together with the third condition concerning smooth contact of neighboring and consecutive segments $\mathbf{x}_j^m$, Theorem 4.5/63 implies

Theorem 4.28 (C_0^k-subdivision algorithm yields C_0^k-spline). *Let (A, G) be a C_0^k-subdivision algorithm. Then, for any choice of initial data $\mathbf{Q}$,*

$$\mathbf{x} : \mathbf{S}_n \ni \mathbf{s} \mapsto \begin{cases} G(2^m \mathbf{s})A^m\mathbf{Q} & \textit{if} \quad \mathbf{s} \in \mathbf{S}_n^m \\ \mathbf{p}_0 & \textit{if} \quad \mathbf{s} = \mathbf{0} \end{cases}$$

is a C_0^k-spline in subdivision form, where $\mathbf{p}_0$ is the fist eigencoefficient of $\mathbf{Q}$ according to Lemma 4.17/75.

Bibliographic Notes

1. The characterization of subdivision surfaces as a nested sequence of rings (4.6/62) is central to an analytic approach. This idea first appeared in the doctoral thesis of Reif [Rei93] and was inspired by thoughts of his advisor Klaus Höllig. The parts of the thesis relevant to subdivision were published later on in [Rei95c].

2. The representation of a subdivision surface as a finite linear combination of functions and coefficients, $\mathbf{x} = B\mathbf{Q}$, is useful for computational purposes. For instance, it was used in [HKD93] to compute subdivision surfaces that minimize a certain fairness functional while interpolating a given set of points.

3. The notion of ineffective eigenvectors (4.19/76), first developed in [Rei99], highlights the equal contribution of functions and coefficients to the final surface.

4. The importance of a strictly dominant eigenvalue $\lambda_0 = 1$ for the convergence of subdivision algorithms at extraordinary points was already observed in [DS78, BS86, BS88]. For a long time, it was taken for granted that this condition was necessary and sufficient. However, it was pointed out in [Rei99] that ineffective eigenvectors have to be excluded to verify such a result.

5. An example showing that requiring linear independence of the generating rings implies a loss of generality can be found in [RP05]. The same reference also provides a constructive procedure, similar to the one described in the proof of Theorem 4.20/77, to efficiently compute from a given matrix $\tilde{A}$ a subdivision matrix A without ineffective eigenvectors, which is similar to the one suggested here.

6. By its nature, the Jordan decomposition (4.22/73) of the subdivision matrix A provides an efficient way to compute powers of A: $A^m = VJ^mV^{-1}$. If the matrix A is dense, or if m is large, it is efficient to map the coefficients $\mathbf{Q}$ into the eigenspace, apply J^m and recover the coefficients before applying standard spline evaluation with the local coefficients of the ring [Sta98c, Sta98a]. However, since the matrices A are typically sparse, but V and its inverse are dense, a rough computation shows that the method pays off for Catmull–Clark subdivision when $m \approx 16$ or larger. For rendering purposes, when many points have to be evaluated at many consecutive rings, it is typically much more efficient to proceed step by step using the sparse matrix product by A.

7. The rows of V^{-1} are the left eigenvectors of the subdivision matrix A. In particular, the rows corresponding to the leading three eigenvalues, w_0^t, w_1^t and w_2^t, yield position $w_0^t\mathbf{Q}$ (4.32/75) and tangent directions $w_j^t\mathbf{Q}$, $j = 1, 2$ (5.2/85) at the central point. Explicit formulas for the central limit point and tangent plane of Catmull–Clark subdivision are given, for example, in [HKD93], and for the Butterfly algorithm in [SDL99].

Chapter 5
C_1^k-Subdivision Algorithms

In the last chapter, we have defined a C_0^k-subdivision algorithm as a pair (A, G) consisting of a subdivision matrix A and a C^k-system G of generating rings. The conditions given in Definition 4.27/80 guarantee that the generated splines are consistent at the center. Such algorithms are easy to construct, but of course, they do not live up to the demands arising in applications, where smoothness is required also at extraordinary knots. In this chapter, we consider subdivision algorithms in more detail with the goal to find conditions for normal continuity and single-sheetedness. First, in Sect. 5.1/84, we define 'generic' sets of initial data $\mathbf{Q}$. Restriction to generic data is necessary to exclude degenerate configurations which, even for impeccable algorithms, yield non-smooth surfaces. Section 5.2/84 defines standard algorithms. This class of algorithms, which is predominant in applications, is characterized by a double positive subdominant eigenvalue. Here, the characteristic ring, which is a planar ring built from the subdominant eigenfunctions, plays a key role in the analysis. With a careful generalization of terms, Sect. 5.3/89 yields a complete classification of all C_1^k-subdivision algorithms. Because we will mostly focus on standard algorithms throughout the book, this part, which is quite technical, may be skipped on a first reading. In Sect. 5.4/95, we consider shift invariant algorithms. Shift invariant algorithms have the property that the shape of the generated splines is independent of the starting point which we choose for labeling the segments $\mathbf{x}_j, j \in \mathbb{Z}_n$. The subdivision matrix of shift invariant algorithms is block-circulant and can be transformed to block-diagonal form by means of the Discrete Fourier Transform. This process is of major importance in applications, as well as for the further development of the theory. Typically, subdivision algorithms are not only shift invariant, but also indifferent with respect to a reversal of orientation of segment labels. Such symmetric algorithms are discussed in Sect. 5.5/103. We show that symmetric algorithms necessarily have a pair of real subdominant eigenvalues, justifying our focus on such schemes. Further, we specify easy-to-verify conditions for the characteristic ring which guarantee normal continuity and single-sheetedness of the generated spline surfaces.

5.1 Generic Initial Data

Since degenerate cases are unavoidable in any linear setting, we cannot expect a subdivision algorithm to generate geometrically smooth spline surfaces for all choices of initial data. In the extreme, subdivision will not even generate a surface: if all coefficients $\mathbf{q}_0 = \cdots = \mathbf{q}_{\bar{\ell}}$ coincide, it generates a sequence of rings that are all shrunk to a single point. The following definition allows us to discard degenerate constellations of coefficients so that we can focus on situations that have practical meaning.

Definition 5.1 (Generic initial data). A vector $\mathbf{Q} = [\mathbf{q}_0; \ldots; \mathbf{q}_{\bar{\ell}}]$ of initial data $\mathbf{q}_\ell \in \mathbb{R}^3$, and equally the corresponding vector $\mathbf{P} = V^{-1}\mathbf{Q}$ of eigencoefficients $\mathbf{p}_\ell \in \mathbb{C}^3$, is called *generic*, if any triple of eigencoefficients has full rank,

$$\det(\mathbf{p}_{r_1}^{i_1}, \mathbf{p}_{r_2}^{i_2}, \mathbf{p}_{r_3}^{i_3}) \neq 0, \quad (r_1, i_1) \neq (r_2, i_2) \neq (r_3, i_3) \neq (r_1, i_1).$$

Imposing conditions on all triples of eigencoefficients is more than needs to be required in the following. For instance, in the next section on standard algorithms, it is sufficient to assume that the eigencoefficients $\mathbf{p}_1, \mathbf{p}_2$ are linearly independent. However, since the set of non-generic initial data as introduced above has measure zero in $\mathbb{R}^{(\bar{\ell}+1)\times 3}$, anyway, we choose the simple, more stringent form of the requirement that will cover all cases of interest.

To classify subdivision algorithms, we regard smoothness of the generated surfaces for generic initial data only.

Definition 5.2 (C_r^k-subdivision algorithm). A C_0^k-subdivision algorithm is called

- C_r^k, respectively
- *normal continuous*, respectively
- *single-sheeted*,

if it generates spline surfaces that are

- C_r^k in the sense of Definition 3.12/51, respectively
- normal continuous in the sense of Definition 3.11/51, respectively
- single-sheeted in the sense of Definition 3.11/51

for any generic vector $\mathbf{Q}$ of initial data.

5.2 Standard Algorithms

Most subdivision algorithms of practical relevance have a double subdominant eigenvalue that is real. As will be explained in Sect. 5.5/103, double subdominant eigenvalues arise from symmetry properties of the algorithms.

Definition 5.3 (Standard algorithm, subdominant eigenvalue λ). A C_0^k-subdivision algorithm (A, G) is called a *standard algorithm*, if the subdivision matrix A has eigenvalues according to

$$1 > \lambda_1 = \lambda_2 > |\lambda_3|, \quad \ell_1 = \ell_2 = 0.$$

Moreover,

$$\lambda := \lambda_1 = \lambda_2$$

is called the *subdominant eigenvalue* of A.

This definition means that the second largest eigenvalue λ of the subdivision matrix is positive and has geometric and algebraic multiplicity 2. According to (4.26/74) and (4.29/74), with w_1^t, w_2^t denoting the second and third row of the matrix V^{-1} of left eigenvectors, the equations

$$Av_i = \lambda v_i, \quad f_i = Gv_i, \quad \mathbf{p}_i = w_i^t \mathbf{Q}, \quad i \in \{1,2\}, \tag{5.1}$$

characterize the corresponding pairs of *subdominant eigenvectors, eigenrings, and eigencoefficients*, respectively. With $f_0 = Gv_0 = 1$, we obtain the asymptotic expansion

$$\mathbf{x}^m = FJ^m\mathbf{P} \stackrel{*}{=} \mathbf{p}_0 + \lambda^m(f_1\mathbf{p}_1 + f_2\mathbf{p}_2) \tag{5.2}$$

for the sequence of rings generated by a standard algorithm. That is, first order behavior of $\mathbf{x}^m$ is completely determined by the user-given data $\mathbf{p}_0, \mathbf{p}_1, \mathbf{p}_2$ and the eigenrings f_1, f_2, which depend only on the algorithm. Together, f_1 and f_2 form a planar ring whose properties are crucial for understanding first order differentiability properties of subdivision surfaces.

Definition 5.4 (Characteristic ring ψ and spline χ, standard). Let (A,G) be a standard algorithm with Jordan decomposition $A = VJV^{-1}$ of A according to (4.22/73) and subdominant eigenrings f_1, f_2 according to (5.1/85). The planar ring

$$\psi := [f_1, f_2] = F[v_1, v_2] \in C^k(\mathbf{S}_n^0, \mathbb{R}^2, G)$$

is called the *characteristic ring* of the algorithm corresponding to V. Accordingly, with the subdominant eigensplines e_1, e_2 of Definition 4.24/78,

$$\chi := [e_1, e_2] = B[v_1, v_2] \in C_0^k(\mathbf{S}_n, \mathbb{R}^2) \tag{5.3}$$

is called the *characteristic spline*.

Since $A^m[v_1, v_2] = \lambda^m[v_1, v_2]$, the rings of the characteristic spline are scaled copies of the characteristic ring,

$$\chi^m = \lambda^m \psi. \tag{5.4}$$

For standard algorithms, the characteristic spline χ inherits from equation (4.17/69) the scaling property

$$\chi(2^{-m}\mathbf{s}) = \lambda^m \chi(\mathbf{s}), \quad \mathbf{s} \in \mathbf{S}_n, \; m \in \mathbb{N}_0. \tag{5.5}$$

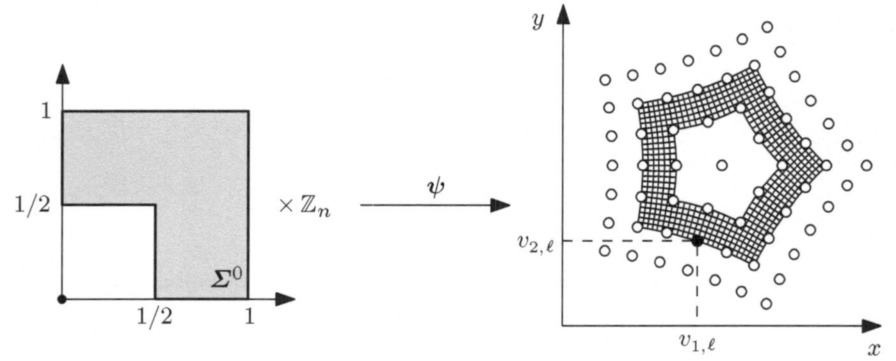

Fig. 5.1 Illustration of Definition 5.4/85: Characteristic ring ψ and its coefficients ∘, which are given by the components of the subdominant eigenvectors v_1, v_2.

The coefficients of the characteristic ring are points in $\mathbb{R}^2$ given by the rows of the matrix $[v_1, v_2]$ of subdominant eigenvectors v_1, v_2 (cf. Fig. 5.1/86):

$$\psi = G[v_1, v_2].$$

These eigenvectors are not uniquely defined, and hence also the matrix V used for Jordan decomposition is ambiguous. However, any two admissible pairs are related by a regular (2×2)-matrix L according to $[\tilde{v}_1, \tilde{v}_2] = [v_1, v_2]L$. The corresponding characteristic rings satisfy $\tilde{\psi} = \psi L$. That is, ψ and $\tilde{\psi}$ are related by a regular linear map. By this relation, the set of all possible characteristic rings becomes an equivalence class. The basic properties of characteristic rings that are employed in the sequel, namely regularity and induced winding numbers, are invariant under that relation. In this regard, any choice of V is as good as any other. Therefore, we omit the suffix "corresponding to V" when talking about characteristic rings or characteristic splines (Fig. 5.2/86).

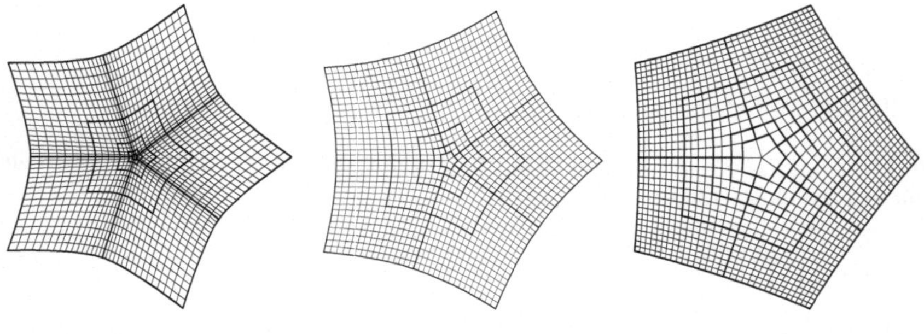

Fig. 5.2 Illustration of Definition 5.4/85: Characteristic spline χ of a standard algorithm for subdominant eigenvalues (*left*) $\lambda = 3/8$, (*middle*) $\lambda = 1/2$, and (*right*) $\lambda = 5/8$.

Now, (5.2/85) reads

$$\mathbf{x}^m \stackrel{*}{=} \mathbf{p}_0 + \lambda^m \psi[\mathbf{p}_1; \mathbf{p}_2]. \tag{5.6}$$

In order to compute normal vectors, we use the operator ${}^{\times}\!D$, as introduced in (2.1/17). By (2.2/17),

$${}^{\times}\!D\mathbf{x}^m = D_1\mathbf{x}^m \times D_2\mathbf{x}^m \stackrel{*}{=} \lambda^{2m}\, {}^{\times}\!D\psi\, (\mathbf{p}_1 \times \mathbf{p}_2), \tag{5.7}$$

where we recall that ${}^{\times}\!D\psi = D_1 f_1 D_2 f_2 - D_2 f_1 D_1 f_2$ is the Jacobian determinant of the characteristic ring.

Definition 5.5 (Regularity of ψ). The characteristic ring ψ is called *regular*, if ${}^{\times}\!D\psi$ has no zeros.

The following theorem shows that regularity of the characteristic ring is sufficient for a standard algorithm to be normal continuous, and, moreover, discards algorithms with ${}^{\times}\!D\psi$ changing sign.

Theorem 5.6 (Regularity of ψ and normal continuity, standard). A standard algorithm with characteristic ring ψ is

- normal continuous with central normal

$$\mathbf{n}^{\mathrm{c}} = \operatorname{sign}({}^{\times}\!D\psi)\, \frac{\mathbf{p}_1 \times \mathbf{p}_2}{\|\mathbf{p}_1 \times \mathbf{p}_2\|},$$

  if ψ is regular;
- not normal continuous, if ${}^{\times}\!D\psi$ changes sign.

Proof. We assume generic initial data, hence $\mathbf{p}_1 \times \mathbf{p}_2 \neq \mathbf{0}$, for both parts of the proof. First, let us assume that ψ is regular. Since, by Theorem 4.7/64, ${}^{\times}\!D\psi$ is continuous on the compact domain $\mathbf{S}_n^0$, the absence of zeros implies that $\operatorname{sign}({}^{\times}\!D\psi)$ is continuous, and that $1/|{}^{\times}\!D\psi|$ is bounded. Hence, we obtain

$$\frac{{}^{\times}\!D\mathbf{x}^m}{\lambda^2\, |{}^{\times}\!D\psi|} \stackrel{*}{=} \operatorname{sign}({}^{\times}\!D\psi)\, (\mathbf{p}_1 \times \mathbf{p}_2) \neq \mathbf{0}$$

and see that $\mathbf{x}^m$ is regular for almost all m. Further, the normal vectors $\mathbf{n}^m$ are convergent according to

$$\mathbf{n}^m = \frac{{}^{\times}\!D\mathbf{x}^m}{\|{}^{\times}\!D\mathbf{x}^m\|} \stackrel{*}{=} \operatorname{sign}({}^{\times}\!D\psi)\, \frac{\mathbf{p}_1 \times \mathbf{p}_2}{\|\mathbf{p}_1 \times \mathbf{p}_2\|} = \mathbf{n}^{\mathrm{c}}.$$

Theorem 4.7/64 implies normal continuity, as stated. Second, let us assume that ${}^{\times}\!D\psi(\mathbf{s}_1)\, {}^{\times}\!D\psi(\mathbf{s}_2) < 0$ for some arguments $\mathbf{s}_1, \mathbf{s}_2 \in \mathbf{S}_n^0$. Here, we obtain

$$\mathbf{n}^m(\mathbf{s}_i) \stackrel{*}{=} \operatorname{sign}({}^{\times}\!D\psi(\mathbf{s}_i))\, \frac{\mathbf{p}_1 \times \mathbf{p}_2}{\|\mathbf{p}_1 \times \mathbf{p}_2\|}, \quad i \in \{1, 2\},$$

and see that $\mathbf{n}^m$ cannot converge to a constant limit since $\|\mathbf{n}^m(\mathbf{s}_1) - \mathbf{n}^m(\mathbf{s}_2)\| \stackrel{*}{=} 2$. □

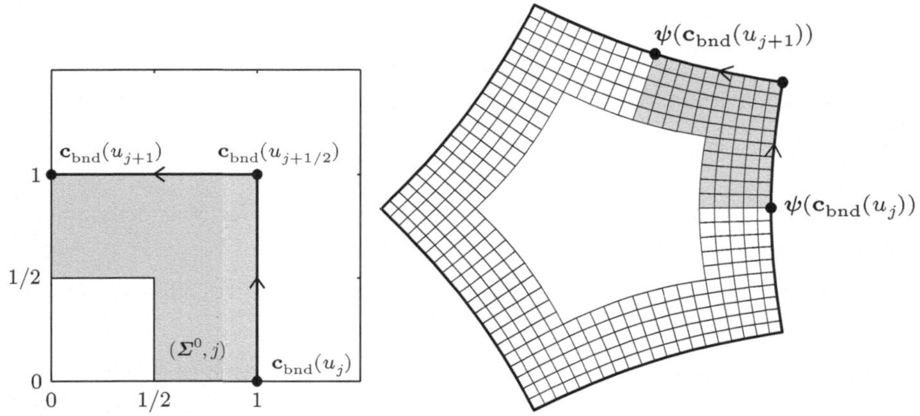

Fig. 5.3 Illustration of Definition 5.7/88: (*left*) Curve $\mathbf{c}_{\mathrm{bnd}}$ in $\mathbf{S}_n^0$ and (*right*) curve $\psi \circ \mathbf{c}_{\mathrm{bnd}}$ in $\mathbb{R}^2$.

Theorem 5.6/87 covers all but the case where ${}^{\times}\!D\psi$ has zeros without changing sign. Here, the behavior of ${}^{\times}\!D\mathbf{x}^m$ depends on higher order eigencoefficients and cannot be determined a priori.

Now, the issue of single-sheetedness has to be addressed, and again, properties of the characteristic ring are crucial. We consider the curve $\mathbf{c}_{\mathrm{bnd}} : U = [0, 1] \to \mathbf{S}_n^0$ in the domain of ψ which parametrizes the outer boundary: With $u_j := j/n$, let

$$\mathbf{c}_{\mathrm{bnd}}(t) := \begin{cases} (1, 2n(u - u_j), j) & \text{if } u_j \le u \le u_{j+1/2} \\ (2n(u_{j+1} - u), 2, j) & \text{if } u_{j+1/2} \le u \le u_{j+1}, \end{cases}$$

see Fig. 5.3/88 *left*. As shown in Example 3.10/50, its winding number is $\nu(\mathbf{c}_{\mathrm{bnd}}, \mathbf{0}) = 1$.

Definition 5.7 (Winding number of ψ). The *winding number* of the characteristic ring $\psi \in C^k(\mathbf{S}_n^0, \mathbb{R}^2)$ is defined as

$$\nu(\psi) := \nu(\psi \circ \mathbf{c}_{\mathrm{bnd}}, \mathbf{0}),$$

see Fig. 5.3/88. We say that ψ is *uni-cyclic* if $|\nu(\psi)| = 1$.

We are now able to prove an easy-to-verify criterion for the single-sheetedness of subdivision algorithms in terms of the winding number of the characteristic ring.

Theorem 5.8 (Winding number of ψ and single-sheetedness, standard). Consider a standard algorithm (A, G) with a regular characteristic ring $\psi \in C^k(\mathbf{S}_n^0, \mathbb{R}^2)$. Then the following assertions are equivalent:

- (A, G) is a C_1^k-subdivision algorithm.
- The characteristic ring ψ is uni-cyclic.
- The characteristic ring ψ is injective.

Proof. First, we prove equivalence of the first and the second assertion. We consider the rings $\boldsymbol{\xi}_*^m$ of the tangential component $\boldsymbol{\xi}_*$ of $\mathbf{x}_*$ according to (4.11/64). By (5.6/87),

$$\boldsymbol{\xi}_*^m = (\mathbf{x}^m - \mathbf{x}^c) \cdot \mathbf{T}^c \overset{*}{=} \lambda^m \psi[\mathbf{p}_1; \mathbf{p}_2] \cdot \mathbf{T}^c.$$

For generic initial data, the (2×2)-matrix $L := [\mathbf{p}_1; \mathbf{p}_2] \cdot \mathbf{T}^c$ is invertible. On one hand, by (2.5/17),

$$^{\times}\!D\boldsymbol{\xi}_*^m \overset{*}{=} \lambda^{2m}\, {}^{\times}\!D\psi \det L.$$

This implies that $\boldsymbol{\xi}_*$ and hence also $\mathbf{x}$ is locally almost regular. On the other hand, let

$$\tilde{\boldsymbol{\xi}}^m := \lambda^{-m} \boldsymbol{\xi}_*^m L^{-1},$$

then $\tilde{\boldsymbol{\xi}}^m \overset{*}{=} \psi$. By continuity and affine invariance of the winding number,

$$\lim_{m\to\infty} \nu(\boldsymbol{\xi}_*^m \circ \mathbf{c}_{\text{bnd}}, 0) = \lim_{m\to\infty} \nu(\tilde{\boldsymbol{\xi}}^m \circ \mathbf{c}_{\text{bnd}}, 0) = \nu(\psi).$$

Combining the two observations, we see that there exists $m_0 \in \mathbb{N}_0$ such that $\boldsymbol{\xi}_*$ is m_0-almost regular and $\nu(\boldsymbol{\xi}_*^{m_0} \circ \mathbf{c}_{\text{bnd}}, 0) = \nu(\psi)$. Hence, by Theorem 4.8/64, $\boldsymbol{\xi}_*$ is single-sheeted if and only if $|\nu(\psi)| = 1$, i.e., if ν is uni-cyclic.

Second, we prove equivalence of the second and the third assertion. We define the spline surface $\mathbf{x} := [\boldsymbol{\chi}, 0]$. Using (5.4/85), we have $\|{}^{\times}\!D\mathbf{x}^m\| = |{}^{\times}\!D\boldsymbol{\chi}^m| = \lambda^{2m}|{}^{\times}\!D\psi|$, showing that $\mathbf{x}$ is almost regular. Further, $\mathbf{x}$ is normal continuous with $\mathbf{n}^c = [0, 0, 1]$ and $\mathbf{x}^c = \mathbf{0}$. Hence, $\boldsymbol{\xi}_* = \boldsymbol{\chi}$, and

$$\nu(\boldsymbol{\xi}_* \circ \mathbf{c}_{\text{bnd}}, 0) = \nu(\boldsymbol{\chi} \circ \mathbf{c}_{\text{bnd}}, 0) = \nu(\psi \circ \mathbf{c}_{\text{bnd}}, 0) = \nu(\psi).$$

If ψ is uni-cyclic, then $\mathbf{x}$ is single-sheeted by Theorem 3.15/53. This implies that $\boldsymbol{\chi}$ and hence also $\psi = \boldsymbol{\chi}^0$ is injective. Conversely, if ψ is injective, then the curve $\psi \circ \mathbf{c}_{\text{bnd}}$ is injective and can be deformed continuously to a circle with winding number ± 1. □

In applications, it is *much* easier to check if ψ is uni-cyclic than to consider global injectivity. Since the conditions given above are sufficient and (almost) necessary for generating C_1^k-surfaces, we conclude with the following definition:

Definition 5.9 (Standard C_1^k-algorithm). A standard algorithm is called a *standard C_1^k-algorithm*, if its characteristic ring ψ is regular with ${}^{\times}\!D\psi > 0$ and uni-cyclic.

Assuming positivity of ${}^{\times}\!D\psi$ is not restrictive. If ${}^{\times}\!D\psi = {}^{\times}\!D(G[v_1, v_2]) < 0$ then interchanging v_1 and v_2 readily yields the desired sign.

5.3 General Algorithms

Standard algorithms cover most cases of practical relevance. Yet, there are legitimate algorithms, for example "simplest subdivision" in Sect. 6.3/120, that have a different eigenstructure-structure; and, certainly, identifying and characterizing more classes of C_1^k-subdivision algorithms is of interest in its own right. This section shows how a careful extension of the concepts used in the standard case yields results of very similar flavor and identical wording also in more general settings.

Specifically, we give six families of possible subdivision matrices that, in principle, are suitable to generate C_1^k-surfaces. We focus on showing that membership in each family implies smoothness; completeness is proven in [Rei99].

Let us consider the sequence $\{\mathbf{x}^m\}_m$ of rings according to (4.26/74) forming the spline $\mathbf{x}$. By (2.3/17), the cross product of partial derivatives has the form

$$^{\times}D\mathbf{x}^m = \sum_i a_i^m h_i \mathbf{c}_i. \tag{5.8}$$

As specified later, the $\{a_i^m\}_m$ form decaying sequences of scaling factors, the $h_i = h_i(\mathbf{s})$ are real-valued rings, and the $\mathbf{c}_i$ are cross products of pairs of eigencoefficients. If the above sum has a single dominant term, i.e.,

$$^{\times}D\mathbf{x}^m \stackrel{*}{=} a_1^m h_1 \mathbf{c}_1, \tag{5.9}$$

and if in addition $a_1^m h_1(\mathbf{s})$ has constant sign $s = \pm 1$ for all $(m, \mathbf{s}) \in \mathbb{N} \times \mathbf{S}_n^0$, then normalization yields normal continuity according to

$$\mathbf{n}^m \stackrel{*}{=} s\,\mathbf{c}_1/\|\mathbf{c}_1\|.$$

It is easy to see that alternating behavior of the sequence with elements a_1^m, or sign changes of h_1 destroy convergence.

Now, we consider the case of a multiple dominant term in (5.8/90). For simplicity, we assume that it is double, and write

$$^{\times}D\mathbf{x}^m \stackrel{*}{=} |a_1^m|\big(s_1^m h_1 \mathbf{c}_1 + s_2^m h_2 \mathbf{c}_2\big), \quad |s_1^m| = |s_2^m| = 1.$$

Choosing a subsequence (for simplicity we reuse the index m) such that $s_1^m \to s_1$, $s_2^m \to s_2$, we obtain

$$\mathbf{n}^m \stackrel{*}{=} \frac{s_1 h_1 \mathbf{c}_1 + s_2 h_2 \mathbf{c}_2}{\|s_1 h_1 \mathbf{c}_1 + s_2 h_2 \mathbf{c}_2\|}.$$

This expression can only converge to a constant limit either if the vectors $\mathbf{c}_1, \mathbf{c}_2$ or the functions h_1, h_2 are linearly dependent. The simple argument is left to the reader. The first case is possible only for non-generic data, while the second one corresponds to the exceptional situation that two functions, which are not interrelated by deeper principles, happen to be linearly dependent.

Consequently, we will search algorithms for which the sum in (5.8/90) has a single dominant term. Using (4.26/74) and (4.27/74), we obtain

$$\begin{aligned}
\mathbf{x}^m = \mathbf{p}_0 &+ \lambda_1^{m,\ell_1} f_1^0 \mathbf{p}_1^{\ell_1} + \lambda_1^{m,\ell_1-1}(f_1^0 \mathbf{p}_1^{\ell_1-1} + f_1^1 \mathbf{p}_1^{\ell_1}) \\
&+ \lambda_1^{m,\ell_1-2}(f_1^0 \mathbf{p}_1^{\ell_1-2} + f_1^1 \mathbf{p}_1^{\ell_1-1} + f_1^2 \mathbf{p}_1^{\ell_1}) \\
&+ \lambda_2^{m,\ell_2} f_2^0 \mathbf{p}_2^{\ell_2} + \mathbf{r}^m,
\end{aligned}$$

where the remainder term satisfies, with the notation for asymptotics of sequences in Sect. 4.5/71,

$$\mathbf{r}^m \preccurlyeq \lambda_1^{m,\ell_1-3} + \lambda_2^{m,\ell_2-1} + \lambda_3^{m,\ell_3}.$$

The cross product of partial derivatives can be computed with the aid of (2.3/17). We obtain

$$^{\times}\!D\mathbf{x}^m = \tilde{\mathbf{n}}_1^m + \tilde{\mathbf{n}}_2^m + \tilde{\mathbf{r}}^m \tag{5.10}$$

with

$$\tilde{\mathbf{n}}_1^m := \lambda_1^{m,\ell_1}\lambda_2^{m,\ell_2}\, {}^{\times}\!D[f_1^0, f_2^0](\mathbf{p}_1^{\ell_1} \times \mathbf{p}_2^{\ell_2})$$
$$\tilde{\mathbf{n}}_2^m := \left((\lambda_1^{m,\ell_1-1})^2 - \lambda_1^{m,\ell_1}\lambda_1^{m,\ell_1-2}\right) {}^{\times}\!D[f_1^0, f_1^1](\mathbf{p}_1^{\ell_1} \times \mathbf{p}_1^{\ell_1-1})$$

and a remainder term $\tilde{\mathbf{r}}^m$ which can be bounded by

$$\tilde{\mathbf{r}}^m \preccurlyeq \lambda_1^{m,\ell_1}(\lambda_1^{m,\ell_1-3} + \lambda_2^{m,\ell_2-1} + \lambda_3^{m,\ell_3}) + \lambda_1^{m,\ell_1-1}(\lambda_1^{m,\ell_1-2} + \lambda_2^{m,\ell_2}).$$

Recalling our convention that $\lambda^{m,\ell} = 0$ for $\ell < 0$, we find $\tilde{\mathbf{n}}_2^m = 0$ if $\ell_1 = 0$, and

$$\tilde{\mathbf{n}}_2^m \stackrel{*}{=} \ell_1^{-1}\,(\lambda_1^{m,\ell_1-1})^2\, {}^{\times}\!D[f_1^0, f_1^1](\mathbf{p}_1^{\ell_1} \times \mathbf{p}_1^{\ell_1-1}) \quad \text{if} \quad \ell_1 > 0.$$

The order of decay of the three summands is easily determined using (4.19/72),

$$\tilde{\mathbf{n}}_1^m \sim (\lambda_1\lambda_2)^m\, m^{\ell_1+\ell_2}$$
$$\tilde{\mathbf{n}}_2^m \sim \begin{cases} 0 & \text{if } \ell_1 = 0 \\ \lambda_1^{2m}\, m^{2\ell_1-2} & \text{if } \ell_1 > 0 \end{cases}$$
$$\tilde{\mathbf{r}}^m \preccurlyeq \begin{cases} (\lambda_1\lambda_3)^m\, m^{\ell_3} & \text{if } \ell_1 = 0 \\ \lambda_1^{2m}\, m^{2\ell_1-3} + (\lambda_1\lambda_2)^m\, m^{\ell_1+\ell_2-1} + (\lambda_1\lambda_3)^m\, m^{\ell_1+\ell_3} & \text{if } \ell_1 > 0. \end{cases}$$

Now, we are prepared to determine a list of cases where in the representation (5.10/91) either $\tilde{\mathbf{n}}_1^m$ or $\tilde{\mathbf{n}}_2^m$ is the strictly dominant term. In the following,

$$\mathcal{A} := \{A : \lambda_1 \neq 0\}$$

denotes the set of all subdivision matrices according to Definition 4.27/80 excluding the trivial case $\lambda_1 = 0$. We distinguish the following cases:

Case 1: $(\lambda_1, \ell_1) \sim (\lambda_2, \ell_2)$, i.e., there is a multiple subdominant eigenvalue. In this case, $\tilde{\mathbf{n}}_2^m \prec \tilde{\mathbf{n}}_1^m$, and $\tilde{\mathbf{r}}^m \prec \tilde{\mathbf{n}}_2^m$ if $(\lambda_2, \ell_2) \succ (\lambda_3, \ell_3)$. We distinguish two sub-cases:

1-1: $\lambda_1 \in \mathbb{R}$. Here, λ_2 is also real. If $\lambda_1 = -\lambda_2$, then $\tilde{\mathbf{n}}_2^m$ is alternating and the algorithm cannot be normal continuous. The case $\lambda_1 = \lambda_2$ does not lead to such problems, and the corresponding class of subdivision matrices is denoted by

$$\mathcal{A}_1^1 := \{A \in \mathcal{A} : (\lambda_1, \ell_1) = (\lambda_2, \ell_2) \succ (\lambda_3, \ell_3),\ \lambda_1 \in \mathbb{R}\}.$$

We note that standard algorithms, as introduced in the last section, belong to this class with $\ell_1 = \ell_2 = 0$.

1-2: $\lambda_1 \notin \mathbb{R}$. The complex sub-case yields the class

$$\mathcal{A}_1^2 := \{A \in \mathcal{A} : (\lambda_1, \ell_1) = (\overline{\lambda_2}, \ell_2) \succ (\lambda_3, \ell_3),\ \lambda_1 \notin \mathbb{R}\}.$$

Case 2: $|\lambda_1| = |\lambda_2|$, $\ell_1 > \ell_2$, i.e., we have equal modulus, but differing multiplicities of the first and second eigenvalue. We distinguish three sub-cases:

2-1: $\ell_1 = \ell_2 + 1$. Here, $\tilde{\mathbf{n}}_2^m \prec \tilde{\mathbf{n}}_1^m$, and $\tilde{\mathbf{r}}^m \prec \tilde{\mathbf{n}}_1^m$ if $(\lambda_2, \ell_2) \succ (\lambda_3, \ell_3)$. In that case, λ_1 and λ_2 are both real, and their product has to be positive to avoid alternating behavior of $\tilde{\mathbf{n}}_1^m$. We obtain the class

$$\mathcal{A}_2^1 := \{A \in \mathcal{A} : \lambda_1 = \lambda_2,\ \ell_1 = \ell_2 + 1,\ (\lambda_2, \ell_2) \succ (\lambda_3, \ell_3)\}.$$

2-2: $\ell_1 > \ell_2 + 2$. Here, $\tilde{\mathbf{r}}^m \preccurlyeq \tilde{\mathbf{n}}_1^m \prec \tilde{\mathbf{n}}_2^m$, and we denote

$$\mathcal{A}_2^2 := \{A \in \mathcal{A} : |\lambda_1| = |\lambda_2|,\ \ell_1 > \ell_2 + 2\}.$$

2-3: $\ell_1 = \ell_2 + 2$. Here, $\tilde{\mathbf{n}}_1^m \sim \tilde{\mathbf{n}}_2^m$ implies decay at equal rates of both terms, and normal continuity cannot be expected by the argument similar to the one ruling out multiple dominant terms in (5.8/90).

Case 3: $|\lambda_1| > |\lambda_2|$. We distinguish two sub-cases:

3-1: $\ell_1 = 0$. Here, $\tilde{\mathbf{n}}_2^m \prec \tilde{\mathbf{n}}_1^m$, and $\tilde{\mathbf{r}}^m \prec \tilde{\mathbf{n}}_1^m$ if $(\lambda_2, \ell_2) \succ (\lambda_3, \ell_3)$. Further, the sign of λ_1 and λ_2 has to be equal to avoid alternating behavior. This sub-case is denoted by

$$\mathcal{A}_3^2 := \{A \in \mathcal{A} : |\lambda_1| > |\lambda_2|,\ \ell_1 = 0,\ \lambda_1\lambda_2 > 0,\ (\lambda_2, \ell_2) \succ (\lambda_3, \ell_3)\}.$$

3-2: $\ell_1 \geq 1$. Here, $\tilde{\mathbf{n}}_1^m \prec \tilde{\mathbf{n}}_2^m$ and also $\tilde{\mathbf{r}}^m \prec \tilde{\mathbf{n}}_2^m$. We denote

$$\mathcal{A}_3^1 := \{A \in \mathcal{A} : |\lambda_1| > |\lambda_2|,\ \ell_1 \geq 1\}.$$

Summarizing, $\tilde{\mathbf{n}}_1^m$ is dominant if the subdivision matrix lies in $\mathcal{A}_1^1, \mathcal{A}_1^2, \mathcal{A}_2^1$ or $\mathcal{A}_3^1$, and $\tilde{\mathbf{n}}_2^m$ is dominant if it lies in $\mathcal{A}_2^2$ or $\mathcal{A}_3^2$.

The, off-hand heuristic, partition of cases into six families will turn out to simplify the analysis that we start by extending the definition of the characteristic ring.

Definition 5.10 (Characteristic ring, general). Let (A, G) be a subdivision algorithm with subdivision matrix $A \in \mathcal{A}_p^q$, $p \in \{1,2,3\}$, $q \in \{1,2\}$. We define the *characteristic ring* $\psi \in C^k(\mathbf{S}_n^0, \mathbb{R}^2)$ by

$$\psi := \begin{cases} [f_1^0, f_2^0] & \text{if } A \in \mathcal{A}_1^1 \cup \mathcal{A}_2^1 \cup \mathcal{A}_3^1 \\ [\operatorname{Re} f_1^0, \operatorname{Im} f_1^0] & \text{if } A \in \mathcal{A}_1^2 \\ [f_1^0, f_1^1] & \text{if } A \in \mathcal{A}_2^2 \cup \mathcal{A}_3^2, \end{cases}$$

and the (2×3)-matrix $\mathbf{P}^*$ of eigencoefficients by

$$\mathbf{P}^* := [\mathbf{p}_1^*; \mathbf{p}_2^*] := \begin{cases} [\mathbf{p}_1^{\ell_1}; \mathbf{p}_2^{\ell_2}] & \text{if } A \in \mathcal{A}_1^1 \cup \mathcal{A}_2^1 \cup \mathcal{A}_3^1 \\ [\operatorname{Re} \mathbf{p}_1^{\ell_1}; -\operatorname{Im} \mathbf{p}_1^{\ell_1}] & \text{if } A \in \mathcal{A}_1^2 \\ [\mathbf{p}_1^{\ell_1}; \mathbf{p}_1^{\ell_1 - 1}] & \text{if } A \in \mathcal{A}_2^2 \cup \mathcal{A}_3^2. \end{cases}$$

ψ is called *regular*, if its Jacobian determinant ${}^\times\!D\psi$ has no zeros.

The following result on normal continuity is completely analogous to Theorem 5.6/87 and the conclusion is verbatim the same.

Theorem 5.11 (Regularity of ψ and normal continuity, general). *A subdivision algorithm (A, G) with $A \in \mathcal{A}_p^q$ and characteristic ring ψ according to the preceding definition is*

- *normal continuous with limit*

$$\mathbf{n}^c = \operatorname{sign}({}^\times\!D\psi) \frac{\mathbf{p}_1^* \times \mathbf{p}_2^*}{\|\mathbf{p}_1^* \times \mathbf{p}_2^*\|},$$

  if the characteristic ring is regular,
- *not normal continuous; if ${}^\times\!D\psi$ changes sign.*

Proof. With the scaling factor

$$a_m := \begin{cases} \lambda_1^{m,\ell_1} \lambda_2^{m,\ell_2} & \text{if } A \in \mathcal{A}_1^1 \cup \mathcal{A}_2^1 \cup \mathcal{A}_3^1 \cup \mathcal{A}_1^2 \\ \ell_1^{-1} (\lambda_1^{m,\ell_1 - 1})^2 & \text{if } A \in \mathcal{A}_2^2 \cup \mathcal{A}_3^2, \end{cases}$$

the cross product of the partial derivatives of the rings is

$${}^\times\!D\mathbf{x}^m \stackrel{*}{=} a_m \, {}^\times\!D\psi \, (\mathbf{p}_1^* \times \mathbf{p}_2^*).$$

In the complex case $A \in \mathcal{A}_1^2$ the relations $f_1^0 = \overline{f}_2^0$ and $\mathbf{p}_1^{\ell_1} = \overline{\mathbf{p}}_2^{\ell_2} = \mathbf{p}_1^* - i\mathbf{p}_2^*$ are used to obtain the real representation. Now, the proof proceeds exactly as for Theorem 5.6/87. □

The wording of the theorem below in the general case is verbatim the same as in the standard case, i.e., as for Theorem 5.8/88 on C_k^1-regularity.

Theorem 5.12 (Winding number of ψ and single-sheetedness, general). *Consider a subdivision algorithm with matrix $A \in \mathcal{A}_p^q$ with a regular characteristic ring $\psi \in C^k(\mathbf{S}_n^0, \mathbb{R}^2)$. Then the following assertions are equivalent:*

- *(A, G) is a C_1^k-subdivision algorithm.*
- *The characteristic ring ψ is uni-cyclic.*
- *The characteristic ring ψ is injective.*

Proof. In all six cases, we will specify sequences $\{\Lambda^m\}$ of invertible (2×2)-matrices such that the rings can be written as

$$\mathbf{x}^m = \mathbf{p}_0 + \psi \Lambda^m \mathbf{P}^* + \mathbf{r}^m$$

with a suitable remainder term $\mathbf{r}^m$. Let us denote the LQ-decomposition of $\mathbf{P}^*$ by $\mathbf{P}^* = L\mathbf{T}$. That is, L is a lower triangular matrix, and $\mathbf{T}$ consists of two orthonormal row-vectors spanning the same plane as the rows of $\mathbf{P}^*$. For generic data, $\mathbf{P}^*$ has full rank, and hence L is invertible. The projection to the tangent plane at the center is

$$\boldsymbol{\xi}^m = (\mathbf{x}^m - \mathbf{p}_0)\cdot \mathbf{T} = \psi \Lambda^m L + \mathbf{r}^m \cdot \mathbf{T}.$$

We define $\tilde{\boldsymbol{\xi}}^m := \boldsymbol{\xi}^m L^{-1}(\Lambda^m)^{-1}$ and obtain

$$\tilde{\boldsymbol{\xi}}^m = \psi + \boldsymbol{\rho}^m, \quad \boldsymbol{\rho}^m := (\mathbf{r}^m \cdot \mathbf{T}) L^{-1}(\Lambda^m)^{-1}.$$

We will show that in all cases the remainder term satisfies $\boldsymbol{\rho}^m \prec 1$, i.e., it converges to 0. Thus, $\tilde{\boldsymbol{\xi}}^m \stackrel{*}{=} \psi$, and all the rest of the conclusion proceeds exactly as in Theorem 5.8/88. It remains to add the details for the six cases. Throughout, we omit the subscript of the first eigenvalue, $(\lambda, \ell) := (\lambda_1, \ell_1)$.

Case 1-1: For $(\lambda_2, \ell_2) = (\lambda, \ell)$, the leading terms are

$$\mathbf{x}^m \stackrel{*}{=} \mathbf{p}_0 + \lambda^{m,\ell} f_1^0 \mathbf{p}_1^\ell + \lambda^{m,\ell} f_2^0 \mathbf{p}_2^\ell.$$

With $\psi = [f_1^0, f_2^0]$, $\mathbf{P}^* = [\mathbf{p}_1^\ell; \mathbf{p}_2^\ell]$, and

$$\Lambda^m := \begin{bmatrix} \lambda^{m,\ell} & 0 \\ 0 & \lambda^{m,\ell} \end{bmatrix}, \quad (\Lambda^m)^{-1} \preccurlyeq 1/\lambda^{m,\ell},$$

the remainder terms satisfy

$$\mathbf{r}^m \prec \lambda^{m,\ell}, \quad \boldsymbol{\rho}^m \sim \mathbf{r}^m (\Lambda^m)^{-1} \prec 1.$$

Case 1-2: For $(\lambda_2, \ell_2) = (\bar{\lambda}, \ell)$, we have

$$\mathbf{x}^m \stackrel{*}{=} \mathbf{p}_0 + 2\,\mathrm{Re}(\lambda^{m,\ell} f_1^0 \mathbf{p}_1^\ell)$$
$$\psi = [\mathrm{Re}\, f_1^0, \mathrm{Im}\, f_1^0], \quad \mathbf{P}^* = [\mathrm{Re}\,\mathbf{p}_1^\ell; -\mathrm{Im}\,\mathbf{p}_1^\ell]$$
$$\Lambda^m := 2\begin{bmatrix} \mathrm{Re}\,\lambda^{m,\ell} & \mathrm{Im}\,\lambda^{m,\ell} \\ -\mathrm{Im}\,\lambda^{m,\ell} & \mathrm{Re}\,\lambda^{m,\ell} \end{bmatrix}, \quad (\Lambda^m)^{-1} = \frac{1}{2}(\Lambda^m)^{\mathrm{t}}/|\lambda^{m,\ell}|^2 \preccurlyeq 1/\lambda^{m,\ell}$$
$$\mathbf{r}^m \prec \lambda^{m,\ell}, \quad \boldsymbol{\rho}^m \sim \mathbf{r}^m(\Lambda^m)^{-1} \prec 1$$

Case 2-1: For $(\lambda_2, \ell_2) = (\lambda, \ell - 1)$, the leading terms of $\mathbf{x}^m$ are

$$\mathbf{x}^m \stackrel{*}{=} \mathbf{p}_0 + \lambda^{m,\ell} f_1^0 \mathbf{p}_1^\ell + \lambda^{m,\ell-1}(f_2^0 \mathbf{p}_2^{\ell-1} + f_1^0 \mathbf{p}_1^{\ell-1} + f_1^1 \mathbf{p}_1^\ell),$$

and the characteristic ring is $\psi = [f_1^0, f_2^0]$. With a vector $\mathbf{n}^c$ perpendicular to $\mathbf{P}^* = [\mathbf{p}_1^\ell; \mathbf{p}_2^{\ell-1}]$ we decompose $\mathbf{p}_1^{\ell-1} = [a, b]\mathbf{P}^* + c\mathbf{n}^c$. Setting

$$\Lambda^m := \begin{bmatrix} \lambda^{m,\ell} & b\lambda^{m,\ell-1} \\ 0 & \lambda^{m,\ell-1} \end{bmatrix}, \quad (\Lambda^m)^{-1} \sim (\lambda^{m,\ell})^{-1} \begin{bmatrix} 1 & -b \\ 0 & m \end{bmatrix},$$

we obtain the remainder term $\mathbf{r}^m \overset{*}{=} \lambda^{m,\ell-1}(cf_1^0\mathbf{n}^c + af_1^0\mathbf{p}_1^\ell + f_1^1\mathbf{p}_1^\ell)$. Using $\mathbf{n}^c\mathbf{P}^* = 0$ and $\mathbf{p}_1^\ell\mathbf{P}^* \sim [1, 0]$, we find

$$\boldsymbol{\rho}^m \sim [1/m, 0] \begin{bmatrix} 1 & -b \\ 0 & m \end{bmatrix} \prec 1.$$

Case 2-2: For $|\lambda_2| = |\lambda|$ and $\ell_2 < \ell - 2$, we have

$$\begin{aligned} \mathbf{x}^m &\overset{*}{=} \mathbf{p}_0 + \lambda^{m,\ell} f_1^0 \mathbf{p}_1^\ell + \lambda^{m,\ell-1}(f_1^0\mathbf{p}_1^{\ell-1} + f_1^1\mathbf{p}_1^\ell) \\ \psi &= [f_1^0, f_1^1], \quad \mathbf{P}^* = [\mathbf{p}_1^\ell; \mathbf{p}_1^{\ell-1}] \\ \Lambda^m &:= \begin{bmatrix} \lambda^{m,\ell} & \lambda^{m,\ell-1} \\ 0 & \lambda^{m,\ell-1} \end{bmatrix}, \quad (\Lambda^m)^{-1} \preccurlyeq (\lambda^{m,\ell-1})^{-1} \\ \mathbf{r}^m &\prec \lambda^{m,\ell-1}, \quad \boldsymbol{\rho} \sim \mathbf{r}^m(\Lambda^m)^{-1} \prec 1. \end{aligned}$$

Case 3-1: For $|\lambda_2| < |\lambda_1|$ and $\ell = 0$, we have

$$\begin{aligned} \mathbf{x}^m &\overset{*}{=} \mathbf{p}_0 + \lambda^{m,\ell} f_1^0 \mathbf{p}_1^\ell + \lambda_2^{m,\ell_2} f_2^0 \mathbf{p}_2^{\ell_2} \\ \psi &= [f_1^0, f_2^0], \quad \mathbf{P}^* = [\mathbf{p}_1^\ell; \mathbf{p}_2^{\ell_2}] \\ \Lambda^m &:= \begin{bmatrix} \lambda^{m,\ell} & 0 \\ 0 & \lambda_2^{m,\ell_2} \end{bmatrix}, \quad (\Lambda^m)^{-1} \preccurlyeq 1/\lambda_2^{m,\ell_2} \\ \mathbf{r}^m &\prec \lambda_2^{m,\ell_2}, \quad \boldsymbol{\rho}^m \sim \mathbf{r}^m(\Lambda^m)^{-1} \prec 1. \end{aligned}$$

Case 3-2: For $|\lambda_2| < |\lambda_1|$ and $\ell > 0$, we have

$$\begin{aligned} \mathbf{x}^m &\overset{*}{=} \mathbf{p}_0 + \lambda^{m,\ell} f_1^0 \mathbf{p}_1^\ell + \lambda^{m,\ell-1}(f_1^0\mathbf{p}_1^{\ell-1} + f_1^1\mathbf{p}_1^\ell) \\ \psi &= [f_1^0, f_1^1], \quad \mathbf{P}^* = [\mathbf{p}_1^\ell; \mathbf{p}_1^{\ell-1}] \\ \Lambda^m &:= \begin{bmatrix} \lambda^{m,\ell} & \lambda^{m,\ell-1} \\ 0 & \lambda^{m,\ell-1} \end{bmatrix}, \quad (\Lambda^m)^{-1} \preccurlyeq (\lambda^{m,\ell-1})^{-1} \\ \mathbf{r}^m &\prec \lambda^{m,\ell-1}, \quad \boldsymbol{\rho} \sim \mathbf{r}^m(\Lambda^m)^{-1} \prec 1. \end{aligned}$$

□

5.4 Shift Invariant Algorithms

A subdivision algorithm is shift invariant if the *shape* generated by the sequence of rings remains unchanged regardless which segment is labeled first when numbering them. Most subdivision algorithms currently in use have this property. It allows analyzing the spectrum of the subdivision matrix with the help of the *Discrete Fourier Transform (DFT)*. We will show that shift invariance is possible only for subdivision matrices with a pair of – either real or complex conjugate – subdominant Jordan blocks. Further, the characteristic ring is symmetric in the sense that neighboring segments are related by a $2\pi/n$-rotation.

Corresponding to the partition of a ring $\mathbf{x}^m = GA^m\mathbf{Q}$ into segments $\mathbf{x}_j^m = \mathbf{x}^m(\cdot, j), j \in \mathbb{Z}_n$, the coefficients $\mathbf{Q}$ can typically be partitioned into n blocks $\mathbf{Q} = [\mathbf{Q}_0; \dots; \mathbf{Q}_{n-1}]$, where all blocks[1] $\mathbf{Q}_j$ have equal size $\tilde{\ell} := (\bar{\ell}+1)/n$. This grouping of coefficients into blocks with equal structure is a natural process; by contrast, assigning the label $j = 0$ to one of these blocks is a random choice, unless the blocks are intentionally treated differently. We expect from a shift invariant algorithm that this choice determines the labelling of segments, but not their shape. To make this precise, let us consider two possible representations $\mathbf{Q}$ and $\tilde{\mathbf{Q}}$ of a given set of initial data, differing only by the labelling of blocks, i.e., $\tilde{\mathbf{Q}}_j = \mathbf{Q}_{j-i}$ for some $i \in \mathbb{Z}_n$. Then the corresponding rings $\mathbf{x}^m$ and $\tilde{\mathbf{x}}^m$ should have segments related by an equal shift of labels, i.e., $\tilde{\mathbf{x}}_j^m = \mathbf{x}_{j-i}^m$. Let us investigate the consequences of this requirement. With $\mathbb{1}$ the identity matrix of size $\tilde{\ell}$, let

$$S := \begin{bmatrix} 0 & 0 & \cdots & 0 & \mathbb{1} \\ \mathbb{1} & 0 & \cdots & 0 & 0 \\ 0 & \mathbb{1} & \cdots & 0 & 0 \\ & & \ddots & & \\ 0 & 0 & \cdots & \mathbb{1} & 0 \end{bmatrix}, \quad \mathbb{1} := \begin{bmatrix} 1 & 0 & \cdots & 0 & 0 \\ 0 & 1 & \cdots & 0 & 0 \\ & & \ddots & & \\ 0 & 0 & \cdots & 1 & 0 \\ 0 & 0 & \cdots & 0 & 1 \end{bmatrix},$$

denote the n-block *shift matrix*. Then $\tilde{\mathbf{Q}} = S^i\mathbf{Q}$, and shift invariance formally reads

$$\tilde{\mathbf{x}}_j^m = G(\cdot, j)A^m S^i\mathbf{Q} = G(\cdot, j-i)A^m\mathbf{Q} = \mathbf{x}_{j-i}^m. \tag{5.11}$$

For $m = 0$, we obtain $G(\cdot, j)S^i = G(\cdot, j-i)$, and hence, for arbitrary m, $G(\cdot, j)A^m S^i = G(\cdot, j)S^i A^m$. Disregarding possible linear dependence of the generating system G, the latter equality suggests that A^m and S^i commute. These considerations give rise to the following definition:

Definition 5.13 (Shift invariance). A subdivision algorithm (A, G) is called *shift invariant*, if the generating system satisfies

$$G(\cdot, j)S = G(\cdot, j-1), \quad j \in \mathbb{Z}_n,$$

and if A and S commute,

$$AS = SA.$$

[1] The partition of vectors of coefficients and functions into n similar blocks must not be confused with the partition into Jordan blocks, see Sect. 4.6/72.

The two conditions imply $G(\cdot, j)S^i = G(\cdot, j-i)$ and $A^m S^i = S^i A^m$ for all $j, i \in \mathbb{Z}_n$ and $m \in \mathbb{N}_0$. Following (5.11/96), we obtain

$$\tilde{\mathbf{x}}_j^m = \mathbf{x}_{j-i}^m \quad \text{if} \quad \tilde{\mathbf{Q}}_j = \mathbf{Q}_{j-i}, \quad i, j \in \mathbb{Z}_n,$$

for a shift invariant algorithm, as intended.

According to the partitioning of the coefficients, the subdivision matrix of a shift-invariant algorithm can be represented by $(n \times n)$ blocks $A_{j,i}$ of size $(\tilde{\ell} \times \tilde{\ell})$,

$$A = \begin{bmatrix} A_{0,0} & \cdots & A_{0,n-1} \\ \vdots & & \vdots \\ A_{n-1,0} & \cdots & A_{n-1,n-1} \end{bmatrix}.$$

If A and S commute, we obtain for the blocks

$$(AS)_{j,i} = A_{j,i+1} = A_{j+1,i} = (SA)_{j,i}, \quad i, j \in \mathbb{Z}_n.$$

Hence, the matrix A is completely determined by the blocks $A_j := A_{j,0}$ of the first column via $A_j = A_{j+i,i}$. We say that A is *block-circulant* and write

$$A = \operatorname{circ}(A_0, \ldots, A_{n-1}) := \begin{bmatrix} A_0 & A_{n-1} & \cdots & A_1 \\ A_1 & A_0 & \cdots & A_2 \\ \vdots & \vdots & \ddots & \vdots \\ A_{n-1} & A_{n-2} & \cdots & A_0 \end{bmatrix}.$$

The given conditions for shift invariance are more general than might appear at first sight. This is best explained by example.

Example 5.14 (Catmull–Clark algorithm in circulant form). In its standard form, the Catmull–Clark algorithm uses $\bar{\ell} + 1 = 12n + 1$ coefficients, arranged as shown in Fig. 6.3/111 to describe a ring. There is one central coefficient $\tilde{\mathbf{q}}_0$, and n blocks $\tilde{\mathbf{Q}}_0, \ldots, \tilde{\mathbf{Q}}_{n-1}$ with always 12 elements. The corresponding subdivision matrix and the generating system have the structure

$$\tilde{A} := \begin{bmatrix} \tilde{a}_0 & \tilde{a}_1 & \tilde{a}_1 & \cdots & \tilde{a}_1 \\ \tilde{a}_2 & \tilde{A}_0 & \tilde{A}_{n-1} & \cdots & \tilde{A}_1 \\ \tilde{a}_2 & \tilde{A}_1 & \tilde{A}_0 & \cdots & \tilde{A}_2 \\ \vdots & & \ddots & \ddots & \\ \tilde{a}_2 & \tilde{A}_{n-1} & \tilde{A}_{n-2} & \cdots & \tilde{A}_0 \end{bmatrix}, \quad \tilde{\mathbf{Q}} := \begin{bmatrix} \tilde{\mathbf{q}}_0 \\ \tilde{\mathbf{Q}}_0 \\ \tilde{\mathbf{Q}}_1 \\ \vdots \\ \tilde{\mathbf{Q}}_{n-1} \end{bmatrix}, \quad \tilde{G} := [\tilde{g}_0, \tilde{G}_1, \ldots, \tilde{G}_{n-1}],$$

where $\tilde{a}_0$ is a real number, $\tilde{a}_1$ is row-vector, and $\tilde{a}_2$ is a column vector with always 12 elements. Of course, one can adapt the notion of shift invariance to cover also such situations, but we want to show now that this is actually not necessary if we slightly modify the structure of the coefficients. The trick is to artificially extend each block $\tilde{\mathbf{Q}}_j$ by a copy $\mathbf{q}_j := \tilde{\mathbf{q}}_0$ of the central coefficient to obtain the arrangement

$$\mathbf{Q} := [\mathbf{Q}_0; \ldots; \mathbf{Q}_{n-1}], \quad \mathbf{Q}_j := [\mathbf{q}_j; \tilde{\mathbf{Q}}_j]$$

with $13n$ coefficients, see Fig. 6.3/111. Accordingly, the subdivision matrix yields the desired circulant structure,

$$A := \operatorname{circ}(A_0, \ldots, A_{n-1}), \quad A_j := \begin{bmatrix} \tilde{a}_0/n & \tilde{a}_1 \\ \tilde{a}_2/n & \tilde{A}_j \end{bmatrix}.$$

Division by n is applied to ensure that also the rows of A sum to 1. Further, all points $\mathbf{q}_0^m = \cdots = \mathbf{q}_{n-1}^m$ remain equal throughout the iteration. The new system of generating rings is

$$G := [G_0, \ldots, G_{n-1}], \quad G_j := [\tilde{g}_0/n, \tilde{G}_j],$$

where division by n retains partition of unity. It is easily shown that the original algorithm and its variant are equivalent in the sense that

$$GA^m\mathbf{Q} = \tilde{G}\tilde{A}^m\tilde{\mathbf{Q}}$$

for any choice of initial data. Unlike the original generating rings, the new system G is linearly dependent. However, no ineffective eigenvectors are introduced. □

The key tool for handling circulant matrices is the *Discrete Fourier Transform (DFT)*. We denote the imaginary unit and the primitive n-th root of unity by

$$\mathbf{i} := \sqrt{-1}, \quad w_n := c_n + \mathbf{i}s_n := \exp(2\pi\mathbf{i}/n). \tag{5.12}$$

With $\mathbb{1}$ the identity matrix of size $\tilde{\ell}$ as above, we define the *Fourier block matrix* $\mathcal{W}$ by

$$\mathcal{W} := (w_n^{-ji}\mathbb{1})_{j,i\in\mathbb{Z}_n} = \begin{bmatrix} \mathbb{1} & \mathbb{1} & \mathbb{1} & \cdots & \mathbb{1} \\ \mathbb{1} & w_n^{-1}\mathbb{1} & w_n^{-2}\mathbb{1} & \cdots & w_n^{1}\mathbb{1} \\ \mathbb{1} & w_n^{-2}\mathbb{1} & w_n^{-4}\mathbb{1} & \cdots & w_n^{2}\mathbb{1} \\ \vdots & \vdots & \vdots & \ddots & \vdots \\ \mathbb{1} & w_n^{1}\mathbb{1} & w_n^{2}\mathbb{1} & \cdots & w_n^{-1}\mathbb{1} \end{bmatrix}.$$

It is easily verified by inspection that the inverse of $\mathcal{W}$ is given by

$$\mathcal{W}^{-1} = \frac{1}{n}(w_n^{+ji}\mathbb{1})_{j,i\in\mathbb{Z}_n} = \frac{1}{n}\overline{\mathcal{W}}.$$

In particular, the i-th block column of $\mathcal{W}^{-1}$ is

$$\mathcal{W}_i^{-1} := \frac{1}{n}\begin{bmatrix} \mathbb{1} \\ w_n^i\mathbb{1} \\ \vdots \\ w_n^{(n-1)i}\mathbb{1} \end{bmatrix}. \tag{5.13}$$

The DFT of the matrix A is defined by $\hat{A} := \mathcal{W} A \mathcal{W}^{-1}$, and a standard computation shows that

$$\hat{A} = \operatorname{diag}(\hat{A}_0, \ldots, \hat{A}_{n-1}) = \begin{bmatrix} \hat{A}_0 & & 0 \\ & \ddots & \\ 0 & & \hat{A}_{n-1} \end{bmatrix} \tag{5.14}$$

is block-diagonal with entries obtained by applying the Fourier matrix to the first block column of A,

$$\begin{bmatrix} \hat{A}_0 \\ \vdots \\ \hat{A}_{n-1} \end{bmatrix} = \mathcal{W} \begin{bmatrix} A_0 \\ \vdots \\ A_{n-1} \end{bmatrix}, \quad \text{that is} \quad \hat{A}_i = \sum_{j \in \mathbb{Z}_n} w_n^{-ji} A_j.$$

By definition, A and $\hat{A}$ are similar, and in particular, they have equal eigenvalues. More precisely, the Jordan decompositions of A and $\hat{A}$ are related by

$$A = VJV^{-1}, \quad \hat{A} = \hat{V} J \hat{V}^{-1}, \quad \hat{V} = \mathcal{W} V.$$

Since $\hat{A}$ is block-diagonal, its Jordan decomposition is obtained from the respective decompositions of the blocks,

$$\hat{V} = \operatorname{diag}(\hat{V}_0, \ldots, \hat{V}_{n-1}), \quad J = \operatorname{diag}(\hat{J}_0, \ldots, \hat{J}_{n-1}), \quad \hat{A}_i = \hat{V}_i \hat{J}_i \hat{V}_i^{-1}.$$

This means that with the help of the DFT, Jordan decomposition of the subdivision matrix A, which typically is quite large, boils down to decomposing the n much smaller blocks $\hat{A}_0, \ldots, \hat{A}_{n-1}$ individually. More specifically, let $\hat{v}$ be a (generalized) eigenvector of $\hat{A}_i$. Then $\hat{v}$ is the i-th block of a column of $\hat{V}$, all other blocks of this column are zero. Hence, using the Kronecker symbol $\delta_{j,i}$ and $V = \mathcal{W}^{-1} \hat{V}$, the corresponding (generalized) eigenvector v of A is

$$v = \mathcal{W}^{-1} \begin{bmatrix} \delta_{0,i} \hat{v} \\ \delta_{1,i} \hat{v} \\ \vdots \\ \delta_{n-1,i} \hat{v} \end{bmatrix} = \frac{1}{n} \begin{bmatrix} w_n^0 \hat{v} \\ w_n^i \hat{v} \\ \vdots \\ w_n^{(n-1)i} \hat{v} \end{bmatrix},$$

or, with (5.13/98), briefly

$$v = \mathcal{W}_i^{-1} \hat{v}. \tag{5.15}$$

Moreover, v is always an eigenvector of S to the eigenvalue w_n^{-i}, i.e.,

$$Sv = w_n^{-i} v. \tag{5.16}$$

This implies for the segments of the corresponding eigenring $f := Gv$

$$f_j = G(\cdot, j) v = w_n^i G(\cdot, j) S v = w_n^i G(\cdot, j-1) v = w_n^i f_{j-1}. \tag{5.17}$$

The observation that every Jordan block of A corresponds to a Jordan block of one of the diagonal blocks leads to the following

Definition 5.15 (Fourier index). For a complex number λ, the set of all indices i with the property that λ is eigenvalue of $\hat{A}_i$ is called the *Fourier index* of λ and denoted by

$$\mathcal{F}(\lambda) := \{i \in \mathbb{Z}_n : \lambda \text{ is eigenvalue of } \hat{A}_i\}.$$

Equally, the *Fourier index* of a Jordan block J, see (4.20/72), is

$$\mathcal{F}(J) := \{i \in \mathbb{Z}_n : J \text{ is Jordan block of } \hat{A}_i\}.$$

It is easily shown that the unique dominant eigenvalue $\lambda_0 = 1$ of a subdivision matrix has the Fourier index

$$\mathcal{F}(1) = \{0\}.$$

Since A is real, the blocks of $\hat{A}$ and also their Jordan decompositions come in complex conjugate pairs,

$$\hat{A}_{n-i} = \overline{\hat{A}_i}, \quad \hat{V}_{n-i} = \overline{\hat{V}}_i, \quad \hat{J}_i = \overline{\hat{J}}_i.$$

In particular, if J is a Jordan block of $\hat{A}_i$, then $\overline{J}$ is a Jordan block of $\hat{A}_{n-i}$,

$$i \in \mathcal{F}(J) \quad \Leftrightarrow \quad n - i \in \mathcal{F}(\overline{J}). \tag{5.18}$$

Together with (5.17/99), this pairing allows us to discard shift invariant subdivision algorithms without a pair of real or complex subdominant Jordan blocks.

Theorem 5.16 (Shift invariant algorithms). *Consider a shift invariant subdivision algorithm (A, G) with $A \in \mathcal{A}_p^q$ according to Sect. (5.3/89) and a regular characteristic ring ψ. Then (A, G) can be a C_1^k-algorithm only if $A \in \mathcal{A}_1^1 \cup \mathcal{A}_1^2$.*

Proof. The excluded cases $A \in \mathcal{A}_p^q, p \geq 2$, are characterized by the fact that the first eigenvalue dominates the second one, $(\lambda_1, \ell_1) \succ (\lambda_2, \ell_2)$. λ_1 has to be real, since otherwise there would exist a similar, but different, eigenvalue $(\overline{\lambda}_1, \ell_1)$. Since the Jordan block J_1 corresponding to (λ_1, ℓ_1) appears only once, its Fourier index contains exactly one element, $\mathcal{F}(J_1) = \{i_1\}$. However, by (5.18/100), $n - i_1$ is also in the Fourier index of $\overline{J}_1 = J_1$, what implies $i_1 = n - i_1 \mod n$. This condition has at most two solutions. Either $i_1 = 0$ or, if n is even, $i_1 = n/2$. In both cases, $2i_1 = 0 \mod n$. Hence, by (5.17/99), any eigenring f_1^i corresponding to J_1 has coinciding segments $f_1^i(\cdot, 2) = f_1^i(\cdot, 0)$. Now, we show that in all excluded cases the characteristic ring ψ is *not* injective, and hence, in view of Theorem 5.12/93, the algorithm is not C_1^k.

If $A \in \mathcal{A}_2^2 \cup \mathcal{A}_3^2$, then both components of the characteristic ring correspond to the first Jordan block, $\psi = [f_1^0, f_1^1]$. Hence, $\psi(\cdot, 2) = \psi(\cdot, 0)$, and ψ is not injective.

If $A \in \mathcal{A}_2^1 \cup \mathcal{A}_3^1$, then the second eigenvalue dominates the third one, $(\lambda_2, \ell_2) \succ (\lambda_3, \ell_3)$. By the same arguments as above, J_2 is real, the single element i_2 of the Fourier index $\mathcal{F}(J_2)$ satisfies $2i_2 = 0 \mod n$, and the segments $f_2^i(\cdot, 2) = f_2^i(\cdot, 0)$ of the eigenring f_2^i coincide. Hence, also in this case, the characteristic ring $\psi = [f_1^0, f_2^0]$ is not injective. □

We now focus on the two remaining classes of algorithms. If $A \in \mathcal{A}_1^1$, then we have a double subdominant Jordan block $J_1 = J_2$ with Fourier index $\mathcal{F}(J_1) = \mathcal{F}(J_2) = \{i, n-i\}$. If $A \in \mathcal{A}_1^2$, then we have a complex conjugate pair of subdominant Jordan blocks $J_1 = \overline{J_2}$ with Fourier indices $\mathcal{F}(J_1) = \{i\}$ and $\mathcal{F}(J_2) = \{n-i\}$. In both cases, we call

$$\mathcal{F}_{\mathrm{sub}} := \{i, n-i\}$$

the *subdominant Fourier index* of the algorithm. With $\hat{v}$ the eigenvector of $\hat{A}_i$ to λ_1 and $v = \mathcal{W}_i^{-1}\hat{v}$, the two subdominant eigenrings $f = Gv_1$ and $\overline{f} = G\overline{v}$ are complex-valued. For $A \in \mathcal{A}_1^2$, λ_1 is complex, and this is just the situation that we expect. We set $v_1^0 := v$, $v_2^0 := \overline{v}$ to obtain $f_1^0 = f$, $f_2^0 = \overline{f}$ and the characteristic ring $\psi := [\operatorname{Re} f_1^0, \operatorname{Im} f_1^0] = [\operatorname{Re} f, \operatorname{Im} f]$. For $A \in \mathcal{A}_1^1$, λ is real, and $v_1^0 := \operatorname{Re} v$, $v_2^0 := \operatorname{Im} v$ are real eigenvectors of A. Hence, $f_1^0 = \operatorname{Re} f$, $f_2^0 = \operatorname{Im} f$ are real subdominant eigenring, and again, the characteristic ring is $\psi := [f_1^0, f_2^0] = [\operatorname{Re} f, \operatorname{Im} f]$. Thus, the case distinction made in Definition 5.10/92 is resolved using the complex-valued eigenring f.

Definition 5.17 (Characteristic ring, complex). Let (A, G), $A \in \mathcal{A}_1^1 \cup \mathcal{A}_1^2$, be a shift invariant C_0^k-subdivision algorithm with subdominant Fourier index $\mathcal{F}_{\mathrm{sub}} = \{i, n-i\}$ and a subdominant eigenvector

$$v := \mathcal{W}_i^{-1}\hat{v}, \quad \hat{A}_i\hat{v} = \lambda_1\hat{v}. \tag{5.19}$$

Then the *characteristic ring in complex form* of the algorithm is defined as the complex-valued ring

$$f = Gv \in C^k(\mathbf{S}_n^0, \mathbb{C}, G).$$

If clear from the context, the suffix "in complex form" is omitted.

As explained above, f is just the complexification of the formerly defined real characteristic ring,

$$\psi = [\operatorname{Re} f, \operatorname{Im} f].$$

Due to the relation (5.17/99), the complex version is sometimes more convenient for analytical purposes than the real form. For instance, it is helpful when proving the following theorem on the Fourier index of the subdominant eigenvalue. Its claim is illustrated by Fig. 5.4/102. On the left hand side, it shows the characteristic ring of the standard Doo–Sabin algorithm for $n = 5$ with weights according to (6.15/116). Here, the subdominant eigenvalue $\lambda = 1/2$ has the correct Fourier index $\mathcal{F}_{\mathrm{sub}} = \{1, 4\}$. On the right hand side, the modified weights $a = [1, 0, 1, 1, 0]/3$ are used, which yield the subdominant eigenvalue $\lambda = (1 + \sqrt{5})/6 \approx 0.54$ with the inappropriate Fourier index $\mathcal{F}_{\mathrm{sub}} = \{2, 3\}$.

Theorem 5.18 (Winding number of ψ and Fourier index). *Let (A, G) be a shift invariant subdivision algorithm with $A \in \mathcal{A}_1^1 \cup \mathcal{A}_1^2$. If the characteristic ring f is uni-cyclic then the subdominant Fourier index is $\mathcal{F}_{\mathrm{sub}} = \{1, n-1\}$.*

Proof. Following Definition 5.7/88, we define the curve $z := f \circ \mathbf{c}_{\mathrm{bnd}}$, which parametrizes the outer boundary of the image of the complex characteristic ring f. Let

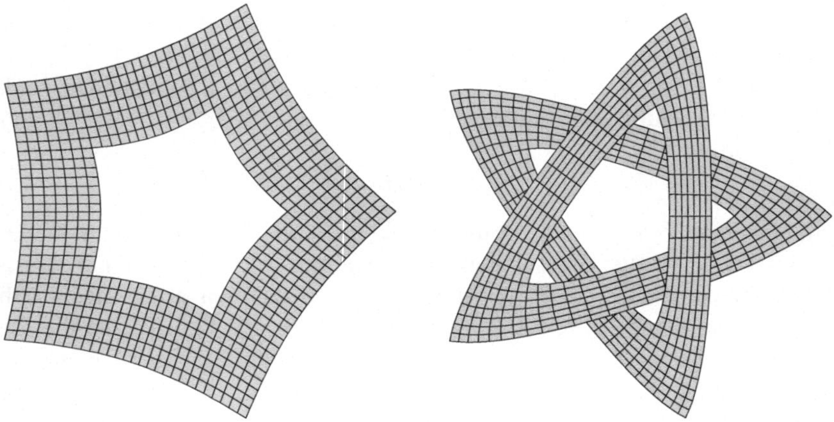

Fig. 5.4 Illustration of Theorem 5.18/101: Characteristic ring ψ of an algorithm for $n = 5$ using (*left*) standard Doo–Sabin weights so that the Fourier index is $\mathcal{F}(\lambda) = \{1, 4\}$ and (*right*) intentionally modified weights so that $\mathcal{F}(\lambda) = \{2, 3\}$. The figure shows and Theorem 5.18/101 proves that ψ is *not* uni-cyclic in the latter case.

us assume that $\mathcal{F}_{\text{sub}} = \{i, n-i\}$, then (5.17/99) implies

$$\frac{z'(u+j/n)}{z(u+j/n)} = \frac{w_n^{ij} z'(u)}{w_n^{ij} z(u)}, \quad u \in [0, 1/n]$$

for all $j \in \mathbb{Z}_n$. We obtain

$$2\pi \mathbf{i}\, \nu(\psi) = \int_0^1 \frac{z'(u)}{z(u)}\, du = n \int_0^{1/n} \frac{z'(u)}{z(u)}\, du = n \ln \frac{z(1/n)}{z(0)},$$

where the imaginary part of the logarithm is only determined up to an integer multiple of 2π. By consistency of neighboring segments according to (4.9/62) and by (5.17/99),

$$z(1/n) = f(0,1,0) = f(1,0,1) = w_n^i z(0). \tag{5.20}$$

Hence, for some $\ell \in \mathbb{Z}$,

$$2\pi\, \nu(f) = n(2\pi i/n + 2\pi\ell),$$

implying that

$$1 = |\nu(f)| = |i + \ell n|.$$

The only solutions to this equation are given by $|i| = 1 \mod n$, as stated. □

Summarizing, a shift invariant C_1^k-algorithm must have a double subdominant eigenvalue, either real or complex, corresponding to the Fourier blocks 1 and $n-1$. The following definition removes some of the ambiguities in choosing the characteristic ring by fixing the index $i = 1$ in (5.19/101) and requiring $f(1,1,0)$ to be real and positive.

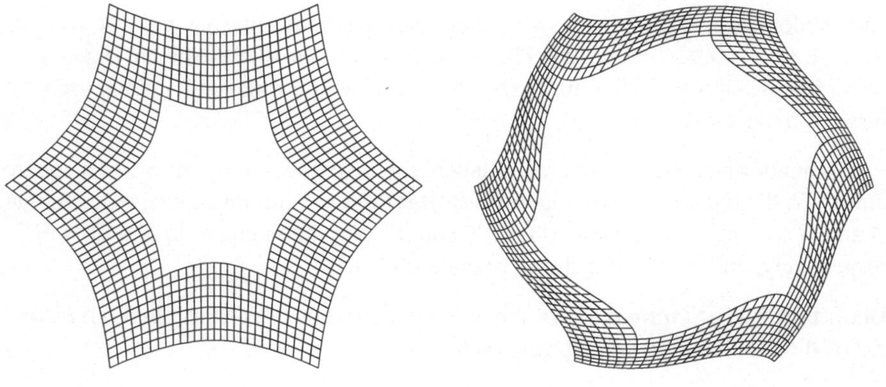

Fig. 5.5 Illustration of Example 5.20/103: Characteristic ring of algorithms for $n = 6$ using (*left*) standard Doo–Sabin weights and (*right*) asymmetric weights.

Definition 5.19 (Characteristic ring, normalized). The characteristic ring $f = Gv$ of a shift-invariant subdivision algorithm (A, G) is called *normalized*, if

$$v = \mathcal{W}_1^{-1}\hat{v}, \quad \hat{A}_1\hat{v} = \lambda_1\hat{v}$$

and the value

$$f(1,1,0) \in \mathbb{R}_{>0}$$

is a positive real number.

It is easily shown that normalization is always possible for a C_1^k-algorithm. In this case, by Theorem 5.12/93, f is injective. Further, $\mathcal{F}_{\text{sub}} = \{1, n-1\}$, and a subdominant eigenvector v can be defined as above. By (5.17/99), the characteristic ring $f = Gv$ satisfies

$$f_j = w_n^j f_0, \quad j \in \mathbb{Z}_n. \tag{5.21}$$

Hence, because f is injective, we have $f(1,1,1) = w_n f(1,1,0) \neq f(1,1,0)$ implying that $f(1,1,0) \neq 0$. Now, the rescaled eigenvector $\tilde{v} := rv$ yields the normalized complex characteristic ring $\tilde{f} = G\tilde{v}$ if we set, e.g., $r := 1/f(1,1,0)$.

5.5 Symmetric Algorithms

Now, we consider subdivision algorithms that are not only invariant under shift but also invariant under reversal of orientation when labelling the initial data. We call the reversal operation 'flipping'. The following example illustrates lack of flip invariance:

Example 5.20 (Flip symmetry). On the left hand side, Fig. 5.5/103 shows the characteristic ring of the standard Doo–Sabin algorithm with weights according to (6.15/116) and subdominant eigenvalue $\lambda = 1/2$. On the right hand side, asymmetric weights $a = [4, 1, 0, 0, 0, 3]/8$ are used. These yield the complex subdominant eigenvalue $\lambda = (6 + \sqrt{3}i)/8 \approx 0.75 + \mathrm{i}0.22$ and the characteristic ring is not symmetric with respect to the x-axis. □

Orientation reversal of coefficient labels can be expressed by means of a square matrix R, the *flip matrix*. Analogous to shift invariance, flip invariance requires that A and R commute and that the rings $\mathbf{x}^m$ and $\tilde{\mathbf{x}}^m$ corresponding to $\mathbf{Q}$ and $\tilde{\mathbf{Q}} = R\mathbf{Q}$, respectively, differ only by a flip of orientation.

Definition 5.21 (Symmetry). A subdivision algorithm (A, G) is called *flip invariant*, if the system of generating rings satisfies

$$G(s, t, j) = G(t, s, -j)R, \quad (s, t, j) \in \mathbf{S}^0_n,$$

for some matrix R commuting with A,

$$AR = RA.$$

The algorithm is called *symmetric*, if it is both shift and flip invariant.

We observe that if the generating rings in G are linearly independent, then R must be an involution, $R = R^{-1}$.

The spectrum of the asymmetric case in Example 5.20/103 included a complex subdominant eigenvalue. This case is ruled out by symmetry.

Theorem 5.22 (Symmetry requires real subdominant eigenvalues). *The symmetric subdivision algorithm (A, G) can be C^k_1 only if $A \in \mathcal{A}^1_1$, i.e., the subdominant Jordan block is double and real,*

$$(\lambda, \ell) := (\lambda_1, \ell_1) = (\lambda_2, \ell_2) \succ (\lambda_2, \ell_3), \quad \lambda \in \mathbb{R}.$$

Proof. According to Theorem 5.16/100, $A \in \mathcal{A}^1_1$ or $A \in \mathcal{A}^2_1$, where we recall form Sect. 5.3/89 that the class $\mathcal{A}^2_1$ contains algorithms with a pair of complex conjugate subdominant eigenvalues. We assume $A \in \mathcal{A}^2_1$ and derive a contradiction:

From $AR = RA$ and $Av = \lambda_1 v$, we conclude $ARv = RAv = \lambda_1 Rv$, i.e., Rv is either 0 or an eigenvector of A to λ_1. Since the eigenvector to λ_1 is unique up to scaling, $Rv = av$ for some $a \in \mathbb{C}$. Using the definition of flip invariance, we obtain

$$f(1, 1, 0) = G(1, 1, 0)v = G(1, 1, 0)Rv = aG(1, 1, 0)v = af(1, 1, 0).$$

As explained in the sequel of Definition 5.19/103, we may assume that f is injective if (A, G) is a C^k_1-algorithm. In particular, we have $f(1, 1, 0) \neq 0$ so that $a = 1$. Further, by (5.21/103),

$$f(1, 0, 0) = G(1, 0, 0)v = G(1, 0, 0)Rv = G(0, 1, 0)v = f(0, 1, 0)$$

contradicting injectivity of f. □

This theorem explains why most subdivision algorithms of practical importance are standard algorithms according to Definition 5.3/84. Shift and flip invariance necessarily lead to a double subdominant Jordan block $J_1 = J_2 = J(\lambda, \ell)$ and, typically, this block is reduced to the trivial case $\ell = 0$, where the Jordan block is a singleton λ. An algorithm with non-trivial Jordan blocks is given in Sect. 6.3/120.

Consider the characteristic spline $h = Bv$ in complex form, where v is the subdominant eigenvector according to Definition 5.17/101. The m-th ring of h is

$$h^m = GA^m v = \lambda^m f.$$

That is, h is built from complex multiples of the characteristic ring. In the real case, applying the factor λ^m simply amounts to scaling, while in the complex case $\lambda = |\lambda| \exp(\mathbf{i}\phi)$. Hence, there is an additional rotation, $\boldsymbol{\xi}^m = |\lambda|^m \exp(\mathbf{i}m\phi) f$. This rotation is illustrated by Fig. 5.5/103 (*right*).

The following theorem establishes an additional symmetry property for the characteristic ring of a symmetric subdivision algorithm.

Theorem 5.23 (Symmetry of the characteristic ring). *Let $f = Gv$ be the normalized characteristic ring of a symmetric subdivision algorithm (A, G) with $A \in \mathcal{A}_1^1$. Then*

$$f(s,t,j) = \overline{f(t,s,-j)}, \quad (s,t,j) \in \mathbf{S}_n^0.$$

Proof. Here, the subdominant eigenvalue $\lambda := \lambda_1 = \lambda_2$ is double. As before, one can show that Rv is 0 or an eigenvector of A to λ. Hence, $Rv = av + b\overline{v}$ for some constants $a, b \in \mathbb{C}$ and by (5.16/99), $Sv = w_n^{-1} v$ and $S\overline{v} = w_n \overline{v}$. This implies $S^j R S^j v = a w_n^{-2j} v + b\overline{v}$. Let us assume without loss of generality that $f(1,1,0) = 1$. By symmetry, we obtain

$$1 = G(1,1,0)v = G(1,1,0) S^j R S^j v = a w_n^{-2j} G(1,1,0)v + bG(1,1,0)\overline{v}$$

for any $j \in \mathbb{Z}_n$. Since G is real, it follows $G(1,1,0)\overline{v} = G(1,1,0)v = 1$, and

$$1 = a w_n^{-2j} + b, \quad j \in \mathbb{Z}.$$

This implies $a = 0, b = 1$ and $Rv = \overline{v}$. Hence,

$$f(s,t,j) = G(s,t,j)v = G(t,s,-j)Rv = G(t,s,-j)\overline{v} = \overline{f(t,s,-j)}.$$

□

In case of symmetry, C_1^k-subdivision algorithms can be detected using significantly simplified criteria, which involve only properties of the upper half of the segment f_0 of the characteristic ring. In particular, the appropriate winding number $\nu(f) = 1$ can be proven by showing that one arc of the outer boundary of f_0 does not intersect the non-positive part of the real axis.

Theorem 5.24 (Conditions for symmetric C_1^k-algorithms). *Let (A, G) be a symmetric C_0^k-subdivision algorithm with $A \in \mathcal{A}_1^1$ and $\mathcal{F}(\lambda) = \{1, n-1\}$, and assume* *that the characteristic ring f is normalized. Then f is regular if and only if the first segment f_0 satisfies*

$$^{\times}\mathcal{D}f_0(s,t) \neq 0 \quad \textit{for all} \quad (s,t) \in \boldsymbol{\Sigma}^0 \textit{ with } s \leq t.$$

Further, if f is regular, then (A, G) is a C_1^k-subdivision algorithm if and only if all real points on the curve $c(u) := f_0(u, 1), u \in [0, 1]$, are positive, i.e.,

$$c(u) \in \mathbb{R} \quad \Rightarrow \quad c(u) > 0.$$

Proof. By Theorem 5.23/105, $^{\times}\mathcal{D}f_0(s,t) = {}^{\times}\mathcal{D}f_0(t,s)$. Further, by (5.21/103), $^{\times}\mathcal{D}f_j(s,t) = {}^{\times}\mathcal{D}f_0(s,t)$, what proves the first part of the theorem.

To prove the second part, let us assume that $c(u_*) = f_0(u_*, 1)$ is a negative real number. Then $u_* \neq 1$ because normalization requires $f_0(1,1) > 0$. By Theorem 5.23/105, $f(u_*, 1, 0) = f(1, u_*, 0)$, showing that f is not injective. Hence, by Theorem 5.12/93, the algorithm is not C_1^k. Conversely, let the condition given in the theorem be satisfied. Following Definition 5.7/88, the winding number of f is

$$\nu(f) := \nu(f \circ \mathbf{c}_{\text{bnd}}, \mathbf{0}).$$

The curves c and $z := f \circ \mathbf{c}_{\text{bnd}}$ are related as follows: Let $u_j := j/n$. The curves c and $\bar{c}$ combine to the outer boundary of the segment f_0,

$$z_0(u) := \begin{cases} \bar{c}(2nu) & \text{if } u_0 \leq u < u_1/2 \\ c(2 - 2nu) & \text{if } u_1/2 \leq u \leq u_1, \end{cases}$$

and the segments of z are rotated copies of z_0,

$$z(u) = w_n^j z_0(u - u_j), \quad u_j \leq u \leq u_{j+1}.$$

Now, we apply Lemma 2.20/36. The disjoint half-lines are given by $h_j := -w_n^{j-1}$. Further, by (5.20/102) and Theorem 5.23/105,

$$c_0(u_1) = w_n c_0(u_0) = \overline{c_0}(u_0).$$

Hence, $\arg(z_1) = -\arg(z_0) = \pi/n$, and therefore

$$\arg(z_j/h_j) - \arg(z_{j-1}/h_j) = (1 + 1/n)\pi - (1 - 1/n)\pi = 2\pi/n.$$

Finally, we obtain the winding number

$$\nu(f) = \nu(z, 0) = \frac{1}{2\pi} \sum_{j=1}^{n} 2\pi/n = 1,$$

showing that f is uni-cyclic. By Theorem (5.12/93), (A, G) is a C_1^k-algorithm. □

In some cases, an even simpler sufficient condition is applicable:

Theorem 5.25 (More conditions for symmetric C_1^k-algorithms). *Let (A, G) be a symmetric C_0^k-subdivision algorithm with $A \in \mathcal{A}_1^1$ and $\mathcal{F}(\lambda) = \{1, n-1\}$, and assume that the characteristic ring f is normalized. Then (A, G) is a C_1^k-subdivision algorithm if both components of $D_2 f_0$ are positive,*

$$\operatorname{Re}(D_2 f_0) > 0, \quad \operatorname{Im}(D_2 f_0) > 0. \tag{5.22}$$

Proof. Symmetry implies $\operatorname{Re}(D_1 f_0(s,t)) = \operatorname{Re}(D_2 f_0(t,s)) > 0$ and $\operatorname{Im}(D_1 f_0 (s,t)) = -\operatorname{Im}(D_2 f_0(t,s)) < 0$. Hence,

$$^{\times}\!D f_0 = \operatorname{Re}(D_1 f_0) \operatorname{Im}(D_2 f_0) - \operatorname{Im}(D_1 f_0) \operatorname{Re}(D_2 f_0) > 0,$$

showing that hat f_0 is regular. Further,

$$\int_u^1 D_1 f_0(\tau, 1)\, dt = f_0(1,1) - f_0(u,1) = f_0(1,1) - c(u).$$

$f_0(1,1)$ is real, and the imaginary part of the integrand is negative so that

$$\operatorname{Im}(c(u)) > 0 \quad \text{for} \quad u \in (0,1].$$

Hence, $c(1) = 1$ is the only real point in the image of c, and the argument is complete. □

Bibliographic Notes

1. Early attempts at verifying normal continuity of subdivision surfaces [DS78, BS86, BS88] were based on considering discrete normals derived from the control net and did not take into account the properties of the generating rings. The incompleteness of the attempts was exposed in [Rei93].

2. The concept of the characteristic ring was introduced by Reif in [Rei93, Rei95c] under the name 'characteristic map'. Earlier, Ball and Storry [BS86] introduced the related notion of *natural configuration* for the geometric layout of the control points of a characteristic ring.

3. The relevance of the Discrete Fourier Transform for the analysis of circulant subdivision matrices was already recognized by Doo and Sabin [DS78], and the approach was used by Ball and Storry [BS86, BS88]. The symmetry properties of characteristic rings and their relation to its Fourier index were highlighted in [PR98]. Further results can be found in [Kob98b].

4. The analysis of general algorithms of Sect. 5.3/89 was first derived in Reif's habilitation [Rei99] and was confirmed by Zorin's results [Zor00a]. Both sources list all possible leading eigenvalues compatible with C_1^k-algorithms. Given the tedium and complexity of the computation this provides welcome agreement.

5. The development of shift-invariant and symmetric algorithms in Sects. 5.4/95 and 5.5/103 follows [PR98]. As pointed out in that paper, shift and flip invariance of a C^k_1-algorithm imply a double subdominant Jordan block.

6. For a long time, it was taken for granted that a double subdominant eigenvalue $\lambda < 1$ were necessary for normal continuity. So it came as a surprise when the analysis in [PR97] revealed an algebraically eightfold, yet innocent, normal continuity preserving, subdominant eigenvalue λ for Simplest subdivision with $n = 3$, see also Sect. 6.3/120.

7. Reif [Rei93, Rei95c] established that regularity and injectivity of the characteristic ring are sufficient for smoothness. Necessity was proven in [PR98].

8. Regularity tests for the characteristic ring can be based on properties of the regular spline. For example, if the directional derivative can be expressed as a difference of spline coefficients of generating splines then the convex hull property can be used to confine directional derivatives to cones whose non-intersection establishes regularity. This geometric approach has been used, for example, by Umlauf and Ginkel [Uml99, Uml04, GU07a]. Zorin [Zor97, Zor00a] proposes to test regularity and injectivity via interval arithmetic. Injectivity for polynomial patches can also be checked by a technique of Goodman and Unsworth [GU96].

9. The simple condition of Theorem 5.24/105 for injectivity of the characteristic ring was not discovered before [RP05]. However, the proof given there is pedestrian. The idea of using concepts form algebraic topology is due to Zorin [Zor00b].

10. Conditions for asymmetric algorithms, as depicted in Fig. 5.5/103 *right*, were first discussed in [Rei95b].

Chapter 6
Case Studies of C_1^k-Subdivision Algorithms

In this chapter, we formally introduce and scrutinize three of the most popular subdivision algorithms, namely the *Catmull–Clark algorithm* [CC78], the *Doo–Sabin algorithm* [DS78], and *Simplest subdivision*[1] [PR97]. Besides the algorithms in their original form, it is instructive to consider certain variants. We selectively modify a subset of weights to obtain a variety of algorithms that is rich enough to illustrate the relevance of the theory developed so far. In particular, we show that a double subdominant eigenvalue is neither necessary nor sufficient for a C_1^k-algorithm: First, there are variants of the Doo–Sabin algorithm with a double subdominant eigenvalue, which provably fail to be C_1^1 because the Jacobian determinant ${}^{\times}\mathcal{D}\psi$ of the characteristic ring changes sign. Second, for valence $n = 3$, Simplest subdivision reveals an eightfold subdominant eigenvalue, but due to the appropriate structure of Jordan blocks, it is still C_1^1. In all cases, the algorithms are symmetric so that the conditions of Theorem 5.24/105 can be used for the analysis.

6.1 Catmull–Clark Algorithm and Variants

The Catmull–Clark algorithm (Fig. 6.1/110) is currently the most popular subdivision algorithm due to its close relationship with the tensor-product spline standard. The algorithm generalizes uniform knot insertion for bicubic tensor-product B-splines. Since each n-gon of the original mesh of control points is subdivided into n quadrilaterals the mesh is purely quadrilateral after the first step. Figure 6.2/110 defines the rules of the subdivision algorithm in terms of stencils. A *stencil* is an intuitive representation of a row of the local subdivision matrix A. In the regular case, when $n = 4$, the control points have the structure of a regular planar grid. In analogy, near an extraordinary vertex, the control points can be arranged with n-fold symmetry.

[1] In the literature, Simplest subdivision is sometimes also called *Mid-edge subdivision*.

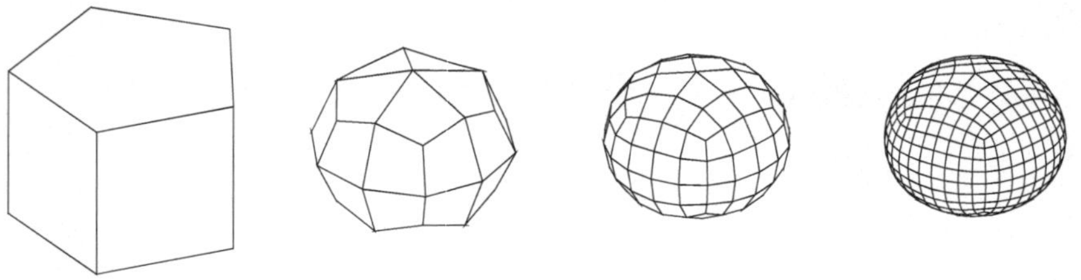

Fig. 6.1 Illustration of Catmull–Clark algorithm: Mesh refinement.

There are three types of stencils for Catmull–Clark subdivision that are inherited from bi-cubic B-spline subdivision and one generalization (see Fig. 6.2/110, *right*) that is expressed in terms of the variables

$$\alpha\,,\ \beta\,,\ \gamma, \quad \alpha+\beta+\gamma=1. \tag{6.1}$$

In [CC78], Catmull and Clark suggest

$$\alpha = 1-\frac{7}{4n},\ \beta=\frac{3}{2n},\ \gamma=\frac{1}{4n}. \tag{6.2}$$

For $n = 4$, this choice coincides with the regular stencil. To establish C_1^2-smoothness for variables α,β,γ summing to 1, we first define an appropriate data structure for the space of rings. Then we determine the characteristic ring ψ and apply Theorem 5.24/105 to obtain necessary and sufficient conditions for smoothness. Let us start with considering a single ring $\mathbf{x}^m$. Each of the n segments $\mathbf{x}_j^m, j \in \mathbb{Z}_n$, consists of three bicubic B-spline patches. The corresponding vector $\mathbf{Q} = [\mathbf{Q}_0;\ldots;\mathbf{Q}_{n-1}]$ of initial data is split into n blocks $\mathbf{Q}_j$ with 13 elements each. We label the coefficients of each block $\mathbf{Q}_j = [\mathbf{q}_{j,1};\ldots;\mathbf{q}_{j,13}]$ as shown in

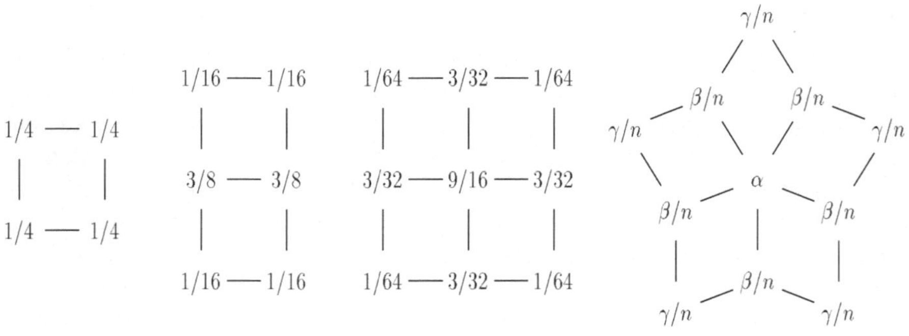

Fig. 6.2 Stencils for the Catmull–Clark algorithm: From *left* to *right*, the weights for generating a new 'face point', a new 'edge point', a new ordinary 'vertex point', and a new extraordinary 'vertex point' of valence n. The scalars α, β and γ are constrained by (6.1/110) and their originally published choice is given in (6.2/110).

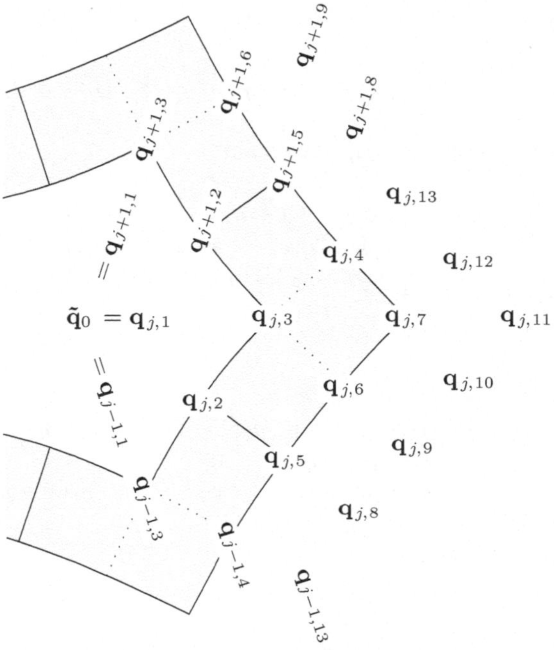

Fig. 6.3 Labelling of the control points $\mathbf{q}_{j,k}$ of the Catmull–Clark algorithm: Following (6.3/111), the center control point is replicated.

Fig. 6.3/111. Following Example 5.14/97, identical copies of the central coefficient $\tilde{\mathbf{q}}_0$ are placed in all blocks to obtain a circulant structure,

$$\tilde{\mathbf{q}}_0 = \mathbf{q}_{0,1} = \cdots = \mathbf{q}_{n-1,1}. \tag{6.3}$$

The corresponding subdivision matrix is block-circulant,

$$A = \operatorname{circ}(A_0, \ldots, A_{n-1}),$$

where the blocks A_j are (13×13)-matrices. Moreover, the algorithm is symmetric in the sense of Definition 5.21/104, and the generated segments satisfy the conditions (4.7/62) and (4.8/62) for $k = 2$. According to (5.14/99), the DFT $\hat{A} = \operatorname{diag}(\hat{A}_0, \ldots, \hat{A}_{n-1})$ of A is block-diagonal. Omitting the details, we find the following: the blocks $\hat{A}_i$ have the form

$$\hat{A}_i = \begin{bmatrix} \hat{A}_i^{0,0} & 0 & 0 \\ \hat{A}_i^{1,0} & \hat{A}_i^{1,1} & 0 \\ \hat{A}_i^{2,0} & \hat{A}_i^{2,1} & 0 \end{bmatrix}. \tag{6.4}$$

Recalling (5.12/98), we set

$$c_n + \mathbf{i}s_n := w_n := \exp(2\pi\mathbf{i}/n), \quad c_{n,i} + \mathbf{i}s_{n,i} := w_n^i = \exp(2\pi i\mathbf{i}/n),$$

for $i \in \mathbb{Z}_n$. With

$$p_1 := 1/64,\ p_2 := 3/32,\ p_3 := 9/16,\ q_1 := 1/16,\ q_2 := 3/8,\ r := 1/4,$$

and using the abbreviation $w := w_n^i$, we obtain

$$\hat{A}_i^{0,0} := \begin{bmatrix} \alpha\delta_{i,0} & \beta\delta_{i,0} & \gamma\delta_{i,0} \\ q_2\delta_{i,0} & 2q_1c_{n,i} + q_2 & q_1(1+\overline{w}) \\ r\delta_{i,0} & r(1+w) & r \end{bmatrix} \tag{6.5}$$

and

$$\begin{bmatrix} \hat{A}_i^{1,0} & \hat{A}_i^{1,1} \\ \hat{A}_i^{2,0} & \hat{A}_i^{2,1} \end{bmatrix} := \left[\begin{array}{ccc|cccc} q_2\delta_{i,0} & q_1 + q_2 w & q_1 & q_1 & q_1 w & 0 & 0 \\ p_2\delta_{i,0} & 2p_1c_{n,i} + p_3 & p_2(1+\overline{w}) & p_1\overline{w} & p_2 & p_1 & 0 \\ q_1\delta_{i,0} & q_1 w + q_2 & q_2 & 0 & q_1 & q_1 & 0 \\ p_1\delta_{i,0} & p_2(1+w) & p_3 & p_2 & p_1(1+w) & p_2 & p_1 \\ \hline 0 & q_2 & q_1(1+\overline{w}) & q_1\overline{w} & q_2 & q_1 & 0 \\ 0 & r & r & 0 & r & r & 0 \\ 0 & q_1 & q_2 & q_1 & q_1 & q_2 & q_1 \\ 0 & 0 & r & r & 0 & r & r \\ 0 & q_1 w & q_2 & q_2 & q_1 w & q_1 & q_1 \\ 0 & rw & r & r & rw & 0 & 0 \end{array}\right].$$

The eigenvalues $1/8, 1/16, 1/32, 1/64$ of the sub-matrix $\hat{A}_i^{1,1}$ are n-fold eigenvalues of A. Other non-zero eigenvalues come only from $\hat{A}_i^{0,0}$. For $i = 0$, we obtain the obligatory eigenvalue $\lambda_0 = 1$ and, with $\gamma := 1 - \alpha - \beta$, the pair

$$\lambda_{1,2}^0 := \left(4\alpha - 1 \pm \sqrt{(4\alpha-1)^2 + 8\beta - 4}\right)/8.$$

Depending on the sign of the discriminant, these two eigenvalues are either both real or complex conjugate. For $i \neq 0$, the non-zero eigenvalues of $\hat{A}_i^{0,0}$ are always real and given by

$$\lambda_{1,2}^i := \left(c_{n,i} + 5 \pm \sqrt{(c_{n,i}+9)(c_{n,i}+1)}\right)/16.$$

Here and in the penultimate display, the subscript 1 refers to the plus sign, and the subscript 2 refers to the minus sign. By Theorem 5.18/101, the subdominant eigenvalue λ must come from the blocks $\hat{A}_1, \hat{A}_{n-1}$. Because the eigenvalue $1/32$ has algebraic multiplicity n, the only candidate is

$$\lambda := \lambda_1^1 = \lambda_1^{n-1} = \left(c_n + 5 + \sqrt{(c_n+9)(c_n+1)}\right)/16. \tag{6.6}$$

Straightforward calculus shows that

$$1 > \lambda > 1/4 > \lambda_2^i > 1/8, \quad i = 1, \ldots, n-1 \tag{6.7}$$

$$\lambda > \lambda_1^i > 1/4, \quad i = 2, \ldots, n-2\,. \tag{6.8}$$

That is, λ is subdominant if α, β, γ are chosen such that

$$\lambda > \max\{|\lambda_1^0|, |\lambda_2^0|\}. \tag{6.9}$$

We will comment on the set of feasible weights at the end of this section, but state already now that the original weights of Catmull–Clark (6.2/110) satisfy the condition.

For computing the characteristic ring, the eigenvector $\hat{v}$ of $\hat{A}_1$ is partitioned into three blocks, $\hat{v} = [\hat{v}_0; \hat{v}_1; \hat{v}_2]$, according to the structure of $\hat{A}_1$ defined in (6.4/111). Then $\hat{A}_1 \hat{v} = \lambda \hat{v}$ is equivalent to

$$\begin{aligned}(\hat{A}_1^{0,0} - \lambda)\hat{v}_0 &= 0\\ (\hat{A}_1^{1,1} - \lambda)\hat{v}_1 &= -\hat{A}_1^{1,0}\hat{v}_0\\ \hat{v}_2 &= (\hat{A}_1^{2,0}\hat{v}_0 + \hat{A}_1^{2,1}\hat{v}_1)/\lambda.\end{aligned} \tag{6.10}$$

Now, $\hat{v}$ can be computed conveniently starting from

$$\hat{v}_0 := [1 + \overline{w}_n, 16\lambda - 2c_n - 6], \tag{6.11}$$

which solves the first eigenvector equation.

By (6.6/112), the characteristic ring depends only on $c_n \in [-1/2, 1)$ and not on the particular choice of weights α, β, γ. For the interval of definition, we can invert the relation to obtain

$$c_n = \frac{16\lambda^2 - 10\lambda + 1}{2\lambda}, \quad \lambda \in \Lambda := \left[(9 + \sqrt{17})/32,\ (3 + \sqrt{5})/8\right), \tag{6.12}$$

and write the characteristic ring in terms of $\lambda \in \Lambda$. After scaling, the eigenvector $\hat{v}$ has the form

$$\hat{v} = ((4\lambda - 1)\hat{v}_{\mathrm{re}} + 2s\lambda(64\lambda - 1)\mathbf{i}\hat{v}_{\mathrm{im}})/13020,$$

where

$$\hat{v}_{\mathrm{re}} := \begin{bmatrix} 0\\ 4\lambda^2(64\lambda-1)(32\lambda-1)(16\lambda-1)(4\lambda-1)\\ 8\lambda^2(64\lambda-1)(32\lambda-1)(16\lambda-1)\\ 4\lambda^2(64\lambda-1)(928\lambda^2+228\lambda-31)\\ 8\lambda^2(64\lambda-1)(16\lambda-1)(4\lambda-1)(4\lambda+13)\\ 4\lambda^2(64\lambda-1)(928\lambda^2+228\lambda-31)\\ 80\lambda^2(1280\lambda^3+2128\lambda^2-56\lambda-13)\\ (64\lambda-1)(16\lambda-1)(4\lambda-1)(100\lambda^2+42\lambda-1)\\ 4\lambda(64\lambda-1)(640\lambda^3+688\lambda^2-82\lambda-1)\\ 20\lambda(2048\lambda^4+11040\lambda^3+812\lambda^2-165\lambda-1)\\ 40\lambda(5248\lambda^3+1568\lambda^2-133\lambda-5)\\ 20\lambda(2048\lambda^4+11040\lambda^3+812\lambda^2-165\lambda-1)\\ 4\lambda(64\lambda-1)(640\lambda^3+688\lambda^2-82\lambda-1) \end{bmatrix}, \quad \hat{v}_{\mathrm{im}} := \begin{bmatrix} 0\\ -4\lambda^2(32\lambda-1)(16\lambda-1)\\ 0\\ 140\lambda^2(8\lambda-1)\\ -8\lambda^2(16\lambda-1)(4\lambda+13)\\ -140\lambda^2(8\lambda-1)\\ 0\\ -(16\lambda-1)(100\lambda^2+42\lambda-1)\\ -4\lambda(160\lambda^2+132\lambda-1)\\ -20\lambda(8\lambda^2+15\lambda+1)\\ 0\\ 20\lambda(8\lambda^2+15\lambda+1)\\ 4\lambda(160\lambda^2+132\lambda-1) \end{bmatrix}.$$

Now, we compute $f(1,1,0)$ to ensure normalization. After reformatting the middle patch according to its tensor product structure and substituting in the parameter

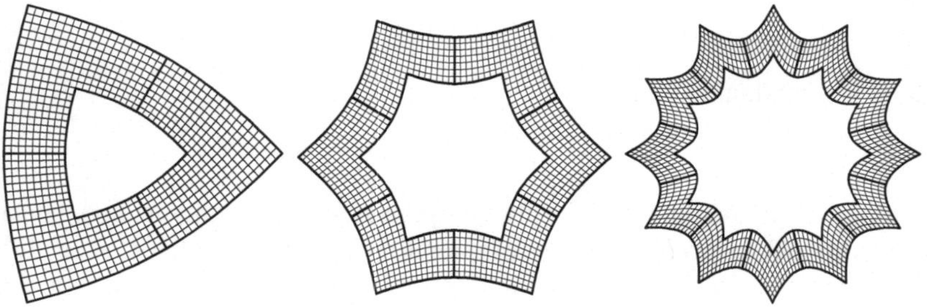

Fig. 6.4 Characteristic ring of the Catmull–Clark algorithm: (*left*) $n = 3$, (*middle*) $n = 6$, and (*right*) $n = 12$.

$\boldsymbol{\sigma} = [1, 1]$, we obtain with $b := [1/6, 2/3, 1/6]$

$$f(1,1,0) = b \begin{bmatrix} \hat{v}^3 & \hat{v}^4 & \hat{v}^{13} \\ \hat{v}^6 & \hat{v}^7 & \hat{v}^{12} \\ \hat{v}^9 & \hat{v}^{10} & \hat{v}^{13} \end{bmatrix} \cdot b = \frac{2}{29295}\lambda(4\lambda-1)\big(139264\lambda^4 + 170496\lambda^3 + 112\,\lambda^2 - 1,476\lambda - 11\big).$$

For $\lambda \in \Lambda$ this value is real and positive. That is, the characteristic ring f is normalized in the sense of Definition 5.19/103.

To establish smoothness, we verify the sufficient conditions (5.22/107) given in Theorem 5.25/107. The derivative of f_0 in t-direction is computed by differencing the Bernstein–Bézier control points of the three bicubic patches. The elements of the resulting three sets of 3×4 coefficients are enumerated $k_1, \ldots, k_{36}$. All k_μ are polynomials in λ with rational coefficients. More precisely,

$$k_\mu(\lambda) = p_\mu(\lambda) + \mathbf{i} s_n q_\mu(\lambda), \quad \mu = 1, \ldots, 36,$$

for certain polynomials p_μ and q_μ of degree ≤ 7 in λ which *are independent* of n or the special weights. Computing the Sturm sequences of all these polynomials on the larger, but more convenient interval $\Lambda' := [0.41, 0.66] \supset \Lambda$, we find either no root or the single root $(3 + \sqrt{5})/8 \notin \Lambda$. Hence, the sign of all polynomials in question is constant and can be determined by evaluation at a single point. At $\lambda = 1/2$, we obtain $p_\mu(\lambda) = q_\mu(\lambda) = 3255/13020 = 1/4$ for all $\mu = 1, \ldots, 36$. Hence, all coefficients k_μ are positive so that, by the convex hull property, $\mathrm{Re}(D_2 f_0) > 0$ and $\mathrm{Im}(D_2 f_0) > 0$. Hence, by Theorem 5.25/107, the algorithm is C_1^2.

Figure 6.4/114 shows the characteristic rings for different values of n. As already mentioned above, it depends only on n, but not on the particular choice of the weights α, β, γ, provided that the conditions, summarized in the following theorem, are satisfied.

Theorem 6.1 (**C_1^2-variants of Catmull–Clark**). *For* $n \geq 3$, $c_n := \cos(2\pi/n)$, *and*

$$\lambda := \left(c_n + 5 + \sqrt{(c_n+9)(c_n+1)}\right)/16 \tag{6.13}$$

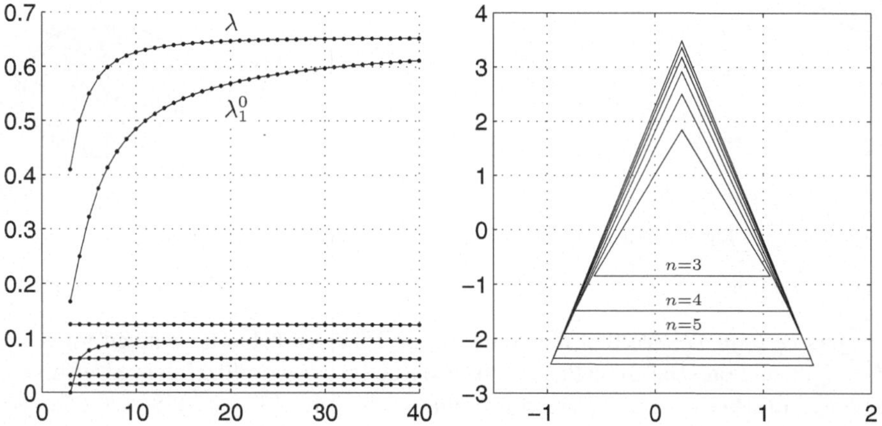

Fig. 6.5 Illustration of Theorem 6.1/114: (*left*) Spectrum of the subdivision matrix of the Catmull–Clark algorithm with standard weights (6.2/110) for $n = 3, \ldots, 40$. The subdominant eigenvalue λ satisfies the condition of the theorem. (*right*) Triangles $\Delta_3, \ldots, \Delta_8$ in the $\alpha\beta$-plane. Choosing (α, β) inside these triangles yields a C_1^2-algorithm.

$$\lambda_{1,2}^0 := \left(4\alpha - 1 \pm \sqrt{(4\alpha-1)^2 + 8\beta - 4}\right)/8, \tag{6.14}$$

the Catmull–Clark algorithm with weights α, β *and* $\gamma = 1 - \alpha - \beta$ *is a standard* C_1^2*-algorithm if and only if* $\lambda > \max\{|\lambda_1^0|, |\lambda_2^0|\}$.

Let us briefly comment on the set of parameters yielding a C_1^2-algorithm. We define

$$\tilde{\alpha} := \frac{\alpha}{2} - \frac{1}{8}, \quad \tilde{\beta} := \frac{\beta}{8} - \frac{1}{16},$$

and obtain the equivalent condition

$$\left|\tilde{\alpha} \pm \sqrt{\tilde{\alpha}^2 + \tilde{\beta}}\right| < \lambda.$$

Distinguishing the cases $\tilde{\alpha}^2 + \tilde{\beta} \geq 0$ and $\tilde{\alpha}^2 + \tilde{\beta} \leq 0$, we find

$$-\lambda^2 < \tilde{\beta} < \lambda(\lambda - 2|\tilde{\alpha}|\lambda).$$

Given the valence n and the corresponding subdominant eigenvalue λ, the set of pairs $(\tilde{\alpha}, \tilde{\beta})$ satisfying this condition forms the interior of a triangle $\tilde{\Delta}_n$. Accordingly, in terms of the original parameters α, β, we obtain a triangle Δ_n in the $\alpha\beta$-plane with corners

$$\left(1/4 \pm 2\lambda, 1/2 - \lambda^2\right), \quad \left(1/4, 1/2 + \lambda^2\right).$$

On the right hand side, Fig. 6.5/115 shows the triangles $\Delta_3, \ldots, \Delta_8$. On the left hand side, we see the complete spectrum of the algorithm when using the standard

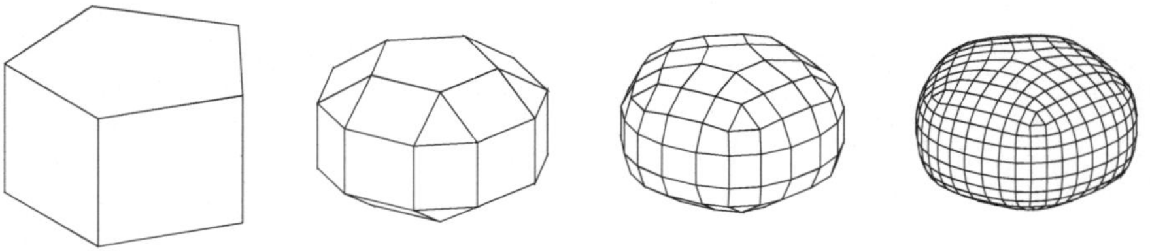

Fig. 6.6 Illustration of Doo–Sabin algorithm: Mesh refinement.

weights (6.2/110). We see that the condition of Theorem 6.1/114 is satisfied for $n = 3, \dots, 40$, and one can show that this is also true for *all* $n > 40$. However, it should be noted that both λ and λ_1^0 are converging to the limit $(3 + \sqrt{5})/8 \approx 0.6545$ as $n \to \infty$. As we will explain in the next chapter, shape may be poor if the ratio of the subdominant eigenvalue λ and the next smaller *subsubdominant eigenvalue* is close to 1.

6.2 Doo–Sabin Algorithm and Variants

The Doo–Sabin algorithm generalizes subdivision of uniform biquadratic tensor-product B-splines. For each n-gon of the original mesh of control points, a new, smaller n-gon is created and connected with its neighbors as depicted in Fig. 6.6/116. Figure 6.7/117 shows the *stencils* for generating a new n-gon from an old one, both for the regular case $n = 4$ (*left*) and the general case (*middle*). For $n = 4$ the weights are those of the biquadratic spline. Doo and Sabin in [DS78] suggested

$$a_j := \frac{\delta_{j,0}}{4} + \frac{3 + 2\cos(2\pi j/n)}{4n} \tag{6.15}$$

for the general case. In the following, we analyze all algorithms that are affine invariant and symmetric:

$$\sum_{j=0}^{n-1} a_j = 1, \quad a_j = a_{n-j}, \quad j \in \mathbb{Z}_n. \tag{6.16}$$

Each of the n segments $\mathbf{x}_j^m, j \in \mathbb{Z}_n$, of the m-th ring generated by the Doo–Sabin algorithm consists of three biquadratic B-spline patches. Accordingly, we can split the control points $\mathbf{Q}^m$ into n groups of nine control points, each, ordered as shown in Fig. 6.7/117 (*right*).

Since the algorithm is symmetric, we can apply DFT as introduced in Sect. 5.4/95 to obtain the block-diagonal form $\hat{A} = \operatorname{diag}(\hat{A}_0, \dots, \hat{A}_{n-1})$ of the subdivision matrix. The non-zeros elements of $\hat{A}_i$ are situated in the first four columns. With

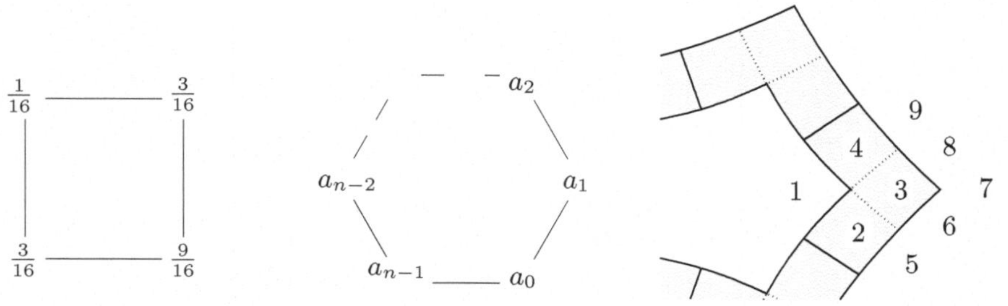

Fig. 6.7 Stencils for the Doo–Sabin algorithm: (*left*) Regular refinement rule, (*middle*) general refinement rule, and (*right*) control point labels of one segment.

$w = w_n^i = \exp(2\pi i\mathbf{i}/n)$, as before, we have

$$\hat{A}_i(:,1:4) = \begin{bmatrix} \hat{\alpha}_i & 0 & 0 & 0 \\ p+\overline{w}q & q & 0 & \overline{w}r \\ p & q & r & q \\ p+wq & wr & 0 & q \\ q+\overline{w}r & p & 0 & \overline{w}q \\ q & p & q & r \\ r & q & p & q \\ q & r & q & p \\ q+wr & wq & 0 & p \end{bmatrix}, \tag{6.17}$$

where $p := 9/16, q := 3/16, r := 1/16$ are the standard weights for quadrilaterals, and

$$\hat{a}_i := \sum_{j=0}^{n-1} w_n^{-ij} a_j$$

are the entries of the DFT of the vector $[a_0, \dots, a_{n-1}]$ of special weights for the inner n-gon. The weights a_j sum to one, i.e., $\hat{a}_0 = 1$, and satisfy $a_j = a_{n-j}$. Hence, $\hat{a}_i = \hat{a}_{n-i}$ is real. The eigenvalues of $\hat{A}_i$ are $\hat{a}_i$, $1/4$, $1/8$, $1/16$, 0. Since each eigenvalue $1/4$ corresponds to a separate eigenspace and also the eigenspace of each $\hat{a}_i$ is spanned by a single vector, and by the requirement on the Fourier index to be $\mathcal{F}(\lambda) = \{1, n-1\}$, the subdominant eigenvalue must be $\lambda := \hat{a}_1 = \hat{a}_{n-1} \in (1/4, 1)$ to generate a C_1^1-algorithm. This yields the inequality

$$1 > \hat{a}_1 > \max\{1/4, |\hat{a}_2|, \dots, |\hat{a}_{n/2}|\}. \tag{6.18}$$

We will see below that this constraint is however *not sufficient*. Using a computer algebra system, one can determine the complex eigenvector $\hat{v}$ of $\hat{A}_1$ corresponding to λ explicitly:

$$\hat{v} = \begin{bmatrix} 2\lambda(16\lambda-1)(8\lambda-1)(4\lambda-1) \\ 6\lambda(16\lambda-1)(6\lambda-1+2\overline{w}_n\lambda) \\ 18\lambda(32\lambda^2-1+4c_n\lambda) \\ 6\lambda(16\lambda-1)(6\lambda-1+2w_n\lambda) \\ (16\lambda-1)\big(12\lambda^2+18\lambda-3+\overline{w}_n(4\lambda^2+12\lambda-1)\big) \\ 6\lambda\big(32\lambda^2+64\lambda-12+c_n(20\lambda+1)-is_n(16\lambda-1)\big) \\ 64\lambda^3+512\lambda^2-46\lambda-8+36c_n\lambda(2\lambda+1) \\ 6\lambda\big(32\lambda^2+64\lambda-12+c_n(20\lambda+1)+is_n(16\lambda-1)\big) \\ (16\lambda-1)\big(12\lambda^2+18\lambda-3+w_n(4\lambda^2+12\lambda-1)\big) \end{bmatrix},$$

where as before $w_n = c_n + \mathbf{i}s_n$.

In particular, for the original Doo–Sabin weights in (6.15/116), we have $\lambda = 1/2$ and, rearranging the entries of $\hat{v}$ in a (3×3)-matrix according to Fig. 6.7/117, *right*,

$$\begin{bmatrix} \hat{v}_5 & \hat{v}_6 & \hat{v}_7 \\ \hat{v}_2 & \hat{v}_3 & \hat{v}_8 \\ \hat{v}_1 & \hat{v}_4 & \hat{v}_9 \end{bmatrix} = 3 \begin{bmatrix} 21+14\overline{w}_n & 28+2w_n+9\overline{w}_n & 35+12c_n \\ 14+7\overline{w}_n & 21+6c_n & 28+2\overline{w}_n+9w_n \\ 7 & 14+7w_n & 21+14w_n \end{bmatrix}.$$

By elementary computations, one can determine the Bernstein–Bézier-form of all three biquadratic patches forming the first segment of the complex characteristic ring f. For $\lambda \in (1/4, 1)$,

$$\begin{aligned} f_0(1,1) &= \frac{\hat{v}_3+\hat{v}_6+\hat{v}_7+\hat{v}_8}{4} = p(\lambda) + c_n q(\lambda) \\ &:= (256\,\lambda^3 + 320\lambda^2 - 52\lambda - 2) + c_n\left(96\lambda^2 + 12\lambda\right). \end{aligned} \tag{6.19}$$

For $n \geq 3$, we have $c_n \geq -1/2$. Furthermore $p(\lambda) > 320\lambda^2 - 54\lambda - 2$ and $q(\lambda) > 0$ for all $\lambda \in (1/4, 1)$ so that

$$\begin{aligned} p(\lambda) + c_n q(\lambda) &> 320\lambda^2 - 54\lambda - 2 - (96\lambda^2 + 12\lambda)/2 \\ &= 272\lambda^2 - 60\lambda - 2 = 2(4\lambda - 1)(34\lambda + 1). \end{aligned} \tag{6.20}$$

That is, $f_0(1,1)$ is real and positive for $\lambda \in (1/4, 1)$. The eigenvector $\hat{v}$ and hence the characteristic ring f depends only on $\lambda = \hat{a}_1 = \hat{a}_{n-1}$ and on the valence n.

For $\lambda \in (1/4, 1)$, the minimum of the real parts of all Bernstein–Bézier coefficients is positive. Hence, by the convex hull property, the condition $c(u) \in \mathbb{R} \Rightarrow c(u) > 0$ in Theorem 5.24/105 is always satisfied. It remains to show regularity of the segment f_0 of the characteristic ring. The Jacobian determinant ${}^{\times}\!\mathcal{D}f_0$ consists of three bicubic patches, which can also be expressed explicitly in Bernstein–Bézier-form. A careful analysis shows that all coefficients are positive if

$$p(\lambda) := 128\lambda^2(1-\lambda) - 7\lambda - 2 + 9\lambda c_n > 0. \tag{6.21}$$

By the convex hull property, this implies regularity of f_0. In particular, for $\lambda = 1/2$, we obtain $p(1/2) = 3/2\,(7 + 3c_n) > 0$ proving that the Doo–Sabin algorithm with standard weights is a C_1^1-algorithm. Figure 6.8/119 illustrates the situation: all Bernstein–Bézier coefficients of the Jacobian ${}^{\times}\!\mathcal{D}f$ depend only on λ and c_n, and

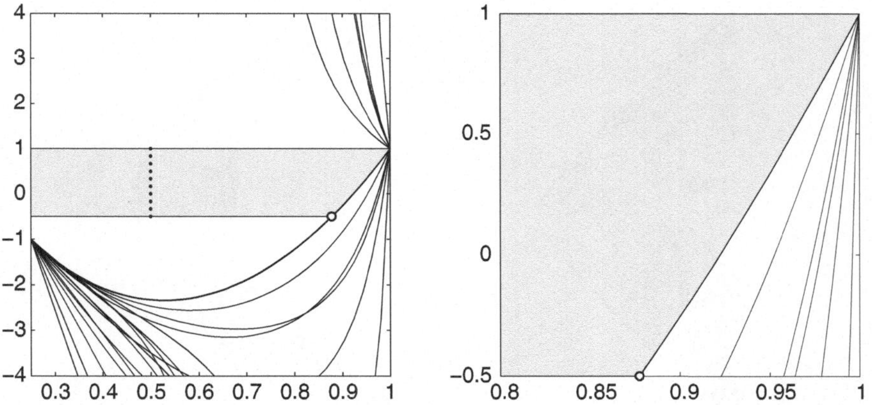

Fig. 6.8 Illustration of Theorem 6.2/119: (*left*) Admissible range of subdominant eigenvalues λ plotted in the (λ, c_n)-plane (see (6.21/118)) and (*right*) magnified detail.

they change sign on the lines plotted in the (λ, c_n)-plane. In particular, $p(\lambda) = 0$ for points on the thick line, and this line is bounding the shaded subset of the interval $(1/4, 1)\times[-1/2, 1)$. For given n, the eigenvalue λ yields a C^1_1-algorithm if and only if the point (λ, c_n) lies in this region. The dotted line, corresponding to the standard case $\lambda = 1/2$, indicates that this value is feasible for all values of n. Surprisingly, there is an upper bound $\lambda_{\text{sup}}(n)$ with $p(\lambda) < 0$ for $1 > \lambda > \lambda_{\text{sup}}(n)$. For such λ, ${}^{\times}\!Df$ actually reveals a change of sign, and the corresponding algorithm cannot be C^1_1. Fortunately, the upper bounds are quite close to 1 so that they do not impose severe restrictions when designing variants on the standard Doo–Sabin algorithm. More precisely, as indicated in Fig. 6.8/119 by the dot, the lowest upper bound occurs for $n = 3$. We have $c_3 = -1/2$ and

$$\lambda_{\text{sup}}(n) \ge \lambda_{\text{sup}}(3) = \frac{\sqrt{187}}{24}\cos\left(\frac{1}{3}\arctan\left(\frac{27\sqrt{5563}}{1576}\right)\right) + \frac{1}{3} \approx 0.8773.$$

The asymptotic behavior for $n \to \infty$ is

$$\lambda_{\text{sup}}(n) \doteq 1 - \frac{\pi^2}{7n^2}.$$

In summary, we have shown the following.

Theorem 6.2 (C^1_1-variants of Doo–Sabin subdivision). *Let $\hat{a}_0, \dots, \hat{a}_{n-1}$ be the Fourier coefficients of a symmetric set of weights for the generalized Doo–Sabin algorithm. Then a standard algorithm is obtained if $\lambda := \hat{a}_1 = \hat{a}_{n-1}$ satisfies the condition*

$$1 > \lambda > \max\{1/4, |\hat{a}_2|, \dots, |\hat{a}_{n-2}|\}.$$

The algorithm is C^1_1 if $p(\lambda) > 0$, and not C^1_1 if $p(\lambda) < 0$. In particular, the algorithm is C^1_1 when choosing the standard weights.

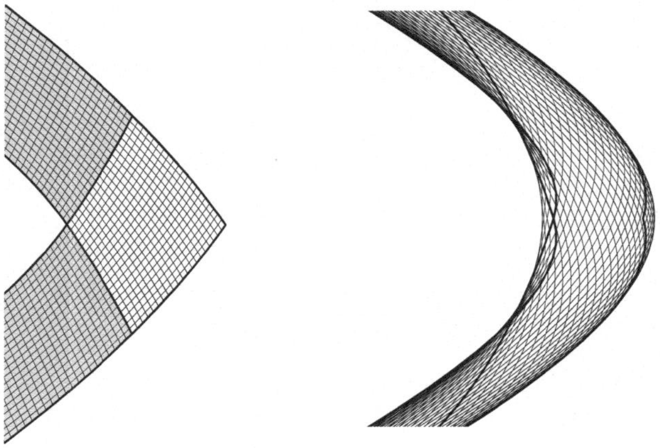

Fig. 6.9 Illustration of Theorem 6.2/119: Corner piece of the characteristic ring for $n = 3$ and subdominant eigenvalue (*left*) $\lambda = 0.5$ and (*right*) $\lambda = 0.95$. On the right hand side, the coordinate axes are scaled differently to clearly visualize non-injectivity.

6.3 Simplest Subdivision

When regarded as an algorithm for refining control meshes, one step of Simplest subdivision connects every edge-midpoint of the given mesh to the four midpoints of the edges that share both a vertex and a face with the current edge. For that reason, Simplest subdivision is sometimes also called *Mid-edge subdivision*. Once all midpoints are linked, the old mesh is discarded, as shown in Fig. 6.10/120. Thus

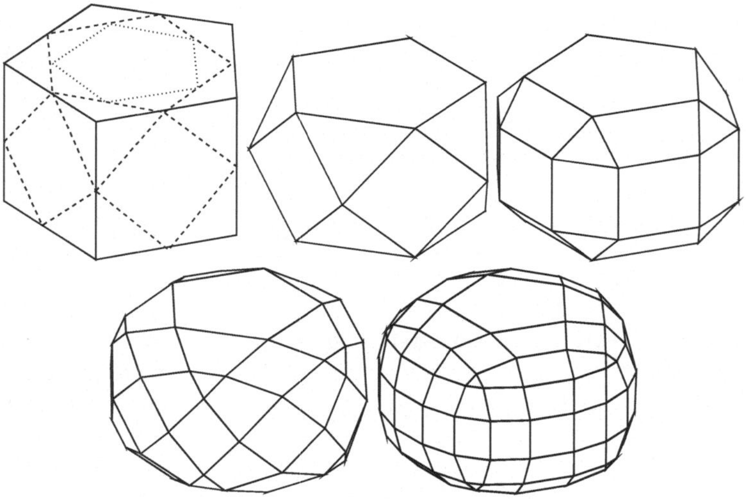

Fig. 6.10 Illustration of Simplest subdivision: Mesh refinement.

every new point, not on the global boundary, has exactly four neighbors; and every mesh point is replaced by a facet that is quadrilateral if the point is new. Each step can be interpreted as cutting off all vertices, along with neighborhoods that stretch half way to the neighbor vertex. The cuts are in general not planar.

The following feature justifies the name of the algorithm: all subdivision stencils are equal and of minimal size 2. The only weight used throughout is $1/2$ so that, in contrast to most other subdivision algorithms, there is no dependence on the valence n. Evidently, this setup is as simple as it can be.

To apply the analysis developed so far, we have to think of the algorithm not in terms of control meshes, but as a recursion for rings. To fit that pattern, we need to combine two steps of mesh refinement to generate a new ring $\mathbf{x}^{m+1}$ from the given ring $\mathbf{x}^m$. For that reason, Simplest subdivision is called a $\sqrt{2}$-algorithm. On a regular, quadrilateral control mesh with 4-valent vertices, a double step of mid-edge mesh refinement coincides with one subdivision step of the 4-direction box spline with directions $\Xi := \left[\begin{smallmatrix} 1 & 0 & -1 & 1 \\ 0 & 1 & -1 & -1 \end{smallmatrix}\right]$. Hence, the resulting limit surface is such a box spline. It is C^1, and each quadrilateral patch consists of four triangles of total degree 2, arranged in a quincunx pattern. From a combinatorial point of view, a double step of mid-edge mesh refinement for a general mesh coincides with one step of Doo–Sabin subdivision. That is, each n-gon is mapped to a smaller one. Using the same arrangement of control points $\mathbf{q}_\ell$ and weights $a = [a_0, \ldots, a_{n-1}]$ as in Fig. 6.7/117, we have

$$a_j = \begin{cases} \frac{1}{2} & \text{for } j = 0, \\ \frac{1}{4} & \text{for } j = 1, n-1, \\ 0 & \text{otherwise.} \end{cases}$$

The decisive point is that these weights are also used in the regular case $n = 4$. The structure of the Fourier blocks $\hat{A}_i$ of the subdivision matrix is the same as for the Doo–Sabin algorithm. But now, the weights in (6.17/117) are $p := 1/2, q := 1/4, r := 0$, and

$$\hat{a}_i = \sum_{j=0}^{n-1} w_n^{-ij} a_j = \frac{1 + \cos(2i\pi/n)}{2}, \quad i \in \mathbb{Z}_n.$$

For $i = 0, \ldots, n-1$, the non-zero eigenvalues of $\hat{A}_i$ are

$$\hat{a}_i \quad \text{and} \quad \frac{1}{4}, \frac{1}{4}.$$

The dominant eigenvalue of the subdivision matrix A is $\hat{a}_0 = 1$. Determining the subdominant eigenvalue is subtle here: If $n \geq 4$, we have $|\hat{a}_i| < |\hat{a}_1|$ for $i = 2, \ldots, n-2$, and also $1/4 < |\hat{a}_1|$ so that

$$\lambda := \hat{a}_1 = \hat{a}_{n-1} = \frac{1 + c_n}{2}, \quad c_n := \cos(2\pi/n),$$

is the double subdominant eigenvalue. That is, we obtain a standard algorithm. However, if $n = 3$, the upper left (4×4)-submatrices $\hat{A}'_i := \hat{A}_i(1:4, 1:4)$ of $\hat{A}_i$ read

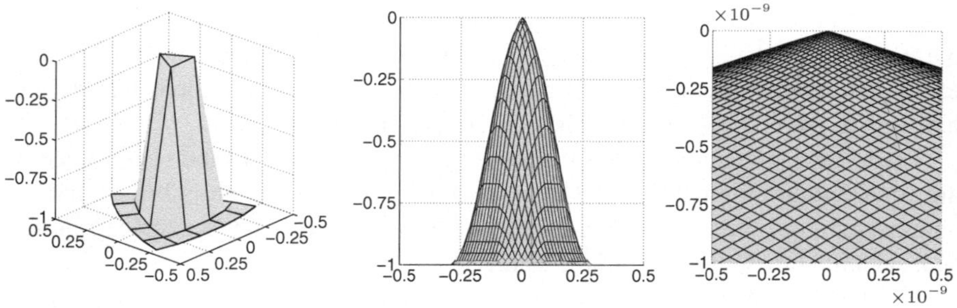

Fig. 6.11 Illustration of Simplest subdivision: (*left*) Input mesh, (*middle*) side view after three double-steps of refinement, and (*right*) close-up of visual cone tip after 20 double-steps of refinement. Nevertheless, from a mathematical point of view, the limit surface is smooth.

$$\hat{A}'_0 = \frac{1}{4}\begin{bmatrix} 4 & 0 & 0 & 0 \\ 3 & 1 & 0 & 0 \\ 2 & 1 & 0 & 1 \\ 3 & 0 & 0 & 1 \end{bmatrix}, \quad \hat{A}'_1 = \overline{\hat{A}'_2} = \frac{1}{8}\begin{bmatrix} 2 & 0 & 0 & 0 \\ 3-\mathbf{i}\sqrt{3} & 2 & 0 & 0 \\ 4 & 2 & 0 & 2 \\ 3+\mathbf{i}\sqrt{3} & 0 & 0 & 2 \end{bmatrix}.$$

The subdominant eigenvalue $\lambda = 1/4$ appears in multitude: its algebraic multiplicity in $\hat{A}'_0, \hat{A}'_1, \hat{A}'_2$ is $2, 3, 3$, respectively, while its geometric multiplicity in all three matrices is 2. Consequently, the Jordan decomposition of the subdivision matrix comprises the non-zero Jordan blocks

$$J_0 = 1, \quad J_1 = J_2 = \begin{bmatrix} 1/4 & 1 \\ 0 & 1/4 \end{bmatrix}, \quad J_3 = \cdots = J_6 = 1/4.$$

Thus, for $n = 3$, the algorithm is non-standard. Here, the more general theory, developed in Sect. 5.3/89, applies. The subdivision matrix has type $A \in \mathcal{A}_1^1$, and the characteristic ring ψ is defined according to Definition 5.10/92. The complex eigenvector defining ψ, arranged in matrix form as shown in Fig. 6.7/117, is

$$\hat{v} := \begin{bmatrix} 6+4\,c_n & 8+3\,c_n & 10+4\,c_n \\ 4+2\,c_n & 6+2\,c_n & 8+3\,c_n \\ 2 & 4+2\,c_n & 6+4\,c_n \end{bmatrix} + \mathbf{i}s_n \begin{bmatrix} 4 & 2 & 0 \\ 2 & 0 & -2 \\ 0 & -2 & -4 \end{bmatrix}.$$

As detailed in [PR97], regularity and uni-cyclicity can be verified using the same techniques as described above for the Doo–Sabin algorithm. Hence, we can state

Theorem 6.3 (Simplest subdivision is C_1^1). *Simplest subdivision is a C_1^1-algorithm for all valences $n \geq 3$.*

It remains to touch on the following two subjects: First, unlike tensor product B-splines, the nodal functions of box spline spaces do not always form bases. In particular, the system G of generating rings of Simplest subdivision is linearly dependent. But fortunately, the matrix A as specified here is a subdivision matrix in the sense that it does not have any ineffective eigenvectors.

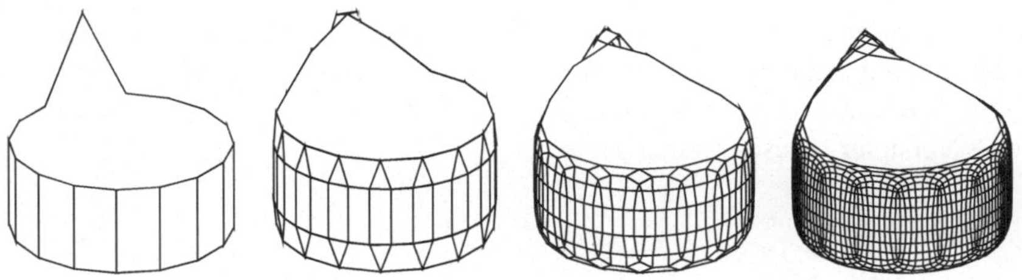

Fig. 6.12 Illustration of Simplest subdivision: Fast convergence for 3-sided facets and slow convergence for large facets.

Second, it must be noted that simplest subdivision, as described here, reveals serious shape artifacts for $n = 3$, and extremely slow convergence for high valences. In Fig. 6.11/122 *left*, we see an input mesh including a triangle, corresponding to an extraordinary point of valence $n = 3$. The first two coordinates correspond to the characteristic spline, while the third one has initial data which are 0 for the three innermost coefficients, and -1 otherwise. At the center, the surface seems to have a cone point, even if we zoom in by a factor of one billion to visualize the control mesh after 20 double steps of refinement (see Fig. 6.11/122, *right*). By the standards of Computer Graphics, this is hardly considered a smooth surface.[2] The apparent inconsistence between the theoretical result and the practical realization can be explained as follows: In the tangent plane, the behavior of rings is governed by the factor $\lambda^{m,1} = m\lambda^{m-1}$, while the component perpendicular to it decays as λ^m. To put it differently, let us consider the subdivision step from ring $\mathbf{x}^m$ to $\mathbf{x}^{m+1}$. Asymptotically, the tangential components are multiplied by $(1 + 1/m)/4$ and the normal component is multiplied by $1/4$, what does not make too much of a difference. As a consequence, the surface locally resembles the tip of a cone. Only for *very* large values of m, the slightly slower decay of the normal component prevails, and forces the rings to approach the tangent plane[3].

Another problem is depicted in Fig. 6.12/123. We see an input mesh which, after the first step, consists of triangles, quadrangles, and a 16-gon. The obvious problem concerns the extremely slow shrinkage of the 16-gon. It is due to the corresponding subdominant eigenvalue $\lambda \approx 0.962$, which is only slightly smaller than 1. As a consequence, a *very* large number of subdivision steps is required to obtain a mesh which is sufficiently dense for, say, visualization.

For practical purposes, Simplest subdivision should be modified, for instance by the Doo–Sabin weights (6.15/116) for $n \neq 4$. The modest increase in complexity is easily compensated for by superior shape properties.

[2] Imagine that the middle part of Fig. 6.11/122 is the size of Mount Everest. Then the detail on the right hand side is smaller than the breadth a hair. But mathematicians think in different categories.

[3] A quite instructive univariate analog of this case, justified by identifying $u = \lambda^m$, is given by the curve $c(u) = [u \ln |u|, |u|]$, $u \in (-1/2, 1/2)$. Although the x-axis is easily verified to be the tangent at the origin, the image of the curve suggests a kink. The reader is encouraged to generate plots of the curve and its curvature at different scales.

Anyway, the latter observations clearly show that it is not sufficient to classify a subdivision algorithm as C_1^k to ensure fairness of the generated surfaces. Rather, an exacting analysis is necessary to scrutinize shape properties of subdivision surfaces. The next chapter focuses on that subject.

Bibliographic Notes

1. In the literature, the terms 'mask' and 'stencil' are not used in a consistent way. By our understanding, the entries in a row of the subdivision matrix, gathering contributions of the given control points to a single new one, can be displayed as a diagram called *stencil*. Conversely, a column of the subdivision matrix indicates how a single given control point influences the different new ones. This 'scattering' of contributions is called *mask*.

2. The spectrum of both the Catmull–Clark algorithm [CC78] and the Doo–Sabin algorithm [DS78] were first analyzed by Doo and Sabin [DS78] in order to assess and improve shape. In [BS86] an (almost correct) characterization of admissible weights α, β, γ for the Catmull–Clark algorithm is given.

3. A complete analysis of variants of the Doo–Sabin algorithm and of the Catmull–Clark algorithm can be found in [PR98]. Additional variants of the Catmull–Clark algorithm that modify rules not only for the central coefficient but also rules influenced by the central coefficient are proposed by Augsdörfer, Dodgson and Sabin [ADS06].

4. Simplest subdivision was published and analyzed by Peters and Reif [PR97] and also by Habib and Warren [HW99]. It is based on the Zwart–Powell element [Zwa73].

5. The smoothness properties of a number of other algorithms have been established. Loop derived the central stencil [Loo87] by analyzing the range of the eigenvalues. The analysis of Loop's algorithm is taken one step further in Schweitzer's thesis [Sch96] and complete analyses can be found in Umlauf's and Zorin's theses [Uml99, Zor97]. The interpolatory *Butterfly algorithm* was defined by Dyn, Gregory and Levin in [DGL90] and analyzed by Zorin in [Zor00a]. A bicubic interpolatory subdivision was introduced by Kobbelt [Kob96a] and analyzed in [Zor00a]. Kobbelt's surprising, since non-polynomial, $\sqrt{3}$-subdivision was derived and analyzed in [Kob00]. Velho and Zorin derived and analyzed the 4-8-algorithm [VZ01], an algorithm based on 4-direction subdivision. Leber's analysis [Leb94] was the first for a non-polynomial, bivariate interpolatory subdivision algorithm. While today, all relevant algorithms have been thoroughly analyzed, the algorithm by Qu and Gregory [QG92, GQ96b] still awaits a detailed treatment.

6. Zorin has developed and implemented automated machinery for proving C_1^k-regularity of subdivision algorithms [Zor00a].

Chapter 7
Shape Analysis and C^k_2-Algorithms

In the preceding chapters, we have studied first order properties of subdivision surfaces in the vicinity of an extraordinary point. Now we look at second order properties, such as the Gaussian curvature or the embedded Weingarten map, which characterize shape. To simplify the setup, we assume $k \geq 2$ throughout. That is, second order partial derivatives of the patches $\mathbf{x}^m_j$ exist and satisfy the contact conditions (4.7/62) and (4.8/62) between neighboring and consecutive segments. However, most concepts are equally useful in situations where the second order partial derivatives are well defined only almost everywhere. In particular, all piecewise polynomial algorithms, such as Doo–Sabin type algorithms or Simplest subdivision, can be analyzed following the ideas to be developed now.

In Sect. 7.1/126, we apply the higher-order differential geometric concepts of Chap. 2/15 to subdivision surfaces and derive asymptotic expansions for the fundamental forms, the embedded Weingarten map, and the principal curvatures. In particular, we determine limit exponents for L^p-integrability of principal curvatures in terms of the leading eigenvalues of the subdivision matrix. The *central ring* will play a key role, just as the characteristic ring for the study for first order properties.

In Sect. 7.2/134, we can leverage the concepts to characterize fundamental shape properties. To this end, the well-known notions of ellipticity and hyperbolicity are generalized in three different ways to cover the special situation in a vicinity of the central point. Properties of the central ring reflect the local behavior, while the Fourier index $\mathcal{F}(\mu)$ of the subsubdominant eigenvalue μ of the subdivision matrix is closely related to the variety of producible shapes. In particular, $\mathcal{F}(\mu) \supset \{0, 2, n-2\}$ is necessary to avoid undue restrictions. Further, we introduce *shape charts* as a tool for summarizing, in a single image, information about the entirety of producible shape.

Conditions for C^k_2-algorithms are discussed in Sect. 7.3/140. Following Theorem 2.14/28, curvature continuity is equivalent to convergence of the embedded Weingarten map. This implies that the subsubdominant eigenvalue μ must be the square of the subdominant eigenvalue λ, and the subsubdominant eigenrings must be quadratic polynomials in the components of the characteristic ring. These extremely restrictive conditions explain the difficulties encountered when trying to construct C_2^k-algorithms. In particular, they lead to a lower bound on the degree of piecewise polynomial schemes, which rules out all schemes generalizing uniform B-spline subdivision, such as the Catmull–Clark algorithm.

Section 7.4/145 presents hitherto unpublished material concerning a general principle for the construction of C_2^k-algorithms, called the *PTER-framework.* This acronym refers to the four building blocks: projection, turn-back, extension, and reparametrization. The important special case of *Guided subdivision*, which inspired that development, is presented in Sect. 7.5/149.

7.1 Higher Order Asymptotic Expansions

We focus on symmetric standard C_1^2-algorithms and assume, for simplicity of exposition, that the subdominant Jordan blocks are singletons, i.e.,

$$1 > \lambda := \lambda_1 = \lambda_2 > |\lambda_3|, \quad \ell_1 = \ell_2 = 0.$$

All subsequent arguments are easily generalized to the case of subdominant Jordan blocks of higher dimension (see Sect. 5.3/89), but the marginal extra insight does not justify the higher technical complexity. We obtain the structure

$$(1,0) \succ (\lambda,0) \sim (\lambda,0) \succ (\lambda_3,\ell_3) \sim \cdots \sim (\lambda_{\bar q},\ell_{\bar q}) \succ (\lambda_{\bar q+1},\ell_{\bar q+1})$$

for the eigenvalues, and denote by μ the common modulus of the *subsubdominant eigenvalues* and by ℓ the size[1] of the corresponding Jordan blocks minus one:

$$\mu := |\lambda_3| = \cdots = |\lambda_{\bar q}|, \quad \ell := \ell_3 = \cdots = \ell_{\bar q}.$$

Consider a subdivision surface $\mathbf{x}$ corresponding to generic initial data $\mathbf{Q}$. Following Definition 2.11/25, we denote by $\mathbf{n}^c$ the central normal, and by $(\mathbf{t}_1^c, \mathbf{t}_2^c, \mathbf{n}^c)$ an orthonormal system defining the central frame $\mathbf{F}^c$,

$$\mathbf{T}^c := \begin{bmatrix} \mathbf{t}_1^c \\ \mathbf{t}_2^c \end{bmatrix}, \quad \mathbf{F}^c := \begin{bmatrix} \mathbf{T}^c \\ \mathbf{n}^c \end{bmatrix}.$$

With (4.28/74), the *second order* asymptotic expansion of the rings $\mathbf{x}^m$ reads

$$\mathbf{x}^m \doteq \mathbf{x}^c + \lambda^m \boldsymbol{\psi}[\mathbf{p}_1;\mathbf{p}_2] + \mu^{m,\ell}\mathbf{d}^m. \tag{7.1}$$

The term

$$\mathbf{d}^m := \sum_{q=3}^{\bar q} d_q^{m-\ell} f_q \mathbf{p}_q$$

summarizes the contribution of the subsubdominant eigencoefficients $\mathbf{p}_q$ and eigenrings f_q. The directions $d_q = \lambda_q/\mu$, as defined in (4.31/75), are numbers on the

[1] Note that the symbol ℓ does not indicate the size of the *subdominant* Jordan block, as in earlier chapters, but the size of the *subsubdominant* Jordan block.

complex unit circle, referring to the angles of the potentially complex subsubdominant eigenvalues $\lambda_3, \ldots, \lambda_{\bar{q}}$.

According to (4.21/73), the scaling factor in (7.1/126) is $\mu^{m,\ell} = \binom{m}{\ell}\mu^{m-\ell}$ provided that $m \geq \ell$. Hence, if $\mu = 0$, the rings $\mathbf{x}^m$ become entirely flat after a few steps. To exclude this trivial situation, we assume $\mu > 0$ throughout. Appropriate asymptotic expansions of the rings of the tangential and the normal component of the transformed spline $\mathbf{x}_* = (\mathbf{x} - \mathbf{x}^c) \cdot \mathbf{F}^c$, as defined in (4.11/64), are given by

$$\boldsymbol{\xi}_*^m = (\mathbf{x}^m - \mathbf{x}^c) \cdot \mathbf{T}^c \overset{*}{=} \lambda^m \psi\, [\mathbf{p}_1; \mathbf{p}_2] \cdot \mathbf{T}^c$$

and

$$z_*^m = (\mathbf{x}^m - \mathbf{x}^c) \cdot \mathbf{n}^c \overset{*}{=} \mu^{m,\ell} \mathbf{d}^m \cdot \mathbf{n}^c, \tag{7.2}$$

respectively. We will focus on algorithms without negative or complex directions d_q. For if, say d_3, is negative or complex then d_3^m oscillates, and if the corresponding coefficient $\mathbf{p}_3 \cdot \mathbf{n}^c$ dominates then z_*^m repeatedly attains positive and negative values as m is growing. In other words, the rings $\mathbf{x}^m$ repeatedly cross the central tangent plane, an undesirable behavior for applications. We therefore focus on algorithms with the following properties:

Definition 7.1 (Algorithm of type (λ, μ, ℓ)). A subdivision algorithm (A, G) is said to be of *type* (λ, μ, ℓ), if

- (A, G) is a symmetric standard C_1^2-algorithm, and
- the *subsubdominant Jordan blocks* have a unique positive eigenvalue,

$$\mu := \lambda_3 = \cdots = \lambda_{\bar{q}} > 0, \quad \ell := \ell_3 = \cdots = \ell_{\bar{q}}, \quad (\mu, \ell) \succ (\lambda_{\bar{q}+1}, \ell_{\bar{q}+1}).$$

Let us briefly discuss some simple consequences of the assumptions made here: In view of Definition 5.3/84, we have a double subdominant eigenvalue,

$$1 > \lambda := \lambda_1 = \lambda_2 > |\lambda_3|, \quad \ell_1 = \ell_2 = 0.$$

Further, by Definition 5.9/89 and Theorem 5.18/101, the Fourier index of λ must be

$$\mathcal{F}(\lambda) = \{1, n-1\}$$

to ensure that the characteristic ring ψ is uni-cyclic.

For an algorithm of type (λ, μ, ℓ),

$$\boldsymbol{\xi}_*^m \overset{*}{=} \lambda^m \bar{\boldsymbol{\xi}}, \quad \bar{\boldsymbol{\xi}} := \psi L, \quad L := [\mathbf{p}_1; \mathbf{p}_2] \cdot \mathbf{T}^c$$

$$z_*^m \overset{*}{=} \mu^{m,\ell} \bar{z}, \quad \bar{z} := \mathbf{d}^m \cdot \mathbf{n}^c = \sum_{q=3}^{\bar{q}} f_q \mathbf{p}_q \cdot \mathbf{n}^c. \tag{7.3}$$

The planar ring $\bar{\boldsymbol{\xi}} = \psi L$ is an affine image of the characteristic ring. By (2.5/17),

$${}^{\times}\!D\bar{\boldsymbol{\xi}} = {}^{\times}\!D\psi \det L, \tag{7.4}$$

i.e., it is regular and injective if and only if L is invertible. From

$$[\mathbf{p}_1; \mathbf{p}_2; \mathbf{n}^c] \cdot \mathbf{F}^c = \begin{bmatrix} L & 0 \\ 0 & 1 \end{bmatrix}$$

we conclude that

$$\det L = \det[\mathbf{p}_1; \mathbf{p}_2; \mathbf{n}^c] = \pm\|\mathbf{p}_1 \times \mathbf{p}_2\|. \tag{7.5}$$

Hence, L is invertible if and only if $\mathbf{p}_1$ and $\mathbf{p}_2$ are linearly independent. In particular, $\bar{\boldsymbol{\xi}}$ is regular and injective for generic initial data. Since, by assumption, $d_q = 1$ for $q = 3, \ldots, \bar{q}$, the factors $d_q^{m-\ell}$ in the definition of $\mathbf{d}^m$ disappear so that the real-valued ring $\bar{z} = \mathbf{d}^m \cdot \mathbf{n}^c$, appearing in the formula for z_*^m, is independent of m. Together, we find the expansion

$$\mathbf{x}_* = (\mathbf{x}^m - \mathbf{x}^c) \cdot \mathbf{F}^c \stackrel{*}{=} \left[\lambda^m \bar{\boldsymbol{\xi}},\, \mu^{m,\ell} \bar{z}\right] = \left[\bar{\boldsymbol{\xi}},\, \bar{z}\right] \operatorname{diag}(\lambda^m, \lambda^m, \mu^{m,\ell}), \tag{7.6}$$

where the asymptotic equivalence of sequences is understood component-wise. That is, the tangential and the normal component are specified exactly up to terms of order $o(\lambda^m)$ and $o(\mu^{m,\ell})$, respectively. Equation (7.6/128) shows that, up to a Euclidean motion, the rings $\mathbf{x}^m$ are asymptotically just scaled copies of the surface $\left[\bar{\boldsymbol{\xi}}, \bar{z}\right]$. For the forthcoming investigation of curvature and shape properties, this surface plays a most important role.

Definition 7.2 (Central ring and central spline). Consider a subdivision surface $\mathbf{x} = B\mathbf{Q} \in C^k(\mathbf{S}_n, \mathbb{R}^3)$ generated by an algorithm of type (λ, μ, ℓ) with central normal $\mathbf{n}^c$, central frame $\mathbf{F}^c$, and eigencoefficients $\mathbf{P} := V^{-1}\mathbf{Q}$. Let $\bar{\mathbf{P}}$ be a vector of points in $\mathbb{R}^3$ with the same block structure as $\mathbf{P}$, see (4.25/74), and all entries zero except for

$$\bar{\mathbf{p}}_0 := \mathbf{0}, \quad [\bar{\mathbf{p}}_1; \bar{\mathbf{p}}_2] := [\mathbf{p}_1; \mathbf{p}_2] \cdot \mathbf{F}^c, \quad \bar{\mathbf{p}}_q^0 := [0,\, \mathbf{p}_q \cdot \mathbf{n}^c], \quad q = 3, \ldots, \bar{q}.$$

The *central ring* $\bar{\mathbf{r}}$ and the *central spline* $\bar{\mathbf{x}}$ corresponding to $\mathbf{x}$ are defined by

$$\bar{\mathbf{r}} := F\bar{\mathbf{P}} \in C^k(\mathbf{S}_n^0, \mathbb{R}^3), \quad \bar{\mathbf{x}} := BV\bar{\mathbf{P}} \in C^k(\mathbf{S}_n, \mathbb{R}^3).$$

Recalling (7.3/127), we find $[\bar{\mathbf{p}}_1; \bar{\mathbf{p}}_2] = [L, 0]$ and

$$\bar{\mathbf{r}} := \left[\bar{\boldsymbol{\xi}},\, \bar{z}\right].$$

Further, we observe the following: According to the structure defined in (4.25/74), $\bar{\mathbf{p}}_q^0$ is the *first* entry in the block $\bar{\mathbf{P}}_q$ of $\bar{\mathbf{P}}$, while $\mathbf{p}_q = \mathbf{p}_q^\ell$ is the *last* entry in the block $\mathbf{P}_q$. Hence, when computing the ring $\bar{\mathbf{x}}^m = FJ^m\bar{\mathbf{P}}$ of the central spline, the summands with index $q = 3, \ldots, \bar{q}$ are $F_q J_q^m \bar{\mathbf{P}}_q = \mu^m f_q \bar{\mathbf{p}}_q^0$. We obtain

$$\bar{\mathbf{x}}^m = \left[\lambda^m \bar{\boldsymbol{\xi}},\, \mu^m \sum_{q=3}^{\bar{q}} f_q \bar{\mathbf{p}}_q^0\right] = \bar{\mathbf{r}} \operatorname{diag}(\lambda^m, \lambda^m, \mu^m)$$

and see that these rings are scaled copies of $\bar{\mathbf{r}}$. The central point and the central normal of $\bar{\mathbf{x}}$ are given by

$$\bar{\mathbf{x}}^{\mathrm{c}} = \bar{\mathbf{p}}_0 = \mathbf{0}, \quad \bar{\mathbf{n}}^{\mathrm{c}} = \frac{\bar{\mathbf{p}}_1 \times \bar{\mathbf{p}}_2}{\|\bar{\mathbf{p}}_1 \times \bar{\mathbf{p}}_2\|} = \mathbf{e}_3 := [0,0,1], \tag{7.7}$$

respectively.

Unlike the characteristic ring, the central ring depends on the initial data via the eigencoefficients $\mathbf{p}_1, \dots, \mathbf{p}_{\bar{q}}$ in (7.3/127). If these data are generic then the central ring is regular, i.e., ${}^{\times}\!\mathcal{D}\bar{\mathbf{r}} \neq \mathbf{0}$. More precisely, using (7.4/127), one easily shows that

$$\|{}^{\times}\!\mathcal{D}\bar{\mathbf{r}}\| \geq |{}^{\times}\!\mathcal{D}\bar{\boldsymbol{\xi}}| = {}^{\times}\!\mathcal{D}\psi \,|\det L|,$$

where we recall that, by definition, ${}^{\times}\!\mathcal{D}\psi > 0$ for a standard algorithm. We start with a lemma concerning the first and second fundamental form.

Lemma 7.3 (Asymptotic expansion of fundamental forms). *For generic initial data consider a subdivision surface* $\mathbf{x} = B\mathbf{Q} \in C^k(\mathbf{S}_n, \mathbb{R}^3)$ *with segments* $\mathbf{x}_j^m$ *generated by a subdivision algorithm of type* (λ, μ, ℓ). *Then we obtain the following asymptotic expansions:*

- *The first fundamental form of* $\mathbf{x}_j^m$ *is a symmetric matrix* $I_j^m \in C^{k-1}(\boldsymbol{\Sigma}^0, \mathbb{R}^{2\times 2})$ *with*

$$I_j^m \stackrel{*}{=} \lambda^{2m} I_j, \quad \text{where} \quad I_j := D\bar{\boldsymbol{\xi}}_j \cdot D\bar{\boldsymbol{\xi}}_j. \tag{7.8}$$

- *There exists* $\bar{m}$ *such that the inverse* $(I_j^m)^{-1}$ *exists for all* $m \geq \bar{m}$, $j \in \mathbb{Z}_n$, *and satisfies*

$$(I_j^m)^{-1} \stackrel{*}{=} \lambda^{-2m} I_j^{-1}. \tag{7.9}$$

- *Let* $\bar{I}_j$ *and* $\bar{I\!I}_j$ *denote the first and second fundamental form of the segments of the central ring* $\bar{\mathbf{r}}$. *The second fundamental form of* $\mathbf{x}_j^m$ *is a symmetric matrix* $I\!I_j^m \in C^{k-2}(\boldsymbol{\Sigma}^0, \mathbb{R}^{2\times 2})$ *with*

$$I\!I_j^m \stackrel{*}{=} \mu^{m,\ell} I\!I_j, \quad \text{where} \quad I\!I_j := \sqrt{\frac{\det \bar{I}_j}{\det I_j}}\, \bar{I\!I}_j. \tag{7.10}$$

Proof. The first formula, (7.8/129), follows immediately from the definition $I_j^m := D\mathbf{x}_j^m \cdot D\mathbf{x}_j^m$ and the expansion

$$D\mathbf{x}_j^m \stackrel{*}{=} \lambda^m D\bar{\boldsymbol{\xi}}_j \mathbf{T}_j^{\mathrm{c}}. \tag{7.11}$$

To compute $(I_j^m)^{-1}$, we note that the inverse of any (2×2)-matrix M with $\det M \neq 0$ can be expressed in the form

$$M^{-1} = \frac{1}{\det M}(C \cdot M) \cdot C, \quad \text{where} \quad C := \begin{bmatrix} 0 & -1 \\ 1 & 0 \end{bmatrix},$$

where we recall that the dot operator transposes its right argument. Now, using (7.4/127),

$$\det I_j^m \stackrel{*}{=} \lambda^{4m} \det I_j = \lambda^{4m} ({}^{\times}\!D\bar{\boldsymbol{\xi}}_j)^2 = \lambda^{4m} ({}^{\times}\!D\boldsymbol{\psi}_j)^2 (\det L)^2. \tag{7.12}$$

By (7.5/128), $(\det L)^2 = \|\mathbf{p}_1 \times \mathbf{p}_2\|^2$ does not vanish for generic initial data, while $({}^{\times}\!D\boldsymbol{\psi}_j)^2 \ge c_j > 0$ for some constant c_j by regularity of $\boldsymbol{\psi}_j$, compactness of the domain $\boldsymbol{\Sigma}^0$, and continuity of ${}^{\times}\!D\boldsymbol{\psi}_j$. Hence, the right hand side in the last display is bounded away from zero so that there exists an integer $\bar{m}$ with $\det I_j^m > 0$ for all $m \ge \bar{m}$, $j \in \mathbb{Z}_n$, and

$$(\det I_j^m)^{-1} \stackrel{*}{=} \lambda^{-4m} (\det I_j)^{-1}. \tag{7.13}$$

As claimed in (7.9/129), we obtain

$$(I_j^m)^{-1} = \frac{1}{\det I_j^m} (C \cdot I_j^m) \cdot C \stackrel{*}{=} \frac{\lambda^{-2m}}{\det I_j} (C \cdot I_j) \cdot C = \lambda^{-2m}\, I_j^{-1}.$$

To prove (7.10/129), we conclude from (7.6/128)

$$\det[D_i D_k \mathbf{x}_j^m; D\mathbf{x}_j^m] \stackrel{*}{=} \lambda^{2m} \mu^{m,\ell} \det[D_i D_k \bar{\mathbf{x}}_j; {}^{\times}\!D\bar{\mathbf{x}}_j].$$

Then, by comparing the components

$$(I\!I_j^m)_{i,k} = \frac{\det[D_i D_k \mathbf{x}_j^m; D\mathbf{x}_j^m]}{\sqrt{\det I_j^m}}, \quad (I\!I_j^{\mathrm{c}})_{i,k} = \frac{\det[D_i D_k \bar{\mathbf{x}}_j; D\bar{\mathbf{x}}_j]}{\sqrt{\det I_j^{\mathrm{c}}}}$$

of $I\!I_j^m$ and $I\!I_j^{\mathrm{c}}$ according to (2.8/19) and using (7.13/130), we obtain the given expansion. □

In the following, we will assume without further notice that, if required, $m \ge \bar{m}$ so that I_j^m is invertible. With the help of the expansions for the fundamental forms, we are now able to derive the expansion for the embedded Weingarten map of the rings.

Theorem 7.4 (Asymptotic expansion of $\mathbf{W}^m$). Under the assumptions of Lemma 7.3/129, the embedded Weingarten maps of the rings $\mathbf{x}^m \in C^k(\mathbf{S}_n^0, \mathbb{R}^3)$ are rings $\mathbf{W}^m \in C^{k-2}(\mathbf{S}_n^0, \mathbb{R}^{3\times 3})$ with

$$\mathbf{W}^m \stackrel{*}{=} \varrho^{m,\ell} \,(\mathbf{T}^{\mathrm{c}})^{\mathrm{t}}\, W \mathbf{T}^{\mathrm{c}}, \quad \varrho := \frac{\mu}{\lambda^2}, \tag{7.14}$$

where W is a symmetric (2×2)-matrix with segments

$$W_j := (D\bar{\boldsymbol{\xi}}_j)^{-1} I\!I_j \cdot (D\bar{\boldsymbol{\xi}}_j)^{-1}, \quad j \in \mathbb{Z}_n. \tag{7.15}$$

Moreover, consecutive rings $\mathbf{W}^m, \mathbf{W}^{m+1}$ join smoothly in the sense that the segments satisfy the contact conditions (4.8/62) up to order $k-2$.

Proof. Recalling Definition 2.4/20, and using (7.11/129) and (7.9/129), the pseudo-inverse of $D\mathbf{x}_j^m$ is

$$(D\mathbf{x}_j^m)^+ \stackrel{*}{=} \lambda^{-m}\big((\mathbf{T}^c)^t \cdot D\bar{\boldsymbol{\xi}}_j\big)(D\bar{\boldsymbol{\xi}}_j \cdot D\bar{\boldsymbol{\xi}}_j)^{-1} = \lambda^{-m}(\mathbf{T}^c)^t(D\bar{\boldsymbol{\xi}}_j)^{-1}.$$

Together with (7.10/129), we find the desired expansion. The C^{k-2}-contact of consecutive and neighboring segments is shown as follows. Using the fractional power embedding $\boldsymbol{\pi}$, as introduced in Example 3.10, we define the reparametrized surface $\tilde{\mathbf{x}} := \mathbf{x} \circ \boldsymbol{\pi}^{-1}$, which is not a spline, but an almost regular standard C_0^k-surface. Its embedded Weingarten map $\tilde{\mathbf{W}}$ is well defined and C^{k-2} away from the origin. Because the images of $\mathbf{x}$ and $\tilde{\mathbf{x}}$ coincide, so do the corresponding embedded Weingarten maps, see Theorem 2.5/22. Hence, smooth contact of the segments $\mathbf{W}_j^m$ follows from smoothness of $\tilde{\mathbf{W}}$. □

Now, using the formulas (2.11/22) and the identities

$$\operatorname{trace}\big((\mathbf{T}^c)^t\, W\mathbf{T}^c\big) = \operatorname{trace} W, \quad \|(\mathbf{T}^c)^t\, W\mathbf{T}^c\|_F = \|W\|_F,$$

we easily find the asymptotic expansions

$$\kappa_M^m \stackrel{*}{=} \frac{\varrho^{m,\ell}}{2} \operatorname{trace} W \tag{7.16}$$

for the mean curvature, and

$$\kappa_G^m \stackrel{*}{=} \frac{(\varrho^{m,\ell})^2}{2}(\operatorname{trace}^2 W - \|W\|_F^2) = (\varrho^{m,\ell})^2 \det W \tag{7.17}$$

for the Gaussian curvature. Let us derive two further asymptotic formulas from these expansions. First, we see immediately that the principal curvatures $\kappa_{1,2}^m$ of $\mathbf{x}^m$ and the eigenvalues $\kappa_{1,2}^W$ of W are related by

$$\kappa_i^m \stackrel{*}{=} \varrho^{m,\ell}\kappa_i^W, \quad i \in \{1, 2\}. \tag{7.18}$$

Second, let $\bar{\kappa}_G := \det \bar{I\!I}/\det \bar{I}$ denote the Gaussian curvature of the central ring. Then, with the definitions (7.10/129) of $I\!I$ and (7.15/130) of W, we further find using $|{}^\times D\bar{\boldsymbol{\xi}}| = \sqrt{\det I}$

$$\kappa_G^m \stackrel{*}{=} \left(\varrho^{m,\ell}\,\frac{\det \bar{I}}{\det I}\right)^2 \bar{\kappa}_G. \tag{7.19}$$

In particular, this formula shows that elliptic and hyperbolic points of the central ring $\bar{\mathbf{r}}$ correspond to elliptic and hyperbolic points of the rings $\mathbf{x}^m$, respectively, for sufficiently large m. Of course, parabolic points of $\bar{\mathbf{r}}$ do not admit a similar conclusion.

The preceding formulas, and in particular (7.18/131), indicate that the ratio $\varrho = \mu/\lambda^2$ together with the dimension ℓ of the subsubdominant Jordan block governs the limit behavior of the principal curvatures of the rings. Clearly, $\varrho < 1$ implies convergence to 0, while $(\varrho, \ell) = (1, 0)$ guarantees boundedness. However, it is not obvious that $(\varrho, \ell) \succ (1, 0)$ necessarily causes divergence since both eigenvalues of W could still be 0. This case is excluded by the following lemma.

Lemma 7.5 (Generically, $W \neq 0$). *For generic initial data* $\mathbf{P}$*, the matrix* W *does not vanish identically.*

Proof. Let us assume that $W_j = (D\bar{\boldsymbol{\xi}}_j)^{-1} II_j \cdot (D\bar{\boldsymbol{\xi}}_j)^{-1} = 0$. Because $\bar{\boldsymbol{\xi}}$ is regular for generic initial data, we have $II_j = \overline{II}_j = 0$ so that the principal curvatures of the central ring vanish identically. This is possible only if the image of $\bar{\mathbf{r}}$ is contained in a plane. Now, we consider the central spline $\bar{\mathbf{x}}$. As we have shown above, its rings $\bar{\mathbf{x}}^m$ are scaled copies of $\bar{\mathbf{r}}$, hence planar, too. Because $\bar{\mathbf{x}}$ is continuous and normal continuous, the image of $\bar{\mathbf{x}}$ must be a subset of a single plane. In view of (7.7/129), this must be the xy-plane,

$$\bar{\mathbf{r}} \cdot \mathbf{e}_3 = \bar{z} = \sum_{q=3}^{\bar{q}} f_q \, \mathbf{p}_q \cdot \mathbf{n}^{\mathrm{c}} = 0.$$

By Lemma 4.22/78, the eigenrings f_q are linearly independent, implying $\mathbf{p}_q \cdot \mathbf{n}^{\mathrm{c}} = 0$ and $\det[\mathbf{p}_1; \mathbf{p}_2; \mathbf{p}_q] = 0$ for all $q = 3, \dots, \bar{q}$. This, however, contradicts the assumption that the initial data $\mathbf{P}$ be generic, see Definition 5.1/84. □

As a consequence of the lemma, we can be sure that the factor $\varrho^{m,\ell}$ in the asymptotic expansion (7.18/131) of the principal curvatures provides not only an upper bound. In fact, it describes the precise asymptotic behavior of at least one out of κ_1^m and κ_2^m since, for generic initial data, at least one eigenvalue of W is non-zero. For that reason, the following critical exponents for L^p-integrability of principal curvatures cannot be improved. We define the L^p-norm $\|\kappa\|_{p,\bar{m}}$ of a spline κ, built from rings κ^m, as the sum of integrals over all surface rings $\mathbf{x}^m$ with index $m \geq \bar{m}$, by

$$\|\kappa\|_{p,\bar{m}}^p := \sum_{m \geq \bar{m}} \int |\kappa|^p \, d\mathbf{x}^m = \sum_{m \geq \bar{m}} \sum_{j \in \mathbb{Z}_n} \int_{\boldsymbol{\Sigma}^0} |\kappa(s,t,j)|^p \, \|{}^{\times}\!D\mathbf{x}^m\| \, dsdt.$$

The space of all functions κ for which $\|\kappa\|_{p,\bar{m}}$ is well defined and finite for sufficiently large $\bar{m}$ is denoted by L^p_{loc}. Then the following theorem holds and is illustrated in Fig. 7.1/133.

Theorem 7.6 (Curvature integrability). *For generic data, let* $\mathbf{x} \in C^k(\mathbf{S}_n, \mathbb{R}^3)$ *be a subdivision surface with principal curvatures* $\kappa_i, i \in \{1,2\}$*. Then, for sufficiently large* $\bar{m}$ *and* $m \geq \bar{m}$*, the rings* κ_i^m *are well-defined. Furthermore,* $\kappa_i \in L^p_{\mathrm{loc}}$ *for all* p *with*

- $p < \frac{2 \ln \lambda}{2 \ln \lambda - \ln \mu}$*, if* $\mu > \lambda^2$*;*
- $p < \infty$*, if* $\mu = \lambda^2$ *and* $\ell > 0$*;*
- $p \leq \infty$*, if* $(\mu, \ell) \preccurlyeq (\lambda^2, 0)$*.*

In any case, $\kappa_i \in L^2_{\mathrm{loc}}$*.*

Proof. The principal curvatures κ_i^m are well-defined and continuous if $\det I^m > 0$. Now, the asymptotic expansion (7.12/130) guarantees the existence of an index $\bar{m}$ such that $\det I^m > 0$ for all $m \geq \bar{m}$. By (7.18/131), both principal curvatures are bounded

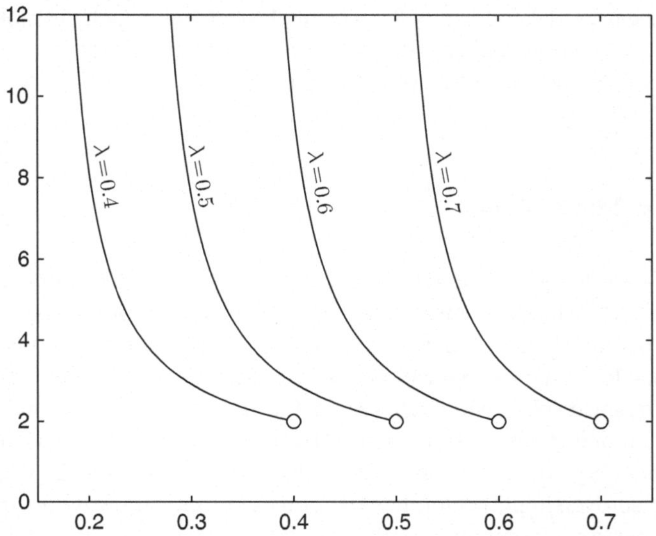

Fig. 7.1 Illustration of Theorem 7.6/132: Limit exponent p of curvature integrability plotted over subsubdominant eigenvalue μ for different values of λ.

if and only if $(\mu, \ell) \preccurlyeq (\lambda^2, 0)$. Hence, it remains to consider the case $p < \infty$. We use (5.7/87), (7.4/127), and (7.5/128) to find

$$\|{}^{\times}\!D\mathbf{x}^m\| \overset{*}{=} \lambda^{2m}\|{}^{\times}\!D\boldsymbol{\psi}\|\, \|\mathbf{p}_1 \times \mathbf{p}_2\| = \lambda^{2m}\|{}^{\times}\!D\bar{\boldsymbol{\xi}}\|.$$

Hence, with (7.18/131),

$$|\kappa_i^m|^p\, \|{}^{\times}\!D\mathbf{x}^m\| \overset{*}{=} (\varrho^{m,\ell})^p |\kappa_i^{\mathrm{w}}|^p \lambda^{2m} |{}^{\times}\!D\bar{\boldsymbol{\xi}}| \overset{*}{=} (m/\varrho)^{\ell p}\, r_p^m\, k_i^p,$$

where we used the abbreviations

$$r_p := \frac{\mu^p}{\lambda^{2(p-1)}} \quad \text{and} \quad k_i^p := |\kappa_i^{\mathrm{w}}|^p\, |{}^{\times}\!D\bar{\boldsymbol{\xi}}|.$$

Denoting the integral of the ring k_i^p by

$$K_i^p := \sum_{j \in \mathbb{Z}_n} \int_{\boldsymbol{\Sigma}^0} k_i^p(s,t,j)\, ds dt,$$

we obtain

$$\|\kappa_i\|_{p,\bar{m}}^p = \sum_{m \geq \bar{m}} \sum_{j \in \mathbb{Z}_n} \int_{\boldsymbol{\Sigma}^0} |\kappa_i^m(s,t,j)|^p\, \|{}^{\times}\!D\mathbf{x}^m\|\, ds dt \overset{*}{=} K_i^p \sum_{m \geq \bar{m}} (m/\varrho)^{\ell p}\, r_p^m.$$

The latter series converges if and only if $r_p < 1$. For $\mu > \lambda^2$, this inequality is equivalent to p being smaller than the bound given in the first item of the theorem, while it is always satisfied for $\mu \le \lambda^2$. The final statement, which guarantees square integrability of the principal curvatures for *any* algorithm of type (λ, μ, ℓ), follows immediately from the above results and $\mu < \ell$. □

7.2 Shape Assessment

As it will be explained in the next section, C_2^k-subdivision algorithms are hard to find, and most schemes currently in use are merely C_1^k. While many popular C_1^k-algorithms live up to the standards of Computer Graphics, they do not satisfy the higher demands arising in applications like car body design. To put it shortly, one could say that most subdivision surfaces are fair from afar, but far from being fair.

When scrutinizing subdivision surfaces by means of shaded images or curvature plots, one possibly encounters an erratic behavior of shape near the central point. It would be an oversimplification to explain these observations by just pointing to the lack of curvature continuity. Rather, it pays off to explore the deeper sources of shape deficiencies. Based on such additional insight, one can develop guidelines for tuning algorithms. Even for families of subdivision algorithms where curvature continuity is beyond reach, this may result in a significant improvement of shape.

As a motivation, consider the following facts regarding Catmull–Clark subdivision, as discussed in the preceding chapter:

- For standard weights and valence $n \ge 5$, the principal curvatures grow unboundedly when approaching the central point.
- For standard weights and valence $n \ge 5$, the generated surfaces are generically not convex.
- Even when tuning the weights α, β, γ carefully to get rid of the latter restriction, the generated surfaces sometimes reveal a *hybrid behavior*, what means that there are both elliptic and hyperbolic points in any neighborhood of the central point.

The first observation can be understood when considering the asymptotic expansion (7.18/131) derived in the preceding section: the ratio $\varrho = \mu/\lambda^2 > 1$ causes divergence of the principal curvatures. Also the second observation can be explained by spectral properties of the subdivision matrix. The subsubdominant eigenvalue μ has Fourier index $\mathcal{F}(\mu) = \{2, n-2\}$, and we will show below that this generically leads to non-convex shape. The third observation is quite subtle, and can be explained only with the help of a so-called shape chart, which summarizes properties of central rings for all possible choices of initial data.

Before we come to that point, let us start with developing concepts for classifying shape at the central point. Because, in general, the Gaussian curvature is not well defined at $\mathbf{x}^c$, we have to generalize the notions of ellipticity and hyperbolicity. We suggest three different approaches, respectively based on:

- The local intersections of the subdivision surface with its tangent plane
- The limit behavior of the Gaussian curvature
- Local quadratic approximation

We will show that in all cases the behavior of the subdivision surface is closely related to the shape of the central surface ring and, in the first and third case, to spectral properties of the subdivision matrix. For simplicity, we continue to consider algorithms of type (λ, μ, ℓ) according to Definition 7.1/127.

We start by introducing an appropriate notion of periodicity for rings.

Definition 7.7 ($\mathcal{P}$-periodicity). Let $\mathcal{P} = \{k_1, \dots, k_q\}$ be a set of indices, which are understood modulo n. A ring $f \in C^k(\mathbf{S}_n^0, \mathbb{K})$ is called *$\mathcal{P}$-periodic*, if there exist functions $g_i, \bar{g}_i \in C^k(\boldsymbol{\Sigma}^0, \mathbb{K})$ such that its segments are given by

$$f(\cdot, j) = \sum_{i=1}^{q} \big(g_i \sin(2\pi k_i j/n) + \bar{g}_i \cos(2\pi k_i j/n)\big).$$

One easily shows that

$$\sum_{j \in \mathbb{Z}_n} f(\cdot, j) = 0 \quad \text{if} \quad 0 \notin \mathcal{P}. \tag{7.20}$$

Further, the space of $\mathcal{P}$-periodic functions is linear. The product of a $\mathcal{P}$-periodic function f and a $\mathcal{Q}$-periodic function g yields an $\mathcal{R}$-periodic function fg, where $\mathcal{R} := \mathcal{P} \pm \mathcal{Q}$ contains all sums and differences of elements of $\mathcal{P}$ and $\mathcal{Q}$.

By (5.17/99) and $\mathcal{F}(\lambda) = \{1, n-1\}$, the tangential component $\bar{\boldsymbol{\xi}} = [f_1, f_2]L$ of the central ring $\bar{\mathbf{r}}$ is $\{1, n-1\}$-periodic, while the third component $\bar{z} = \sum_q f_q\, \mathbf{p}_q \cdot \mathbf{n}^c$ is $\mathcal{F}(\mu)$-periodic.

Now, we introduce three variants on the notion of an elliptic or hyperbolic point, which apply to the special situation at the central point. As a first approach, let us consider a non-parabolic point of a regular C^2-surface. If it is elliptic, then the surface locally lies on one side of the tangent plane. By contrast, if it is hyperbolic, then the surface intersects the tangent plane in any neighborhood. This basic observation motivates the following generalization. It involves the notion of the *central tangent plane* which is the plane perpendicular to $\mathbf{n}^c$ through the point $\mathbf{x}^c$.

Definition 7.8 (Sign-type). The central point $\mathbf{x}^c$ of a subdivision surface $\mathbf{x}$ is called

- *elliptic in sign* if, in a sufficiently small neighborhood of $\mathbf{x}^c$, the subdivision surface intersects the central tangent plane only in $\mathbf{x}^c$;
- *hyperbolic in sign*, if in any neighborhood of $\mathbf{x}^c$ the subdivision surface has points on both sides of the central tangent plane.

This classification defines a minimum standard for subdivision surfaces: any high-quality algorithm should be able to generate both sign-types in order to cover basic shapes. The sign-type can be established by looking at the third component of the central ring.

Theorem 7.9 (Central surface and sign-type). *Let $\bar{\mathbf{r}} = [\bar{\boldsymbol{\xi}}, \bar{z}]$ be the central ring of the subdivision surface* $\mathbf{x}$.

- *If $\bar{z} > 0$ or $\bar{z} < 0$, then $\mathbf{x}^c$ is elliptic in sign.*
- *If $\bar{z}$ changes sign, then $\mathbf{x}^c$ is hyperbolic in sign.*

Proof. Intersections of $\mathbf{x}$ and the central tangent plane correspond to zeros of the normal component z_* of the transformed spline surface $\mathbf{x}_*$. According to (7.2/127) and (7.6/128), its rings z_*^m satisfy

$$z_*^m = (\mathbf{x}^m - \mathbf{x}^c) \cdot \mathbf{n}^c \stackrel{*}{=} \mu^{m,\ell}\overline{z},$$

and the assertion follows easily. □

The last display implies more than is stated in the theorem. We see that the sign map of z_*^m is equivalent to the sign map of $\overline{z}$ in an asymptotic way. Thus, the distribution of signs of the normal component z_* can be studied with the help of the central ring, except for points corresponding to zeros of $\overline{z}$. The next theorem relates the sign-type and the Fourier index of the subsubdominant eigenvalue.

Theorem 7.10 (Fourier index and sign-type). *For generic initial data, the central point $\mathbf{x}^c$ is hyperbolic in sign unless $0 \in \mathcal{F}(\mu)$.*

Proof. The function $\overline{z}$ is $\mathcal{F}(\mu)$-periodic. Hence, if $0 \notin \mathcal{F}(\mu)$, the sum of its segments vanishes, $\sum_{j \in \mathbb{Z}_n} \overline{z}_j = 0$. Since $\overline{z} \neq 0$ for generic initial data, it has to have positive and negative function values. □

The strong consequence of this theorem is that, for any good subdivision algorithm, one of the subsubdominant eigenvalues must correspond to the zero Fourier block of the subdivision matrix. Otherwise, the resulting surfaces will locally intersect the tangent plane at the extraordinary vertex for almost all initial data. For example, the standard Catmull–Clark algorithm reveals this shortcoming for $n \geq 5$: the Fourier index of μ is $\{2, n-2\}$ and the generated subdivision surfaces are, for generic data, not elliptic in sign. In particular, *they are not convex.*

The second approach to a classification of the central point makes use of the fact that the Gaussian curvature is well defined for all rings $\mathbf{x}^m$ with sufficiently large index m.

Definition 7.11 (Limit-type). Wherever it is well defined, denote by κ_{G} the Gaussian curvature of a subdivision surface $\mathbf{x}$. The central point $\mathbf{x}^c$ is called

- *elliptic in the limit* if $\kappa_{\mathrm{G}} > 0$ in a sufficiently small neighborhood of $\mathbf{x}^c$;
- *hyperbolic in the limit* if $\kappa_{\mathrm{G}} < 0$ in a sufficiently small neighborhood of $\mathbf{x}^c$;
- *hybrid,* if κ_{G} changes sign in every neighborhood of $\mathbf{x}^c$.

Again, the limit-type of an extraordinary vertex is closely related to the central ring.

Theorem 7.12 (Central surface and limit-type). *Denote by $\overline{\kappa}_{\mathrm{G}}$ the Gaussian curvature of the central ring $\overline{\mathbf{r}}$. For generic initial data, the central point is*

- *elliptic in the limit, if $\overline{\kappa}_{\mathrm{G}} > 0$;*
- *hyperbolic in the limit, if $\overline{\kappa}_{\mathrm{G}} < 0$;*
- *hybrid, if $\overline{\kappa}_{\mathrm{G}}$ changes sign.*

The proof follows immediately from (7.19/131). Again, this expansion implies more than is stated in the theorem. We see that the sign map of the Gaussian curvature

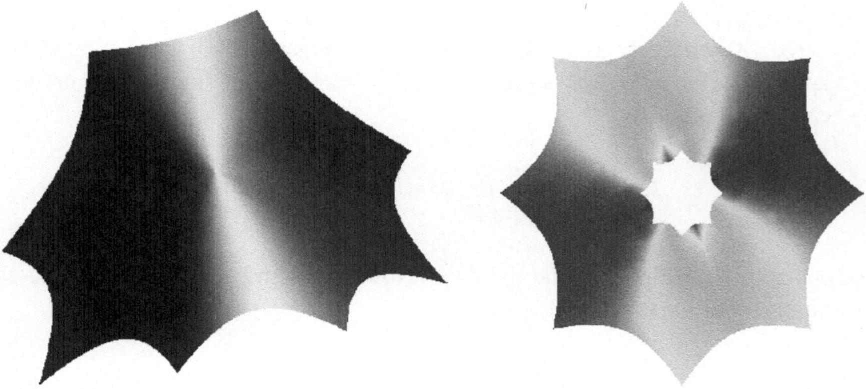

Fig. 7.2 Illustration of hybrid case: Hybrid shape of a subdivision surface generated by a modified Catmull–Clark algorithm. (*left*) Lighted surface with an undesired pinch-off near the central point. (*right*) Part of the surface shaded by Gaussian curvature. *Blue* and *green* colors indicate hyperbolic points, *yellow* and *red* colors indicate elliptic points.

of $\bar{\mathbf{r}}$ is equivalent to that of the rings in an asymptotic way. Thus, the distribution of the sign of the Gaussian curvature in a vicinity of an extraordinary vertex can be studied with the help of the central ring – except at parameters corresponding to parabolic points of the central ring. The study of the Gaussian curvature of the central surface is a basic tool for judging the quality of a subdivision surface since, in applications, fairness requires that the extraordinary point be either elliptic or hyperbolic in sign. The hybrid case leads to shape artifacts (see Fig. 7.2/137). A high quality subdivision algorithm should therefore exclude the hybrid case completely, while facilitating both elliptic and hyperbolic shape in the limit-sense. This is a very strong requirement that is hard to fulfill in practice. To explain the problem, let us consider two sets of initial data: $\mathbf{Q}[0]$ is chosen so that the central ring has positive Gaussian curvature and $\mathbf{Q}[1]$ so that the central ring has negative Gaussian curvature. Now, we consider any continuous transition $\mathbf{Q}[t], t \in [0,1]$, connecting the two cases. The Gaussian curvature of the corresponding central rings is a family $\bar{\kappa}_{\mathrm{G}}[t]$ of functions connecting $\bar{\kappa}_{\mathrm{G}}[0] > 0$ and $\bar{\kappa}_{\mathrm{G}}[1] < 0$. If hybrid behavior is to be excluded then the transition between the positive and the negative case has to be restricted to isolated t-values where $\bar{\kappa}_{\mathrm{G}}[t] \equiv 0$. However, to devise an algorithm with such a property is challenging since the relation between initial data and curvature of the central ring is highly non-linear.

Relating $\bar{\kappa}_{\mathrm{G}}$ to spectral properties is rather difficult and does not promise results beyond Theorem 7.10/136. Since we want to be able to distinguish the desired cup- and saddle-shapes from unstructured local oscillations, we consider a third approach. As we will show in Theorem 7.16/143 of the next section, the subdivision surface $\mathbf{x}$ is C^k_2 if and only if the function $\bar{z}$ is a quadratic polynomial in the subdominant eigenrings f_1, f_2, i.e., there exists a constant symmetric (2×2)-matrix H such that the components of the central ring satisfy

$$\bar{\boldsymbol{\xi}} H \cdot \bar{\boldsymbol{\xi}} - \bar{z} = 0.$$

Then, the Gaussian curvature of the central point is given by $\det(H/2)$. In general, no matrix will satisfy the above identity exactly. But one can still try to determine a best approximation in the least squares sense. To this end, we define an inner product for real-valued rings by

$$\langle f, g\rangle := \sum_{j\in\mathbb{Z}_n} \int_{\boldsymbol{\Sigma}^0} f(s,t,j)g(s,t,j)\,dsdt$$

and denote the corresponding norm by $|\cdot|$. Now, for given $\overline{\boldsymbol{\xi}}$ and $\overline{z}$, we define H as the minimizer of the functional

$$\varphi(H) := \left|\overline{\boldsymbol{\xi}}H\cdot\overline{\boldsymbol{\xi}} - \overline{z}\right|^2.$$

The matrix H provides information on the global shape of the central ring in the sense of averaging, and its determinant is now used to define a third notion of hyperbolicity and ellipticity.

Definition 7.13 (Average-type). The central point $\mathbf{x}^{\mathrm{c}}$ is called

- *elliptic in average*, if $\det H > 0$;
- *hyperbolic in average*, if $\det H < 0$.

The average-type is closely related to the Fourier index of the subsubdominant eigenvalue.

Theorem 7.14 (Central surface and average-type). *For generic initial data, the central point is*

- *not elliptic in average unless* $0 \in \mathcal{F}(\mu)$;
- *not hyperbolic in average unless* $\{2, n-2\} \subset \mathcal{F}(\mu)$.

Proof. We start with a simple observation for periodic functions. Let f be $\mathcal{P}$-periodic and g be $\mathcal{Q}$-periodic. By (7.20/135),

$$\langle f, g\rangle = 0 \qquad \text{if} \qquad 0 \notin \mathcal{P} \pm \mathcal{Q}, \tag{7.21}$$

where we recall that $\mathcal{P} \pm \mathcal{Q}$ contains all sums and differences of elements of $\mathcal{P}$ and $\mathcal{Q}$ modulo n. To put the optimization problem in a more convenient form, we set $p := \psi_1^2 + \psi_2^2$, $q := \psi_1^2 - \psi_2^2$, $r := 2\psi_1\psi_2$, and write

$$\varphi(H) = |\boldsymbol{\psi}(LH\cdot L)\cdot\boldsymbol{\psi} - \overline{z}|^2 = |ap + bq + cr - \overline{z}|^2 \tag{7.22}$$

where the coefficients a, b, c are defined by

$$LH\cdot L =: \begin{bmatrix} a+b & c \\ c & a-b \end{bmatrix}. \tag{7.23}$$

The sign of the determinant of H, which we are going to determine, is given by

$$\operatorname{sign}(\det H) = \operatorname{sign}(\det LH\cdot L) = \operatorname{sign}(a^2 - b^2 - c^2).$$

Minimizing the functional φ according to (7.22/138) is equivalent to solving the Gramian system

$$\begin{bmatrix} \langle p,p\rangle & \langle p,q\rangle & \langle p,r\rangle \\ \langle p,q\rangle & \langle q,q\rangle & \langle q,r\rangle \\ \langle p,r\rangle & \langle q,r\rangle & \langle r,r\rangle \end{bmatrix} \begin{bmatrix} a \\ b \\ c \end{bmatrix} = \begin{bmatrix} \langle p,\overline{z}\rangle \\ \langle q,\overline{z}\rangle \\ \langle r,\overline{z}\rangle \end{bmatrix}. \tag{7.24}$$

Now, we determine the periodicity of the functions p, q, r. With the rotation matrix

$$R := \begin{bmatrix} \cos(2\pi/n) & \sin(2\pi/n) \\ -\sin(2\pi/n) & \cos(2\pi/n) \end{bmatrix}$$

we obtain for the segments

$$\begin{aligned} p_j &= \psi_0 R^j \begin{bmatrix} 1 & 0 \\ 0 & 1 \end{bmatrix} \cdot (\psi_0 R^j) = p_0 \\ q_j &= \psi_0 R^j \begin{bmatrix} 1 & 0 \\ 0 & -1 \end{bmatrix} \cdot (\psi_0 R^j) = \cos(4\pi j/n) q_0 - \sin(4\pi j/n) r_0 \\ r^j &= \psi_0 R^j \begin{bmatrix} 0 & 1 \\ -1 & 0 \end{bmatrix} \cdot (\psi_0 R^j) = \cos(4\pi j/n) q_0 + \sin(4\pi j/n) r_0 \end{aligned}$$

and observe that p is $\{0\}$-periodic, while q and r are $\{2\}$-periodic. Hence, by (7.21/138), the off-diagonal elements of the Gramian matrix in the first row and column vanish, $\langle p,q\rangle = \langle p,r\rangle = 0$. If $0 \notin \mathcal{F}(\mu)$, the function $\overline{z}$ is $\mathcal{P}$-periodic with $0 \notin \mathcal{P}$, and the first entry of the right hand side of (7.24/139) becomes $\langle p,\overline{z}\rangle = 0$. Thus, $a = 0$ and $\operatorname{sign}(\det H) = \operatorname{sign}(-b^2 - c^2) \le 0$. If $\{2, n-2\} \not\subset \mathcal{F}(\mu)$, the function $\overline{z}$ is $\mathcal{P}$-periodic with $\{2, n-2\} \cap \mathcal{P} = \emptyset$, and the second and third entry of the right hand side of (7.24/139) become $\langle q,\overline{z}\rangle = \langle r,\overline{z}\rangle = 0$. Thus, $b = c = 0$ and $\operatorname{sign}(\det H) = \operatorname{sign}(a^2) \ge 0$. □

As a consequence of this theorem, we see that the variety of producible shapes will cover both basic average-types only if the subsubdominant eigenvalue is at least triple with Fourier index $\mathcal{F}(\mu) \supset \{0, 2, n-2\}$. However, it must be emphasized that this spectral property is by no means a sufficient condition for a good subdivision algorithm, but merely a basic requirement.

Deeper insight is provided by the concept of *shape charts*, that classify the space of shapes that can be generated by a subdivision algorithm. Let us consider a subdivision algorithm of type (λ, μ, ℓ) with a triple subsubdominant eigenvalue and Fourier index $\mathcal{F}(\mu) = \{0, 2, n-2\}$. Then the third component of the central ring is

$$\overline{z} = \alpha f_3 + \beta f_4 + \gamma f_5,$$

where the coefficients α, β, γ depend on the initial data. Further, we observe that all three shape types of the central point are invariant with respect to regular linear maps. That is, if $\mathbf{P}$ and $\tilde{\mathbf{P}} = \mathbf{P}M$ are initial data related by an invertible (3×3)-matrix M, then the classifications of the corresponding central points $\mathbf{x}^{\mathrm{c}}$ and $\tilde{\mathbf{x}}^{\mathrm{c}}$ coincide. For that reason, we may assume that the matrix L in (7.23/138) is the identity and that, without loss of generality,

$$\alpha^2 + \beta^2 + \gamma^2 = 1, \quad \gamma \geq 0.$$

This observation implies that we can restrict a basic investigation of possible shapes to the two-parameter family

$$\bar{\mathbf{r}}^{\alpha,\beta} := \left[\psi,\ \alpha f_3 + \beta f_4 + \sqrt{1-\alpha^2-\beta^2} f_5\right]$$

of surface rings, where the parameters vary inside the unit circle,

$$(\alpha, \beta) \in \Gamma := \left\{(\alpha, \beta) \in \mathbb{R}^2 : \alpha^2 + \beta^2 \leq 1\right\}.$$

A *shape chart* $c := \Gamma \to \mathbb{Z}$ is a map which assigns to each α, β an indicator for the shape-type, for instance

$$c_{\text{limit}}(\alpha, \beta) := \begin{cases} 1 & \text{if } \bar{\kappa}_{\text{G}}^{\alpha,\beta} \geq 0 \\ 0 & \text{if } \bar{\kappa}_{\text{G}}^{\alpha,\beta} \text{ changes sign} \\ -1 & \text{if } \bar{\kappa}_{\text{G}}^{\alpha,\beta} \leq 0, \end{cases} \tag{7.25}$$

where $\bar{\kappa}_{\text{G}}^{\alpha,\beta}$ is the Gaussian curvature of $\bar{\mathbf{r}}^{\alpha,\beta}$. By Theorem 7.12/136, the value $c_{\text{limit}}(\alpha, \beta)$ indicates whether the corresponding subdivision surface is elliptic, hybrid, or hyperbolic in the limit. A shape chart thus summarizes, in a single image for all input data, information about the possible shape in a neighborhood of the central point. In particular, the hybrid region, i.e., the set of pairs (α, β) such that $c_{\text{limit}}(\alpha, \beta) = 0$, can be used to assess the quality of a subdivision algorithm: the smaller that region, the better the algorithm.

Shape charts can be visualized by coloring the different regions of Γ and thereby partitioning the unit circle into two or three subsets as in Fig. 7.3/141. When computing shape charts, possible symmetry properties can be exploited to increase efficiency. Variants on the concept include in particular the following:

- Different normalizations of the triple (α, β, γ). For example, $\max\{|\alpha|, |\beta|, |\gamma|\} = 1$ leads to square-shaped plots.
- Continuous variation of values. For example, the variance of trace W, see (7.16/131), shows the deviation of the mean curvature of the rings $\mathbf{x}^m$ from a constant value.

7.3 Conditions for C_2^k-Algorithms

In this section, we derive necessary and sufficient conditions for curvature continuity at the central point. It turns out that the sufficient conditions are extremely restrictive. This explains the failure of many early attempts to construct such algorithms. We start with a necessary condition on the spectrum of the subdivision matrix.

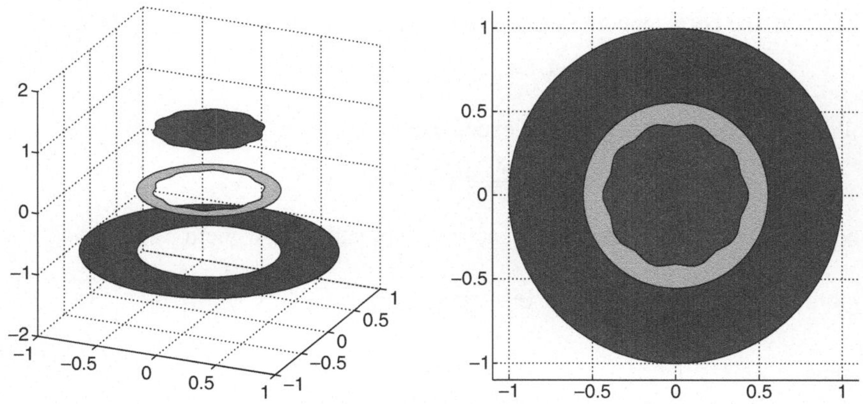

Fig. 7.3 Illustration of (7.25/140): Shape chart for the Catmull–Clark algorithm with $n = 10$ and flexible weights. *(left) Perspective view* and *(right) top view*. Respectively, the colors *red*, *green*, and *blue* indicate elliptic, hybrid, and hyperbolic behavior in the limit.

Theorem 7.15 (Necessity of $\mu \leq \lambda^2$). *A subdivision algorithm of type (λ, μ, ℓ) can be C_2^k only if $(\mu, \ell) \preccurlyeq (\lambda^2, 0)$.*

Proof. Let us recall the expansion (7.14/130),

$$\mathbf{W}^m \overset{*}{=} \varrho^{m,\ell}\,(\mathbf{T}^c)^t\, W \mathbf{T}^c, \quad \varrho = \frac{\mu}{\lambda^2}.$$

In view of Lemma 7.5/132, which states that $W \neq 0$ for generic initial data, we see that pointwise convergence of the sequence $\mathbf{W}^m$, as required by Theorem 2.14/28, is possible only if $\varrho^{m,\ell}$ converges. □

If $\mu < \lambda^2$ then $\varrho < 1$ and $\mathbf{W}^m$ converges to 0. According to Theorem 2.14/28, this guarantees curvature continuity. However, in this case the central point is necessarily a *flat spot*, i.e., the principal curvatures vanish here. For most applications, such a restriction is not acceptable so that we do not elaborate on that case. Rather, we seek conditions for *nontrivial curvature continuity* and assume from now on

$$(\mu, \ell) = (\lambda^2, 0).$$

Then, according to Theorem 2.14/28, a necessary and sufficient condition for curvature continuity is that the limit

$$\mathbf{W}^c := \lim_{m \to \infty} \mathbf{W}^m = \begin{bmatrix} W & 0 \\ 0 & 0 \end{bmatrix}$$

be a constant (3×3)-matrix, i.e., it does not depend on the arguments (s, t, j). Now, we *reparametrize* the rings $\mathbf{x}^m$ via the inverse of the planar ring $\overline{\boldsymbol{\xi}} = \psi L$, which is an embedding for generic data,

$$\tilde{\mathbf{x}}^m(u, v) := \mathbf{x}^m(\mathbf{s}), \quad \mathbf{s} := \overline{\boldsymbol{\xi}}^{-1}(u, v) \in \mathbf{S}_n^0.$$

By (2.5/22), the corresponding embedded Weingarten maps are equal up to sign:

$$\tilde{\mathbf{W}}^m(u,v) = \pm\mathbf{W}^m(\mathbf{s}).$$

Following (7.6/128), the asymptotic expansion of $\tilde{\mathbf{x}}^m$ is

$$(\tilde{\mathbf{x}}^m - \mathbf{x}^c)\cdot\mathbf{F}^c \stackrel{*}{=} \big[\lambda^m u, \lambda^m v, \lambda^{2m}\tilde{z}(u,v)\big],$$

where $\tilde{z}(u,v) := \overline{z}(\mathbf{s})$. Some elementary computations now yield

$$D\tilde{\mathbf{x}}^m \stackrel{*}{=} \lambda^m\mathbf{T}^c, \quad (D\tilde{\mathbf{x}}^m)^+ \stackrel{*}{=} \lambda^{-m}(\mathbf{T}^c)^t, \quad \tilde{\mathbf{n}}^m \stackrel{*}{=} \mathbf{n}^c, \quad \tilde{I\!I}^m \stackrel{*}{=} \lambda^{2m}\begin{bmatrix}\tilde{z}_{uu} & \tilde{z}_{uv}\\ \tilde{z}_{uv} & \tilde{z}_{vv}\end{bmatrix}$$

so that

$$\tilde{\mathbf{W}}^m \stackrel{*}{=} (\mathbf{T}^c)^t\tilde{W}\mathbf{T}^c, \quad \tilde{W} := \begin{bmatrix}\tilde{z}_{uu} & \tilde{z}_{uv}\\ \tilde{z}_{uv} & \tilde{z}_{vv}\end{bmatrix}.$$

Hence, the limit $\tilde{\mathbf{W}}^c = \pm\mathbf{W}^c$ is constant if and only if the three functions $\tilde{z}_{uu}$, $\tilde{z}_{uv}$, and $\tilde{z}_{vv}$ are constant. This holds if and only if

$$\tilde{z} \in \operatorname{span}\{1, u, v, u^2, uv, v^2\}.$$

That is, $\tilde{z}$ is a quadratic polynomial in u, v. Since $[u,v] = \overline{\boldsymbol{\xi}}(\mathbf{s})$, and the components of $\overline{\boldsymbol{\xi}}$ are linear combinations of the subdominant eigenrings f_1 and f_2, we obtain the equivalent condition

$$\overline{z} \in \operatorname{span}\{1, f_1, f_2, f_1^2, f_1 f_2, f_2^2\}.$$

Now, we consider the central spline $\mathbf{x}$ according to Definition 7.2/128. As observed above, its rings satisfy

$$\overline{\mathbf{x}}^0 = \overline{\mathbf{r}} = \big[\overline{\boldsymbol{\xi}}, \overline{z}\big], \quad \overline{\mathbf{x}}^m = \big[\lambda^m\overline{\boldsymbol{\xi}}, \lambda^{2m}\overline{z}\big].$$

We know that $\overline{z}$ is a quadratic polynomial in the components of $\overline{\boldsymbol{\xi}}$ and write $\overline{z} = p(\overline{\boldsymbol{\xi}})$. Being scaled copies of $\overline{\mathbf{x}}^0$, the other rings satisfy similar equations $\lambda^{2m}\overline{z} = p^m(\lambda^m\overline{\boldsymbol{\xi}})$, where the functions $p^m := \lambda^{2m}p(\lambda^{-m}\cdot)$ are also quadratic polynomials. However, because the rings $\overline{\mathbf{x}}^m$ join C^2, all these polynomials must in fact coincide, i.e., $p^m = p$. The resulting relation

$$\lambda^{2m}p = p(\lambda^m\cdot), \quad m \in \mathbb{N}_0,$$

shows that p is a *homogeneous* quadratic polynomial. Hence,

$$\overline{z} \in \operatorname{span}\{f_1^2, f_1 f_2, f_2^2\}.$$

Finally, because

$$\overline{z} = \sum_{q=3}^{\bar{q}} f_q\mathbf{p}_q\cdot\mathbf{n}^c = a_1 f_1^2 + a_2 f_1 f_2 + a_3 f_2^2 \tag{7.26}$$

must hold for any choice of generic initial data, we obtain the following result.

Theorem 7.16 (C_2^k-criterion). *A subdivision algorithm of type $(\lambda, \lambda^2, 0)$ is a C_2^k-algorithm if and only if the subsubdominant eigenrings satisfy*

$$f_q \in \operatorname{span}\{f_1^2, f_1 f_2, f_2^2\}, \quad q = 3, \dots, \bar{q}.$$

Moreover, $\bar{q} \leq 5$

Proof. The first part of the theorem was derived above. The second part, saying that the subsubdominant eigenvalue is at most triple, follows from linear independence of the subsubdominant eigenrings according to Lemma 4.22/78. □

The functional dependence required by the theorem is extremely restrictive and accounts, for instance, for the impossibility of finding C_2^2-variants on the Catmull–Clark algorithm. To see this, we now focus on piecewise polynomial algorithms.

Definition 7.17 ($C_r^{k,q}$-algorithm). Let $\{\boldsymbol{\Sigma}_i\}_i$ be a finite family of intervals forming a partition of the domain $\boldsymbol{\Sigma}^0$ of segments,

$$\boldsymbol{\Sigma}^0 = \bigcup_i \boldsymbol{\Sigma}_i.$$

A ring $\mathbf{x}^m \in C^k(\mathbf{S}_n^0, \mathbb{R}^d, G)$ is said to have *bi-degree q with respect to $\{\boldsymbol{\Sigma}_i\}_i$* if $\mathbf{x}^m$ restricted to $\boldsymbol{\Sigma}_i$ is a polynomial of bi-degree at most q for all i, and a polynomial of bi-degree q for at least one i; we write

$$\deg \mathbf{x}^m = q.$$

Further, a C_r^k-subdivision algorithm (A, G) is called a *$C_r^{k,q}$-algorithm*, if

$$\max_\ell \deg g_\ell = q$$

for the generating rings g_ℓ.

For instance, the Catmull–Clark algorithm is a $C_1^{2,3}$-algorithm, and the Doo–Sabin algorithm is a $C_1^{1,2}$-algorithm. For tensor-product splines with simple knots, the bi-degree q exceeds the smoothness k only by 1. However, non-trivial $C_2^{k,q}$-algorithms require a substantially higher degree. The results in that direction are all based on the following observation:

Lemma 7.18 (Degree estimate for $\boldsymbol{\psi}$). *For $n \neq 4$, the characteristic map $\boldsymbol{\psi}$ of a standard $C_1^{k,q}$-algorithm satisfies*

$$\deg \boldsymbol{\psi} > k.$$

Proof. Let us assume that $\deg \boldsymbol{\psi} \leq k$. Then the segments $\boldsymbol{\psi}_j$ are in fact not piecewise polynomials on a partition, but simply polynomials on $\boldsymbol{\Sigma}^0$. Equally, two neighboring segments $\boldsymbol{\psi}_j$ and $\boldsymbol{\psi}_{j+1}$ differ only by a change of parameters,

$$\boldsymbol{\psi}_{j+1}(s,t) = \boldsymbol{\psi}_j(t, -s).$$

Hence, $\boldsymbol{\psi}_{j+4} = \boldsymbol{\psi}_j$, implying that injectivity is possible only for $n = 4$. □

For $n = 4$, the characteristic ring of the Catmull–Clark-algorithm and of the Doo–Sabin-algorithm have $\deg \psi = 1$. For $n \neq 4$, the lemma and Theorem 7.16/143 suggest, and the following shows, that the generating system G must have at least bi-degree $2k + 2$ to represent subsubdominant eigenfunctions.

Theorem 7.19 (Degree estimate for $C_2^{k,q}$-algorithms). *Let $n \neq 4$. For a non-trivial $C_2^{k,q}$-algorithm with characteristic ring ψ,*

$$q \geq 2 \deg \psi \geq 2k + 2.$$

In particular, the lowest degree for $k = 2$ is $q = 6$.

Proof. By (7.26/142), with the complex characteristic ring $f = f_1 + \mathbf{i} f_2$, the j-th segment of the normal component of the central ring can be written as

$$\begin{aligned} \bar{z}_j &= a_1 f_{1,j}^2 + a_2 f_{1,j} f_{2,j} + a_3 f_{2,j}^2 \\ &= \operatorname{Re}(\alpha f_j^2) + \beta |f_j|^2 = \operatorname{Re}(\alpha w_n^{2j} f_0^2) + \beta |f_0|^2, \end{aligned}$$

where $\alpha := (a_1 - a_3 - \mathbf{i} a_2)/2, \beta := (a_1 + a_3)/2$. The last equality follows from (5.21/103), saying that the segments of f are related by $f_j = w_n^j f_0$. By Lemma 7.18/143, the complex-valued piecewise polynomial f_0 has degree $\deg f_0 \geq k + 1$.

For a bivariate polynomial p of degree $d := \deg p$ we define the leading coefficient $c[p] \neq 0$ and the leading monomial $m[p](s,t) = s^\ell t^{d-\ell}$ by the split

$$p = c[p]\, m[p] + T[p],$$

where the trailing term

$$T[p] := \sum_{i=\ell+1}^{d} c_i s^i t^{d-\ell} + \sum_{i+k<d} c_{i,k} s^i t^k$$

summarizes all terms of degree d which contain at least the factor $s^{\ell+1}$, and all terms of degree $< d$. Obviously, for two polynomials p_1, p_2 with $m[p_1] = m[p_2]$ it is

$$c[p_1 p_2] = c[p_1]\, c[p_2], \quad m[p_1 p_2] = (m[p_1])^2.$$

When restricted to a suitable subset of its domain,

$$f_0 = c[f_0] m[f_0] + T[f_0], \quad \deg m[f_0] \geq k + 1.$$

Because the characteristic ring f can be scaled arbitrarily, we may assume without loss of generality that the leading coefficient is $c[f_0] = 1$. Hence,

$$f_0^2 = (m[f_0])^2 + T\left[f_0^2\right], \quad |f_0|^2 = f_0 \overline{f_0} = (m[f_0])^2 + T\left[|f_0^2|\right],$$

and the coefficient of $\overline{z}_j$ to the monomial $(m[f_0])^2$ is

$$\mathrm{Re}(\alpha w_n^{2j}) + \beta.$$

This expression can vanish for all $j \in \mathbb{Z}_n$ only if $\alpha = \beta = 0$. This implies $a_1 = a_2 = a_3 = 0$ and $\overline{z} = 0$, contradicting the assumption that the initial data be generic. Hence, $m[\overline{z}_j] = (m[f_0])^2$ at least for one j, showing that the degree of $\bar{z}$ is bounded by $\deg \overline{z} \geq \deg \overline{z}_j = 2d[f_0] \geq 2(k+1)$. □

7.4 A Framework for C_2^k-Algorithms

In this section, we provide a framework for constructing C_2^k-algorithms. So far, the algorithm (A, G) was assumed to be given, and ψ was determined as the planar ring corresponding to the subdominant eigenvalues of the subdivision matrix A. By contrast, we now start with a function $\varphi \in C^k(\mathbf{S}_n^0, \mathbb{R}^2, G)$ and then derive a matrix A so that (A, G) defines a C_2^k-algorithm with $\psi := \varphi$ as its characteristic ring. More precisely, we say that the planar ring $\varphi \in C^k(\mathbf{S}_n^0, \mathbb{R}^2, G)$ is a *regular C^k-embedding of $\mathbf{S}_n^0$ with scale factor* λ if it has the two key properties of a characteristic ring, i.e.,

- φ is regular and injective, and
- there exists a real number $\lambda \in (0, 1)$ such that φ and $\lambda\varphi$ join C^2 according to (4.8/62) when regarded as consecutive rings.

For instance, the characteristic ring of the Catmull–Clark algorithm represents a regular C^2-embedding of bi-degree 3, which may be used to construct a $C_2^{2,6}$-algorithm. But as mentioned already above, there is no need to derive φ from an existing algorithm. The image of φ is denoted by

$$\boldsymbol{\Omega} := \varphi(\mathbf{S}_n^0).$$

Now, we define a family of reparametrization operators, taking rings to functions on scaled copies of $\boldsymbol{\Omega}$.

Definition 7.20 (Reparametrization $\mathcal{R}_m$). For $m \in \mathbb{N}_0$, the *reparametrization operator* $\mathcal{R}_m$ maps a ring $\mathbf{p} \in C^k(\mathbf{S}_n^0, \mathbb{R}^d)$ to a C^k-function $\mathbf{q} := \mathcal{R}_m[\mathbf{p}]$ on $\lambda^m \boldsymbol{\Omega} \subset \mathbb{R}^2$,

$$\mathbf{q} : \lambda^m \boldsymbol{\Omega} \ni \boldsymbol{\xi} \mapsto \mathbf{p}(\varphi^{-1}(\lambda^{-m}\boldsymbol{\xi})) \in \mathbb{R}^d.$$

The inverse operator $\mathcal{R}_m^{-1}$ maps a C^k-function $\mathbf{q}$ on $\lambda^m \boldsymbol{\Omega}$ to a ring $\mathbf{p} := \mathcal{R}_m^{-1}[\mathbf{q}] \in C^k(\mathbf{S}_n^0, \mathbb{R}^d)$,

$$\mathbf{p} : \mathbf{S}_n^0 \ni \mathbf{s} \mapsto \mathbf{q}(\lambda^m \varphi(\mathbf{s})).$$

The operator $\mathcal{R}_m$, and equally $\mathcal{R}_m^{-1}$, is linear in the sense that $\mathcal{R}_m[\alpha f + \beta g] = \alpha\mathcal{R}_m[f] + \beta\mathcal{R}_m[g]$. Given φ, we denote the space of bivariate polynomials of total degree 2 restricted to $\lambda^m \boldsymbol{\Omega}$ by $\mathbb{P}_2(\lambda^m \boldsymbol{\Omega})$. The following definition is crucial. It characterizes subdivision algorithms which are able to represent rings corresponding to quadratic polynomials, and generate such quadratic rings from quadratic rings.

Definition 7.21 (Quadratic precision). The subdivision algorithm (A,G) has *quadratic precision* with respect to φ if

- for each quadratic polynomial $p \in \mathbb{P}_2(\boldsymbol{\Omega})$ there exists a real-valued ring $\mathbf{x}^0 \in C^k(\mathbf{S}_n^0, \mathbb{R}, G)$ with
$$\mathcal{R}_0[\mathbf{x}^0] = p,$$
- for consecutive rings $\mathbf{x}^0 = G\mathbf{Q}$ and $\mathbf{x}^1 = GA\mathbf{Q}$,
$$\mathcal{R}_0[\mathbf{x}^0] \in \mathbb{P}_2(\boldsymbol{\Omega}) \quad \text{implies} \quad \mathcal{R}_0[\mathbf{x}^1] \in \mathbb{P}_2(\boldsymbol{\Omega}).$$

First, we observe that for a subdivision algorithm (A,G) with quadratic precision, $\mathcal{R}_0[\mathbf{x}^0] \in \mathbb{P}_2(\boldsymbol{\Omega})$ implies $\mathcal{R}_0[\mathbf{x}^m] \in \mathbb{P}_2(\boldsymbol{\Omega})$ and also $\mathcal{R}_m[\mathbf{x}^m] \in \mathbb{P}_2(\lambda^m\boldsymbol{\Omega})$ for all m. Second, we consider the sequence

$$\mathcal{R}_0[\mathbf{x}^0],\ \mathcal{R}_1[\mathbf{x}^1],\ \mathcal{R}_2[\mathbf{x}^2],\ \ldots,$$

starting from $\mathcal{R}_0[\mathbf{x}^0] \in \mathbb{P}_2(\boldsymbol{\Omega})$. Corresponding to consecutive rings that join C^2, all these polynomials coincide in the sense that they must have the same monomial expansion. However, strictly speaking, they are not equal because the domains are different. To account for that fact, we write

$$\mathcal{R}_0[\mathbf{x}^0] \cong \mathcal{R}_m[\mathbf{x}^m], \quad m \in \mathbb{N}.$$

In particular, if $\mathcal{R}_0[\mathbf{x}^0]$ is a monomial of total degree $\ell \le 2$, we have

$$\mathcal{R}_0[\mathbf{x}^m] = \lambda^{\ell m}\mathcal{R}_0[\mathbf{x}^0]. \tag{7.27}$$

Remarkably, quadratic precision immediately yields an appropriate eigenstructure for (A,G).

Lemma 7.22 (Quadratic precision yields correct spectrum). *Let φ be a regular C^k-embedding of $\mathbf{S}_n^0$ with scale factor λ. If (A,G) has quadratic precision with respect to φ, then there exist eigenvalues λ_i, eigenvectors v_i, and eigenrings $f_i := Gv_i$, satisfying*

$$\begin{gathered}\lambda_0 = 1, \quad \lambda_1 = \lambda_2 = \lambda, \quad \lambda_3 = \lambda_4 = \lambda_5 = \lambda^2, \\ f_0 = 1, \quad [f_1, f_2] = \varphi, \quad f_3 = f_1^2,\ f_4 = f_1 f_2,\ f_5 = f_2^2.\end{gathered}$$

Here, with a slight abuse of notation, we indexed eigenvalues without assuming that the whole sequence is ordered by modulus. In particular, further eigenvalues with modulus greater than λ^2 are not excluded a priori.

Proof. With $\boldsymbol{\xi} = (x,y)$, we define the monomials

$$p_0(\boldsymbol{\xi}) = 1,\ p_1(\boldsymbol{\xi}) = x,\ p_2(\boldsymbol{\xi}) = y,\ p_3(\boldsymbol{\xi}) = x^2,\ p_4(\boldsymbol{\xi}) = xy,\ p_5(\boldsymbol{\xi}) = y^2$$

in $\mathbb{P}_2(\boldsymbol{\Omega})$. For $i = 0,\ldots,5$, we have $\lambda_i = \lambda^{\ell_i}$, where ℓ_i is the total degree of p_i. By definition of quadratic precision, the function $f_i := \mathcal{R}_0^{-1}[p_i] = p_i \circ \varphi$ can be written as $f_i = Gv_i'$ for some vector $v_i' \neq 0$. By (7.27/146), $\mathcal{R}_0[GA^m v_i'] = \lambda^{\ell_i m} p_i$, and hence, applying $\mathcal{R}_0^{-1}$ on both sides,

$$GA^m v_i' = \lambda_i^m f_i = \lambda_i^m G v_i'.$$

If G is linearly independent, it follows immediately that v_i' is an eigenvector of A to λ_i, but we have to show that the same is true in general.

For $k \in \mathbb{N}_0$, let $v_i := \lambda_i^{-k} A^k v_i'$. Then

$$Gv_i = \lambda_i^{-k} GA^k v_i' = f_i$$

shows that v_i is another possible choice of coefficients corresponding to the polynomial p_i. As before,

$$GA^m v_i = \lambda_i^m f_i = \lambda_i^m G v_i. \tag{7.28}$$

With $A = VJV^{-1}$ the Jordan decomposition of A, let

$$w' := V^{-1} v_i', \quad w := V^{-1} v_i = \lambda_i^{-k} J^k w'.$$

Recalling (4.25/74), F and w are partitioned into blocks F_r and w_r corresponding to the Jordan blocks J_r of J. Condition (7.28/147) yields the equivalent system

$$F_r J_r^m w_r = \lambda_i^m F_r w_r, \quad r = 0, \ldots, \bar{r}.$$

When determining solutions w_r, we distinguish two cases: First, if the eigenvalue corresponding to J_r is $\lambda_r = 0$, then $w_r = \lambda_i^{-k} J_r^k w_r' = 0$ is the only solution for k chosen sufficiently large.

Second, if $\lambda_r \neq 0$, then Lemma 4.22/78 guarantees that the eigenfunction f_r^0 does not vanish. Of course, the trivial solution $w_r = 0$ is possible. Otherwise, if $w_r \neq 0$, let ν denote the largest index of a non-vanishing component, i.e., $w_r^i = 0$ for $i > \nu$ and $w_r^\nu \neq 0$. By (4.27/74), we have the asymptotic expansion

$$\lambda_i^m F_r w_r = F_r J_r^m w_r \overset{*}{=} \lambda_r^{m,\nu} f_r^0 w_r^\nu,$$

implying $\lambda_i = \lambda_r$ and $\nu = 0$. Hence, $w_r = [w_r^0; 0; \ldots; 0]$ is an eigenvector of J_r to the eigenvalue λ_i. Summarizing, we have $J_r w_r = \lambda_i w_r$ for all r. Therefore, $Jw = \lambda_i w$ and $Av_i = \lambda_i v_i$.

For $i = 0, \ldots, 5$, we obtain the eigenvalues $\lambda_0 = 1, \lambda_1 = \lambda_2 = \lambda$ and $\lambda_3 = \lambda_4 = \lambda_5 = \lambda^2$, as stated. The corresponding dominant and subdominant eigenrings are $f_0 = 1$, and $[f_1, f_2] = \varphi$. Hence, $f_3 = p_3 \circ \varphi = p_1^2 \circ \varphi = (p_1 \circ \varphi)^2 = f_1^2$, and equally $f_4 = f_1 f_2$, $f_5 = f_2^2$. □

Together, Lemma 7.22/146 and Theorem 7.16/143 show that quadratic precision and scalable embeddings yield promising candidates for C_2^k-algorithms.

Theorem 7.23 (Quadratic precision suggests C_2^k-algorithm). *Let φ be a regular C^k-embedding of $\mathbf{S}_n^0$ with scale factor λ. If the symmetric C^k-subdivision algorithm (A, G) has quadratic precision with respect to φ, and if $|\lambda_i| < \lambda^2$ for all $i > 5$, then (A, G) is of type $(\lambda, \lambda^2, 0)$ and defines a C_2^k-algorithm.*

Now, we describe a four-step procedure which yields subdivision algorithms with quadratic precision. The four steps *reparametrization – extension – turn-back – projection* suggest the acronym *PTER* for the framework, where as usual the concatenation of operators is from right to left.

Let us assume that a C^k-system G of generating rings and a regular C^k-embedding φ with scale factor λ are given and have the following properties:

- The generating rings g_ℓ are piecewise polynomial with maximal degree q in the sense of Definition 7.17/143.
- In view of Theorem 7.19/144, the embedding $\varphi = G[v_1, v_2]$ has degree $\deg \varphi \leq k/2$.
- There exist vectors v_3, v_4, v_5 with

$$Gv_3 = (Gv_1)^2, \quad Gv_4 = (Gv_1)(Gv_2), \quad Gv_5 = (Gv_2)^2$$

  to account for Theorem 7.16/143. In particular, this assumption is fulfilled if G spans the space of *all* piecewise polynomials with respect to the given partition and the given order of continuity.

To simplify notation, we describe how to compute $\mathbf{x}^1 = G\mathbf{Q}^1$ from $\mathbf{x}^0 = G\mathbf{Q}$ for given initial data $\mathbf{Q}$. But the whole procedure is linear and independent of the level m so that it defines a stationary algorithm. The building blocks are characterized as follows:

R – *Reparametrization:* Reparametrize the ring $\mathbf{x}^0$ as a function $\mathbf{y}^0$ on $\boldsymbol{\Omega}$,

$$\mathbf{y}^0 := \mathcal{R}_0[\mathbf{x}^0].$$

E – *Extension:* Extend $\mathbf{y}^0$ to a function $\mathbf{y}^1$ defined on $\lambda\boldsymbol{\Omega}$ such that quadratic polynomials are extended by themselves,

$$\mathbf{y}^0 \cong \mathbf{y}^1 \quad \text{if} \quad \mathbf{y}^0 \in \mathbb{P}_2(\boldsymbol{\Omega}).$$

We note that smooth contact is required only for quadratic polynomials. In general, $\mathbf{y}^0$ and $\mathbf{y}^1$ do not need to join continuously. Since G is not necessarily linear independent, $\mathbf{y}^1$ may depend not only on $\mathbf{y}^0$, but also directly on the initial data $\mathbf{Q}$. We write in terms of the linear *extension operator* $\mathcal{E}$

$$\mathbf{y}^1 := \mathcal{E}[\mathbf{Q}, \mathbf{y}^0].$$

Some examples of $\mathcal{E}$ are as follows:
(i) A projection from the space of functions on $\boldsymbol{\Omega}$ onto some finite dimensional space $\mathbb{P}(\boldsymbol{\Omega})$ of bivariate polynomials containing $\mathbb{P}_2(\boldsymbol{\Omega})$. This projection could be obtained, e.g., by a least squares fit or by an interpolant $\tilde{\mathbf{y}}^0 \in \mathbb{P}(\boldsymbol{\Omega})$ of $\mathbf{y}^0$. Then, the extension is defined by the polynomial $\tilde{\mathbf{y}}^0$, i.e., $\mathbf{y}^1 \cong \tilde{\mathbf{y}}^0$. In the same way, also spaces of piecewise polynomials can be used.
(ii) The minimizer of some positive semi-definite quadratic fairness functional $\mathcal{F}$, acting on functions defined on $\lambda\boldsymbol{\Omega}$, with the property that $\mathcal{F}$ vanishes on $\mathbb{P}_2(\lambda\boldsymbol{\Omega})$. For instance, for functions $\mathbf{y}^1$ joining C^k with $\mathbf{y}^0$, one can consider

$$\mathcal{F}(\mathbf{y}^1) := \int_{\lambda\Omega} \Delta^k \mathbf{y}^1(\boldsymbol{\xi})\, d\boldsymbol{\xi} \to \min$$

or discrete variants thereof. Also here, if $\mathbf{y}^0 \in \mathbb{P}(\Omega)$, then $\mathbf{y}^1 \cong \mathbf{y}^0$ because $\mathcal{F}(\mathbf{y}^1) = 0$ for $\mathbf{y}^1 \in \mathbb{P}_2(\lambda\Omega)$.

T – *Turn-back:* Convert the function $\mathbf{y}^1$ back into a ring,

$$\tilde{\mathbf{x}}^1 := \mathcal{R}_1^{-1}[\mathbf{y}^1].$$

In general, this ring is neither in the span of G, nor does it join smoothly with $\mathbf{x}^0$.

P – *Projection:* Project $\tilde{\mathbf{x}}^1$ into the subspace of $C^k(\mathbf{S}_n^0, \mathbb{R}^d, G)$ consisting of rings that join C^k with $\mathbf{x}^0$. The coefficients $\mathbf{Q}^1$ of the resulting ring $\mathbf{x}^1 = G\mathbf{Q}^1$ are obtained by a linear operator $\mathcal{P}$,

$$\mathbf{Q}^1 := \mathcal{P}[\mathbf{Q}, \tilde{\mathbf{x}}^1],$$

where the first argument provides information to enforce the C^k-condition. Crucially, $\mathcal{P}$ has to be chosen such that $\mathbf{x}^1 = \tilde{\mathbf{x}}^1$ if $\tilde{\mathbf{x}}^1$ is a quadratic polynomial in the components of φ, i.e., if $\mathcal{R}_0[\tilde{\mathbf{x}}^1] \in \mathbb{P}_2(\Omega)$. Thus, $\mathcal{P}$ is typically defined by a constrained least squares fit with respect to some inner product, either continuous or discrete,

$$\|\tilde{\mathbf{x}}^1 - G\mathbf{Q}^1\| \to \min.$$

We note that, if G is linearly dependent, $\mathcal{P}$ is not uniquely determined by the above optimization problem.

Together, the PTER-framework yields the new coefficients

$$\tilde{A}\mathbf{Q} := \mathbf{Q}^1 := \mathcal{P}[\mathbf{Q}, \mathcal{R}_1^{-1}[\mathcal{E}[\mathbf{Q}, \mathcal{R}_0[G\mathbf{Q}]]]].$$

The columns of $\tilde{A}$ are obtained by substituting in unit vectors for the argument $\mathbf{Q}$. Then, any ineffective eigenvectors should be removed from $\tilde{A}$ according to Theorem 4.20/77 to obtain a genuine subdivision matrix A.

Theorem 7.24 (The PTER-framework works). *The PTER-framework yields a $C_2^{k,q}$-algorithm (A, G) if $|\lambda_i| < \lambda^2$ for $i > 5$.*

Proof. Tracing subdivision of the ring $\mathbf{x}^0$ corresponding to a quadratic function $\mathcal{R}[\mathbf{x}^m]$, one easily sees that the so constructed algorithm (A, G) has quadratic precision and the assumptions of Theorem 7.23/147 are satisfied. □

7.5 Guided Subdivision

The framework in the previous section is inspired by and closely related to that of *Guided subdivision*. Guided subdivision aims at controlling the shape by means of a so-called *guide surfaces*, or guide for short. This guide $\mathbf{g}$ serves as an outline of the local shape of the subdivision surface $\mathbf{x}$ to be constructed, and is not changed as subdivision proceeds. The sequence of rings will be defined such that consecutive rings join C^2 and the reparametrization of $\mathbf{x}^m$ approximates the guide on $\lambda^m \boldsymbol{\Omega} = \lambda^m \varphi(\mathbf{S}_n^0)$. For a given regular C^k-embedding φ of $\mathbf{S}_n^0$ with scale factor λ, we expect

$$\mathcal{R}_m[\mathbf{x}^m] \approx \mathbf{g}_{|\lambda^m \boldsymbol{\Omega}},$$

or equivalently

$$\mathbf{x}^m \approx \mathcal{R}_m^{-1}[\mathbf{g}].$$

Thus, the shape of the spline surface approximates the shape of the guide.

It is instructive to explain the concept of Guided subdivision by means of a concrete and actually quite simple setting. Just like the framework, it has many options, generalizations and extensions, such as algorithms for triangular patches or for higher smoothness and precision.

Let $\{\Sigma_i\}_{i=1}^3$ be the natural partition of $\boldsymbol{\Sigma}^0$ into three squares with side length $1/2$, and choose smoothness $k = 2$ and bi-degree $q = 7$. Bi-degree 7 is not minimal, but chosen to simplify the exposition of Hermite sampling below.

Due to the partition, for $0 \le \ell \le k = 2$ and all $j \in \mathbb{Z}_n$, the functions

$$D_1^\ell \mathbf{x}^m(1/2, \cdot, j), \quad D_2^\ell \mathbf{x}^m(\cdot, 1/2, j)$$

defining the inner boundary of $\mathbf{x}^m$ are *polynomials* of degree at most q. Hence, the C^2-contact conditions (4.8/62) imply that the corresponding functions

$$D_1^\ell \mathbf{x}^{m+1}(1, \cdot, j), \quad D_2^\ell \mathbf{x}^{m+1}(\cdot, 1, j),$$

at the outer boundary of $\mathbf{x}^{m+1}$ are also not piecewise polynomial but each a single polynomial of degree $q = 7$ or less. Therefore, we define $\tilde{G} = [\tilde{g}_1, \ldots, \tilde{g}_{\tilde{q}}]$ to be a system of rings spanning the linear subspace of all C^2-rings with bi-degree $q = 7$, and for which

$$D_1^\ell \tilde{g}_\ell(1, \cdot, j), \quad D_2^\ell \tilde{g}_\ell(\cdot, 1, j), \quad 0 \le \ell \le k = 2,$$

are polynomials of degree ≤ 7. Then a ring in $C^k(\mathbf{S}_n^0, \mathbb{R}^d, \tilde{G})$ is uniquely defined by its partial derivatives up to order $(\frac{q-1}{2}, \frac{q-1}{2}) = (3, 3)$ at the $4n$ points

$$\mathbf{s}_j^1 := (1/2, 0, j),\ \mathbf{s}_j^2 := (1, 0, j),\ \mathbf{s}_j^3 := (1/2, 1/2, j),\ \mathbf{s}_j^4 := (1, 1, j), \quad j \in \mathbb{Z}_n, \tag{7.29}$$

see Fig. 7.4/151. To formalize the construction of rings from partial derivatives, we define the tensor-product *Hermite operator* $\mathcal{H}$ of order $(3, 3)$. The operator $\mathcal{H}$ maps a ring $\mathbf{x}^0$ to the (4×4)-matrix

$$\mathcal{H}[\mathbf{x}^0] := \left[D_1^\alpha D_2^\beta \mathbf{x}^0\right]_{(0,0) \le (\alpha,\beta) \le (3,3)}$$

of partial derivatives up to order $(3, 3)$.

For simplicity, we consider polynomial guides only. To represent them in monomial form, let

$$M_r := [m_{\nu,\mu}]_{\nu+\mu \le r}, \quad m_{\nu,\mu}(x, y) := x^\nu y^\mu,$$

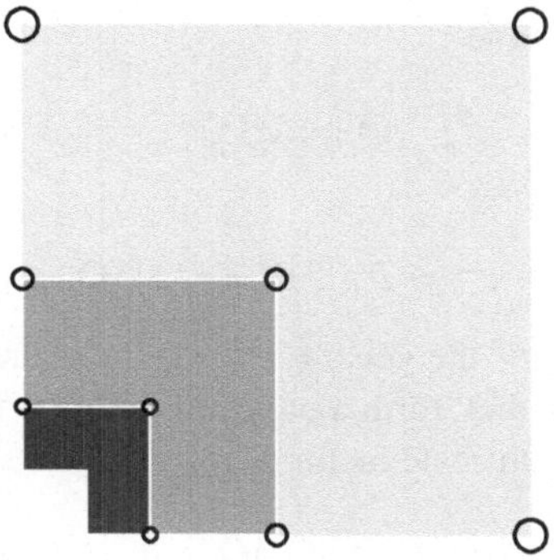

Fig. 7.4 Illustration of (7.29/150): Hermite sampling at the marked points determines the rings.

span the space $\mathbb{P}_r$ of bivariate polynomials of total degree $\leq r$. It is convenient to define the algorithm by means of a diagonal matrix J and a corresponding system F of eigenrings. Let $F = [f_{\nu,\mu}]_{\nu+\mu\leq r}$ be the set of rings $f_{\nu,\mu} \in C^k(\mathbf{S}_n^0, \mathbb{R}, \tilde{G})$ interpolating the reparametrized monomials $\mathcal{R}_0^{-1}[m_{\nu,\mu}]$ up to order $(3,3)$ at the points $\mathbf{s}_j^i$,

$$\mathcal{H}\big[f_{\nu,\mu} - \mathcal{R}_0^{-1}[m_{\nu,\mu}]\big](\mathbf{s}_j^i) = 0, \quad i = 1, \ldots, 4,\ j \in \mathbb{Z}_n.$$

According to the labelling of generating rings $f_{\nu,\mu}$, the vector of initial data has the form $\mathbf{P} := [\mathbf{p}_{\nu,\mu}]_{\nu+\mu\leq r}$ so that

$$\mathbf{x}^0 = F\mathbf{P} = \sum_{\nu=0}^{r}\sum_{\mu=0}^{r-\nu} f_{\nu,\mu}\mathbf{p}_{\nu,\mu}.$$

Since the values of a monomial $m_{\nu,\mu}$ on $\lambda^m\boldsymbol{\Omega}$ and on $\lambda^{m+1}\boldsymbol{\Omega}$ are related by a scale factor $\lambda^{\nu+\mu}$, and since this monomial corresponds to the rings $f_{\mu,\nu}$, Guided subdivision can be defined by a simple scaling process. We define the diagonal matrix

$$J := \operatorname{diag}\big([\lambda^{\nu+\mu}]_{\nu+\mu\leq r}\big)$$

to obtain the recursion

$$\mathbf{x}^m := F\mathbf{P}^m, \quad \mathbf{P}^m := J^m\mathbf{P} = [\lambda^{m(\nu+\mu)}\mathbf{p}_{\nu,\mu}]_{\nu+\mu\leq r}.$$

Although this is needed neither for the analysis nor for an implementation, we briefly discuss a possible conversion of the setup into a subdivision algorithm (A, G) in its genuine form. Let $B_r = [b_{\nu,\mu}]_{\nu+\mu\leq r}$ denote the vector of bivariate *Bernstein polynomials* of total degree $\leq r$ on the unit triangle. Because these Bernstein polynomials are linearly independent, monomials can be represented as linear combinations of them. That is, there exists an invertible matrix V with $M_r = B_r V$ and

$B_r = M_r V^{-1}$. Then we define

$$G := FV^{-1}, \quad \mathbf{Q} := V\mathbf{P}, \quad A := VJV^{-1},$$

to obtain

$$\mathbf{x}^m = FJ^m\mathbf{P} = GA^m\mathbf{Q}.$$

Because the elements $g_{\nu,\mu}$ of the generating system G are Hermite interpolants to the Bernstein polynomials, they form a partition of unity. Further, A represents just de Casteljau's algorithm with scale factor λ,

$$B_r(\lambda\xi)\mathbf{Q} = B_r(\xi)A\mathbf{Q}, \quad \xi = (x, y),$$

showing that the rows of A sum to 1.

Definition 7.25 (Guided $C_{2,r}^{2,7}$-subdivision). For $r \geq 2$, the subdivision algorithm (A, G) with A and G as defined above is called *Guided $C_{2,r}^{2,7}$-subdivision*. The polynomial

$$\mathbf{g} := \bigcup_{m \in \mathbb{N}_0} \lambda^m \boldsymbol{\Omega} \ni \xi \mapsto B_r(\xi)\mathbf{Q} \in \mathbb{R}^d$$

is called the *guide* to the initial data $\mathbf{Q}$.

Although the minimal value $r = 2$ is impeccable from a theoretical point of view, one typically chooses much larger values for r to define a space of rings which covers a sufficiently rich variety of shapes. Let us discuss some implications of the above definition.

First, the subdivision matrix A and the diagonal matrix $J = V^{-1}AV$ are similar so that we can easily read off the common spectrum and see that the structure of the leading eigenvalues is just right.

Second, because J is diagonal, we have

$$\mathbf{x}^m = FJ^m\mathbf{P} = \sum_{\nu=0}^{r} \sum_{\mu=0}^{r-\nu} \lambda^{m(\nu+\mu)} m_{\nu,\mu} \mathbf{p}_{\nu,\mu},$$

showing that $\mathbf{x}^m$ interpolates the reparametrization of $\mathbf{g}_{|\lambda^m \boldsymbol{\Omega}}$,

$$\mathcal{H}\big[\mathbf{x}^m - \mathcal{R}_m^{-1}[\mathbf{g}]\big](\mathbf{s}_j^i) = \sum_{\nu=0}^{r} \sum_{\mu=0}^{r-\nu} \lambda^{m(\nu+\mu)} \mathbf{p}_{\nu,\mu} \mathcal{H}[f_{\nu,\mu} - R_0^{-1} m_{\nu,\mu}](\mathbf{s}_j^i) = 0.$$

Hence, by the chain rule,

$$D_1^\alpha D_2^\beta \mathbf{x}^m(\mathbf{s}_j^1) = 2^{\alpha+\beta} D_1^\alpha D_2^\beta \mathbf{x}^{m+1}(\mathbf{s}_j^2)$$
$$D_1^\alpha D_2^\beta \mathbf{x}^m(\mathbf{s}_j^3) = 2^{\alpha+\beta} D_1^\alpha D_2^\beta \mathbf{x}^{m+1}(\mathbf{s}_j^4)$$

for $(\alpha, \beta) \leq (3, 3)$. Since, for $\ell \leq 2$,

$$D_1^\ell \mathbf{x}^m(1, \cdot, j), \; D_2^\ell \mathbf{x}^m(\cdot, 1, j), \; D_1^\ell \mathbf{x}^{m+1}(2, \cdot, j), \; D_2^\ell \mathbf{x}^{m+1}(\cdot, 2, j)$$

are all polynomials of degree at most 7, we conclude that coincidence of partial derivatives at the points $\mathbf{s}^i_j$ implies

$$D_1^\ell \mathbf{x}^m(1/2,\cdot,j) = 2^\ell D_1^\ell \mathbf{x}^{m+1}(1,\cdot,j)$$
$$D_2^\ell \mathbf{x}^m(\cdot,1/2,j) = 2^\ell D_2^\ell \mathbf{x}^{m+1}(\cdot,1,j).$$

This shows that consecutive rings join C^2 so that Guided subdivision (A,G) is indeed a C^2-algorithm.

Third, the surfaces $\mathbf{x}$ and $\mathbf{g}$ have third-order contact at the points $2^{-m}\mathbf{s}^i_j$. In particular, the points

$$\mathbf{x}(2^{-m}\mathbf{s}^i_j) = \mathbf{g}(\lambda^m \boldsymbol{\xi}^i_j), \quad \boldsymbol{\xi}^i_j := \varphi(\mathbf{s}^i_j)$$

and also the embedded Weingarten maps

$$\mathbf{W}_\mathbf{x}(2^{-m}\mathbf{s}^i_j) = \mathbf{W}_\mathbf{g}(\lambda^m \boldsymbol{\xi}^i_j)$$

coincide. This property accounts for our initial statement, saying that the image of $\mathbf{g}$ yields a good approximation of the image of $\mathbf{x}$. As one approaches the center, the interpolation points become denser and denser so that the shapes are closer and closer.

While the latter observation is relevant for a qualitative assessment of shape, the next theorem verifies analytic smoothness.

Theorem 7.26 (Guided $C^{2,7}_{2,r}$-subdivision works). *For $r \geq 2$, Guided $C^{2,7}_{2,r}$-subdivision (A,G) defines a $C^{2,7}_2$-algorithm.*

Proof. For $\nu+\mu \leq 2$, the ring $\mathcal{R}_0^{-1}[m_{\nu,\mu}]$ lies in $C^2(\mathbf{S}^0_n,\mathbb{R},G)$, and hence $f_{\nu,\mu} = \mathcal{R}_0^{-1}[m_{\nu,\mu}]$. Since J is a diagonal matrix, we can easily read off the non-zero eigenvalues $\lambda^{\nu+\mu}$, and see that the functions $f_{\nu,\mu}$ are the corresponding eigenrings. The eigenring to the dominant eigenvalue $\lambda_0 = \lambda^0 = 1$ is

$$f_{0,0} = \mathcal{R}_0^{-1}[m_{0,0}] = 1,$$

the eigenrings to the subdominant eigenvalue $\lambda_1 = \lambda_2 = \lambda$ are

$$[f_{1,0}, f_{0,1}] = \mathcal{R}_0^{-1}[m_{1,0}, m_{0,1}] = \varphi,$$

and the eigenrings to the subsubdominant eigenvalue $\lambda_3 = \lambda_4 = \lambda_5 = \lambda^2$ are

$$[f_{2,0}, f_{1,1}, f_{0,2}] = \mathcal{R}_0^{-1}[m_{2,0}, m_{1,1}, m_{0,2}] = [f_{1,0}^2, f_{1,0}f_{0,1}, f_{0,1}^2].$$

All other eigenvalues are, by construction, smaller so that the claim follows from Theorem 7.16/143. □

Guided $C^{2,7}_{r,2}$-subdivision fits the pattern of the PTER-framework. The extension process yields the guide $\mathbf{g}$ restricted to the domain $\lambda\boldsymbol{\Omega}$, while the projecting step into the appropriate space is defined via Hermite sampling at the points $\boldsymbol{\xi}^i_j$.

Bibliographical Notes

1. The importance of the ratio $\varrho := \frac{\mu}{\lambda^2}$, as defined in (7.14/130), was already observed in the early days of subdivision, and the search for good algorithms was sometimes guided by aiming at the optimal value $\varrho = 1$ [DS78, Loo87]. It was even conjectured (but of course soon disproven) that $\varrho = 1$ was *sufficient* for curvature continuity [BS90, Sto85]. Necessary conditions for boundedness of curvatures were discussed by Holt in [Hol96]. However, a rigorous proof must exclude degeneracy for generic initial data in the sense of Lemma 7.5/132. This proof appears first in [Rei07].

2. Algorithms generating surfaces with bounded curvature include those of Sabin [Sab91a], Holt [Hol96] and Loop [Loo02a], as well as the C^2-subdivision surfaces with enforced flat spots introduced by Prautzsch and Umlauf [PU98a, PU98b, PU00b], see remark after Theorem 7.15/141. Bounds on the curvature for such algorithms are derived by Peters and Umlauf in [PU01] based on tools for the analysis of the Gauss and mean curvature of subdivision surfaces in [PU00a].

3. The asymptotic expansions in Sect. 7.1/126 and most of the material in Sect. 7.2/134 follow ideas developed in [PR04]. In particular, the concept of the central ring, see Definition 7.2/128, is introduced there.

4. An alternative measure for assessing the variety of shapes that can be generated by a subdivision algorithm is 'r-flexibility', as introduced in [PR99].

5. Amazingly, the inability of the standard Catmull–Clark algorithm to generate convex shape for $n \geq 5$, as explained by Theorem 7.10/136, was not discovered before [PR04]. A detailed study of shape properties and suggestions for tuning Catmull–Clark subdivision can be found in [KPR04].

6. Theorem 7.6/132 is due to Reif and Schröder [RS01]. It is essential for using subdivision surfaces for Finite element simulation of thin shells and plates [CSA$^+$02, COS00].

7. While the original paper by Karčiauskas, Peters and Reif [KPR04] used n-sided *shape charts*, Augsdörfer, Dodgson and Sabin [ADS06] switched to equivalent circular charts. Hartmann [Har05] as well as Ginkel and Umlauf [GU08] took advantage of symmetries to reduce the chart to one sector for even valences and a half-sector for odd valences. Shape charts have been used to improve the curvature behavior at the central point [ADS06, GU06a, GU06b, GU07b]. Ginkel's thesis points out the remarkable fact that a large class of algorithms cannot handle high-valence satisfactorily: no matter how the input data are adjusted, there is no purely convex or a purely hyperbolic neighborhood of the singularity.

8. The C_2^k-criterion of Theorem 7.16/143 appears in similar form in [Pra98]. In [BK04], Barthe and Kobbelt propose tuning subdivision to approximate the conditions of the theorem.

9. Sabin [Sab91a] is the first to argue that Catmull–Clark and other piecewise bicubic algorithms that generate C^2-rings cannot be C^2 at the central point. The precise degree estimate Theorem 7.19/144 is due to Reif [Rei96a]. It is valid not only

for stationary algorithms, but for any scheme generating piecewise polynomial surfaces with the given structure of rings. Generalized degree estimates can be found in [PR99]. A stunning circumvention of the seemingly solid degree barrier has recently been discovered by Karčiauskas and Peters [KP05]. The idea is to increase the number of segments when approaching the singularity.

10. The first C_2^k-subdivision algorithms where constructed by Reif [Rei96b, Rei98] for TURBS, and by Prautzsch [Pra97] for Freeform splines. Although being impeccable from a theoretical point of view, both algorithms failed to gain much popularity. While, for TURBS, this is easily explained by a substantial lack of visual fairness, Freeform splines should have deserved better.

11. The Guided subdivision of Karčiauskas and Peters [KP05, KP07b, KMP06] is based on a fundamental revision and generalization of the ideas behind TURBS and Freeform splines. It combines analytical smoothness with high visual fairness, and should be considered a prototype for a new generation of premium algorithms. The development of the general PTER-framework, at present unpublished, was inspired by the concept of Guided subdivision. In particular, the proof of Lemma 7.22/146 uses ideas appearing already in [KMP06].

Chapter 8
Approximation and Linear Independence

In this chapter, we elaborate on two aspects of subdivision which, besides smoothness, are very important for applications – convergence of sequences of so-called proxy surfaces, such as control polyhedra, and linear independence of generating splines.

In Sect. 8.1/157, we consider a sequence $\{\check{\mathbf{x}}^k\}_k$ of *proxy surfaces* to a subdivision surface $\mathbf{x}$. For example, piecewise linear proxy surfaces arise as 'control polyhedra' in whatever sence, or as a sequence of finer and finer piecewise linear interpolants of $\mathbf{x}$. The analysis to be developed is, however, sufficiently general to cover cases where the proxy surfaces consist of non-linear pieces, for instance, when approximating $\mathbf{x}$ by an increasing, but finite number of polynomial patches. We derive upper bounds on the parametric and geometric distance between $\check{\mathbf{x}}^k$ and $\mathbf{x}$, which are asymptotically sharp up to constants as $k \to \infty$. Our results show that the rate of convergence of the geometric distance, which is crucial for applications in Computer Graphics, depends on the subsubdominant eigenvalue μ.

In Sect. 8.2/169, we consider the question of local and global linear independence of the generating splines $B = [b_0, \ldots, b_{\bar{\ell}}]$. This topic is closely related to the existence and uniqueness of solutions of approximation problems in spaces of subdivision surfaces, such as interpolation or fairing. We show that local linear independence cannot be expected if the valence n is high, and that even global linear independence is lost in special situation, like Catmull–Clark subdivision for a control net with the combinatorial structure of a cube.

8.1 Proxy Splines

Much of the appeal of subdivision lies in the intuitive refinability of a faceted proxy representation that outlines the limit surface ever more closely (see e.g. Fig. 1.1/2): for applications such as rendering on the computer a piecewise linear polyhedron (Fig. 8.1/158, *left*) is displayed in place of the limit surface (Fig. 8.1/158, *right*). Evidently, too few subdivision steps result in a polyhedron that visibly lacks

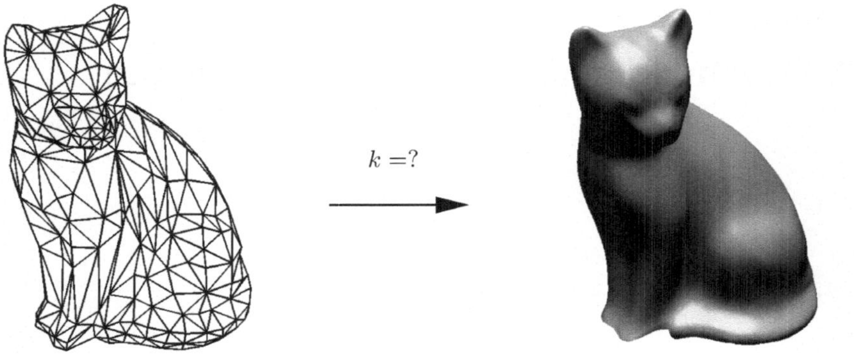

Fig. 8.1 Illustration of approximation by a proxy surface: How many steps k of control mesh refinement are required to ensure an approximation of the subdivision surface with a given accuracy?

smoothness, while too many steps are costly due to exponential growth of the number of facets under uniform refinement. For example, estimating the number to be $k = 10$ when the sharp estimate is $k = 5$ means computing and processing millions instead of thousands of facets for each original facet.

The (maximal) distance between the polyhedron and the limit surface can be measured numerically, after each refinement step using bounds on the generating rings. For many applications, however, it is of interest to obtain tight *a priori* bounds to allocate resources or to guide adaptivity.

In applications, a *control polyhedron* of a subdivision surface $\mathbf{x}$ is a piecewise linear or bi-linear surface that interpolates the control points and thereby approximates $\mathbf{x}$. The following definition addresses this approximation as a special case of a much larger class of *proxy surfaces*. Other instances of proxy surfaces include the piecewise linear or bi-linear interpolant of the subdivision surface $\mathbf{x}$, or the approximation by an increasing, but finite number of polynomial patches.

Let us consider a subdivision surface $\mathbf{x} = B\mathbf{Q} \in C^0(\mathbf{S}_n, \mathbb{R}^3)$. Then a sequence $\check{\mathbf{x}}_k = \check{B}_k\mathbf{Q}$ of *proxy surfaces* is defined by vectors $\check{B}_k = [\check{b}_{0,k}, \dots, \check{b}_{\bar{\ell},k}]$ of *generating proxy splines* $\check{b}_{\ell,k} \in C^0(\mathbf{S}_n, \mathbb{R})$. Typically, the functions $\check{b}_\ell$ are piecewise linear or bi-linear, but this is not assumed in what follows. The index k is called the *level* of the proxy surface and may be thought of as the 'budget' allocated for the approximation. As k increases, the tessellation of the splines $\check{b}_{\ell,k}$ becomes finer and finer. Therefore, incrementing k is also called *refinement* of proxy surfaces. Typically, under refinement, $\check{\mathbf{x}}_k$ rapidly approaches $\mathbf{x}$. Therefore, in applications such as rendering of surfaces, $\mathbf{x}$ is frequently not evaluated pointwise, but replaced by the faceted surface $\check{\mathbf{x}}_k$ for some modest value of the level k. When represented in subdivision form (4.6/62), we denote the rings of the spline $\check{\mathbf{x}}_k$ by

$$\check{\mathbf{x}}_k^m := \check{B}_k^m\mathbf{Q} \in C^0(\mathbf{S}_n^0, \mathbb{R}^d).$$

The following three properties are typical for sequences of control polyhedra, and will serve to define more general systems of generating proxy splines:

- The generating proxy splines form a partition of unity,

$$\sum_{\ell=0}^{\bar\ell} \breve b_{\ell,k} = 1, \quad k \in \mathbb{N}_0.$$

- The $(m+1)$-st ring $\breve{\mathbf{x}}_{k+1}^{m+1}$ of level $k+1$ equals the mth ring of level k corresponding to iterated coefficients $A\mathbf{Q}$, i.e.,

$$\breve{\mathbf{x}}_{k+1}^{m+1} = \breve B_{k+1}^{m+1}\mathbf{Q} = \breve B_k^m A\mathbf{Q}, \quad m,k \in \mathbb{N}_0.$$

- For increasing level k, i.e. repeated refinement, the outermost ring $\breve{\mathbf{x}}_k^0$ converges to $\mathbf{x}^0$ with respect to the maximal Euclidean distance according to

$$\|\mathbf{x}^0 - \breve{\mathbf{x}}_k^0\|_\infty \le cq^k|\mathbf{Q}|$$

  for constants $c > 0$, $q \in (0,1)$, and some semi-norm $|\cdot|$. Estimates of that type are well understood, and in particular, c and q can be determined reliably. In the case of binary refinement, the typical value of q is $q = 1/4$. The semi-norm $|\cdot|$ typically involves certain first or second order differences of coefficients.

Before we proceed, we introduce basic notation required here. For a vector $H = [h_0,\dots,h_{\bar\ell}]$ of splines or rings, let

$$\|H\|_\infty := \sum_{\ell=0}^{\bar\ell} \|h_\ell\|_\infty$$

denote the sum of sup-norms on the respective domain. For vectors $\mathbf{R} = [\mathbf{r}_0;\dots;\mathbf{r}_{\bar\ell}]$ of points in $\mathbb{R}^d$, and $0 \le i \le \bar\ell$, we define the semi-norm

$$|\mathbf{R}|_i := \max_{i\le\ell\le\bar\ell} \|\mathbf{r}_\ell\|$$

as the maximal Euclidean norm of $\mathbf{r}_i,\dots,\mathbf{r}_{\bar\ell}$. If the first i entries of H vanish, i.e., $h_0 = \cdots = h_{i-1} = 0$, then the estimate

$$\|H\mathbf{R}\|_\infty \le \|H\|_\infty\,|\mathbf{R}|_i \tag{8.1}$$

applies. Further, if M is a square matrix with row-sum norm $\|M\|$, then the vector of functions HM is bounded by

$$\|HM\|_\infty \le \|H\|_\infty\,\|M\|. \tag{8.2}$$

Now, we formalize the notion of a proxy spline and show both lower and upper bounds on its convergence to the corresponding subdivision surface. As the name indicates, proxy splines have the same structure as the splines underlying subdivision.

Definition 8.1 (Proxy spline). The splines $\breve B_k = [\breve b_{0,k},\dots,\breve b_{\bar\ell,k}]$, $\breve b_{\ell,k} \in C^0(\mathbf{S}_n,\mathbb{R})$, form a sequence of *generating proxy splines* of the standard subdivision algorithm

(A, G) if there exist constants $c_B > 0$ and $q \in (0,1)$ independent of k so that

$$\sum_{\ell=0}^{\bar\ell} \breve b_{\ell,k} = 1, \quad \breve B_{k+1}^{m+1} = \breve B_k^m A, \quad \|B^0 - \breve B_k^0\|_\infty \le c_B q^k \tag{8.3}$$

for all $m, k \in \mathbb{N}_0$. Then $\breve{\mathbf{x}}_k := \breve B_k \mathbf{Q}$ is a *level k proxy spline* of the spline $\mathbf{x} = B\mathbf{Q}$. Further, we say that the sequence $\breve B_k$ has a *uniformly bounded gradient*, if all rings $\breve b_{\ell,k}^0$ consist of finitely many differentiable pieces, and there exists a constant c' independent of ℓ and k such that $\|D\breve b_{\ell,k}^0\|_\infty \le c'$, wherever $D\breve b_{\ell,k}^0$ is defined.

The third condition in (8.3/160) says that the generating rings $G = B^0$ are approximated with order q^k by the *generating proxy rings*

$$\breve G_k := \breve B_k^0$$

as k tends to infinity. The typical value for binary refinement is $q = 1/4$. The condition concerning uniform boundedness of gradients is satisfied if, for instance, the gradients of generating proxy rings are converging uniformly,

$$\|DG - D\breve G_k\|_\infty \to 0.$$

This property is typically satisfied for control polyhedra, and also for piecewise linear interpolants.

Analogous to the eigenrings $F := GV$, we define *proxy eigenrings* $\breve F_k$, *eigensplines* E, *and proxy eigensplines* $\breve E_k$ by

$$\breve F_k := \breve G_k V, \quad E := BV, \quad \breve E_k := \breve B_k V,$$

respectively, where V is the matrix of eigenvectors of A yielding the Jordan decomposition $A = VJV^{-1}$. Hence, with the eigencoefficients $\mathbf{P} = V^{-1}\mathbf{Q}$,

$$\breve{\mathbf{x}}_k^m = \breve F_k J^m \mathbf{P}, \quad \mathbf{x} = E\mathbf{P}, \quad \breve{\mathbf{x}}_k = \breve B_k \mathbf{P}.$$

By the second relation in (8.3/160), we can express the ring $\breve{\mathbf{x}}_k^m$ either in terms of some proxy on the initial ring, or in terms of the initial proxy of some ring, depending on k being larger or smaller than m. More precisely,

$$\breve{\mathbf{x}}_k^m = \begin{cases} \breve B_{k-m}^0 A^m \mathbf{Q} = \breve E_{k-m}^0 J^m \mathbf{P} & \text{if } m \le k \\ \breve B_0^{m-k} A^k \mathbf{Q} = \breve E_0^{m-k} J^k \mathbf{P} & \text{if } m \ge k. \end{cases} \tag{8.4}$$

First, we consider the *parametric deviation* of a spline $\mathbf{x}$ from its proxy $\breve{\mathbf{x}}_k$,

$$\mathbf{d}_k := \mathbf{x} - \breve{\mathbf{x}}_k,$$

and derive a bound on the maximum of the Euclidean norm,

$$\|\mathbf{d}_k\|_\infty := \max_{\mathbf{s} \in \mathbf{S}_n} \|\mathbf{d}_k(\mathbf{s})\|.$$

This bound is defined in terms of the seminorm $|\mathbf{P}|_1$ in the space of eigencoefficients, and the constants

$$c_E := \|E - \breve{E}_0\|_\infty, \quad c_F^k := q^{-k}\,\|F - \breve{F}_k\|_\infty.$$

By (8.2/159) and the third property in (8.3/160), we have $\|F - \breve{F}_k\|_\infty = \|(B^0 - \breve{B}_k^0)V\|_\infty \le c_B q^k\,\|V\|$ showing that the sequence c_F^k is bounded by

$$c_F := \sup_k c_F^k \le c_B\,\|V\|,$$

where c_B is the constant defined in (8.3/160). While being crucial for the asymptotic behavior, this estimate of the c_F^k is far from being tight. Hence, in applications, these constants, or at least close upper bounds thereof, should be derived independently for some values of k.

Theorem 8.2 (Parametric distance to a proxy spline). *The parametric distance* $\mathbf{d}_k$ *between the spline* $\mathbf{x} = B\mathbf{Q}$ *and the proxy spline* $\breve{\mathbf{x}}_k = \breve{B}_k\mathbf{Q}$ *is bounded by*

$$\|\mathbf{d}_k\|_\infty \le \max\Bigl\{\max_{m\le k} c_F^{k-m} q^{k-m}\,|J^m\mathbf{P}|_1,\; c_E\,|J^k\mathbf{P}|_1\Bigr\}. \tag{8.5}$$

In particular, there exists a constant c such that

$$\|\mathbf{d}_k\|_\infty \le c\,\max\{\lambda^k, q^k\}\,|\mathbf{P}|_1. \tag{8.6}$$

Proof. Since $\breve{B}_k$ forms a partition of unity and the dominant eigenvector v_0 consists of all ones,

$$\breve{e}_{0,k} = \breve{B}_k v_0 = 1 = B v_0 = e_0. \tag{8.7}$$

Hence, the first component of $E - \breve{E}_k$ and also of $F - \breve{F}_k$ always vanishes. With $\mathbf{d}_k^m = (B^m - \breve{B}_k^m)\mathbf{Q}$ denoting the rings of $\mathbf{d}_k$,

$$\|\mathbf{d}_k\|_\infty = \sup_{m\in\mathbb{N}_0} \|\mathbf{d}_k^m\|_\infty.$$

Following (8.4/160), we distinguish two cases: For the outer rings with indices $m \le k$, the approximation is governed by the refinement of the proxy rings, while for the rings close to the central point with indices $m \ge k$, the convergence properties of subdivision are predominant.

If $m \le k$ then, using $B^m = GA^m$ and the estimate (8.1/159), we obtain the first maximand in (8.5/161),

$$\begin{aligned}\|\mathbf{d}_k^m\|_\infty &= \|(G - \breve{G}_{k-m})A^m\mathbf{Q}\|_\infty = \|(F - \breve{F}_{k-m})J^m\mathbf{P}\|_\infty\\ &\le \|F - \breve{F}_{k-m}\|_\infty\,|J^m\mathbf{P}|_1 \le c_F^{k-m} q^{k-m}\,|J^m\mathbf{P}|_1.\end{aligned}$$

To also verify (8.6/161) for $m \le k$, we partition J into blocks J_r as in Sect. 4.6/72, and obtain

$$|J^m\mathbf{P}|_1 \le \max_{1\le r\le \bar{r}} \|J_r^m\|_\infty\,|\mathbf{P}|_1. \tag{8.8}$$

For the subdominant blocks, the row-sum norm is $\|J_1^m\| = \|J_2^m\| = \lambda^m$, while the remaining ones correspond to eigenvalues with modulus $< \lambda$. Hence, there exists a constant c_J^1 so that $\|J_r^m\|_\infty \le c_J^1 \lambda^m$ for all $r \ge 1$. Since $c_F^{k-m} \le c_F$ and $q^{k-m}\lambda^m \le \max\{q^k, \lambda^k\}$, (8.6/161) follows for $m \le k$:

$$\|\mathbf{d}_k^m\|_\infty \le c_F c_J^1 \, q^{k-m}\lambda^m \, |\mathbf{P}|_1 \le c_F c_J^1 \max\{q^k, \lambda^k\} \, |\mathbf{P}|_1.$$

If $m \ge k$ then, using (8.7/161) and (8.1/159) again, we obtain the second maximand in (8.5/161),

$$\begin{aligned}\|\mathbf{d}_k^m\|_\infty &= \|(B^{m-k} - \breve{B}_0^{m-k})A^k\mathbf{Q}\|_\infty = \|(E^{m-k} - \breve{E}_0^{m-k})J^k\mathbf{P}\|_\infty \\ &\le \|E - \breve{E}_0\|_\infty \, |J^k\mathbf{P}|_1 = c_E \, |J^k\mathbf{P}|_1.\end{aligned}$$

Combining the results of the two cases, we obtain (8.5/161). Using the bound (8.8/161) on $|J^k\mathbf{P}|_1$, we also obtain

$$\|\mathbf{d}_k^m\|_\infty \le c_E c_J^1 \, \lambda^k \, |\mathbf{P}|_1.$$

and prove (8.6/161) with $c := \max\{c_F c_J^1, c_E c_J^1\}$. □

Let us briefly discuss the result. The first maximand in (8.5/161) applies to the outer rings of the spline. For fixed m, the term q^{k-m} describes the convergence in the regular setting as $k \to \infty$. The second maximand applies to the inner rings, and in particular to the situation near the center. For its verification, we estimated the deviation $E^{m-k} - \breve{E}_0^{m-k}$ on the $(m-k)$th ring by the constant c_E, which is valid for all rings. When approaching the center, i.e., for $m \gg k$, this is pessimistic. Refined estimates for individual rings $\mathbf{d}_k^m$ can be derived, but here, we are only interested in the maximum over all rings. The second estimate (8.6/161) is in general also not sufficiently tight for applications but is primarily intended to provide insight into the asymptotic decay of the bound as $k \to \infty$. For typical values like $\lambda = 1/2$ and $q = 1/4$, the predicted convergence rate $1/2^k$ is much slower than the q^k-behavior of the proxy surface in the regular setting. The following example shows that, in general, this reduced order of convergence is not an overestimation, but accurately describes the actual deviation.

Example 8.3 (Characteristic proxy spline). According to Definition 5.4/85, the characteristic spline

$$\chi := B[v_1, v_2] = B\mathbf{Q}'$$

of the standard algorithm (A, G) is the planar spline corresponding to the subdominant eigenvectors $\mathbf{Q}' := [v_1, v_2]$. By (5.5/85), its basic scaling property is

$$\chi(\mathbf{s}) = \lambda^{-m}\chi(2^{-m}\mathbf{s}). \tag{8.9}$$

The *characteristic proxy splines* and the *characteristic proxy rings* are defined by

$$\breve{\chi}_k := [\breve{e}_{1,k}, \breve{e}_{2,k}] = \breve{B}_k\mathbf{Q}', \quad \breve{\psi}_k := [\breve{f}_{1,k}, \breve{f}_{2,k}] = \breve{G}_k\mathbf{Q}'. \tag{8.10}$$

Denote the deviation by

$$\boldsymbol{\delta}_k := \boldsymbol{\chi} - \check{\boldsymbol{\chi}}_k = (B - \check{B}_k)\mathbf{Q}'.$$

For $k \geq m$, the deviation on ring m is

$$\begin{aligned}\boldsymbol{\delta}_k^m &= (B^m - \check{B}_k^m)\mathbf{Q}' = (B^0 - \check{B}_{k-m}^0)A^k\mathbf{Q}' \\ &= \lambda^k(G - \check{G}_{k-m})\mathbf{Q}' = \lambda^k(\boldsymbol{\psi} - \check{\boldsymbol{\psi}}_{k-m}).\end{aligned}$$

The vector $\mathbf{P}' = [\delta_1, \delta_2]$ of eigencoefficients consists of two unit vectors and has semi-norm $|\mathbf{P}'|_1 = 1$. Hence, (8.6/161) yields the estimate

$$\|\boldsymbol{\chi} - \check{\boldsymbol{\chi}}_k\| \leq c\max\{\lambda^k, q^k\}. \tag{8.11}$$

In fact, for $k = m$,

$$\boldsymbol{\delta}_k^k = \lambda^k(\boldsymbol{\psi} - \check{\boldsymbol{\psi}}_0) = \lambda^k\boldsymbol{\delta}_0^0.$$

So, unless the characteristic proxy ring $\check{\boldsymbol{\psi}}_0$ reproduces the characteristic ring $\boldsymbol{\psi}$, we have $\boldsymbol{\delta}_0^0 = \boldsymbol{\psi} - \check{\boldsymbol{\psi}}_0 \neq 0$ and $\|\boldsymbol{\delta}_k\|$ converges no faster than λ^k.

On the other hand, for a fixed index m and level k tending to infinity, the convergence cannot be faster than q^k given in Definition 8.1/159. This may not be a tight estimate in general, but for $q < \lambda$ the estimate

$$\|\boldsymbol{\delta}_k^m\|_\infty = \lambda^m \, \|(F - \check{F}_{k-m})\mathbf{P}'\|_\infty \leq \lambda^m \, c_F q^{k-m} \tag{8.12}$$

establishes a much faster decay than λ^k, anyway. □

If $\mathbf{x} \in C^0(\mathbf{S}_n, \mathbb{R}^3)$ is a spline surface, then in many applications the parametric deviation from $\check{\mathbf{x}}_k$ is not important. Instead, the *geometric deviation* is used to judge the quality of approximation. Among the various possibilities to define that quantity, we choose the Hausdorff distance

$$\begin{aligned}d(\mathbf{x}, \check{\mathbf{x}}_k) := \max\bigl\{&\max_{\mathbf{s}\in\mathbf{S}_n}\min_{\mathbf{s}_k\in\mathbf{S}_n}\|\mathbf{x}(\mathbf{s}) - \check{\mathbf{x}}_k(\mathbf{s}_k)\|, \\ &\max_{\mathbf{s}_k\in\mathbf{S}_n}\min_{\mathbf{s}\in\mathbf{S}_n}\|\mathbf{x}(\mathbf{s}) - \check{\mathbf{x}}_k(\mathbf{s}_k)\|\bigr\}.\end{aligned}$$

Obviously, $d(\mathbf{x}, \check{\mathbf{x}}_k) \leq \|\mathbf{d}_k\|_\infty$ so that we can expect tighter bounds than before.

Given a parameter $\mathbf{s}$, it is typically very difficult to find $\mathbf{s}_k$ such that $\|\mathbf{x}(\mathbf{s}) - \check{\mathbf{x}}_k(\mathbf{s}_k)\|$ is actually minimized, and vice versa. Instead, we specify a fixed sequence of relations $\mathbf{s}_k := \mathbf{r}_k(\mathbf{s})$, define the *reparametrized proxy surface* $\bar{\mathbf{x}} := \check{\mathbf{x}}_k \circ \mathbf{r}_k$, and estimate

$$d(\mathbf{x}, \check{\mathbf{x}}_k) \leq \|\bar{\mathbf{d}}_k\|_\infty, \quad \bar{\mathbf{d}}_k := \mathbf{x} - \bar{\mathbf{x}}_k.$$

The reparametrizations $\mathbf{r}_k$ are chosen so that the convergence rate near the center is raised from λ^k to μ^k, where μ is the subsubdominant eigenvalue of the algorithm.

To avoid discussing degenerate cases, such as the image of the characteristic proxy spline $\check{\boldsymbol{\chi}}_k$ shrunk to a point, we make the following assumption: there exists

$\bar{m} \in \mathbb{N}_0$ such that

$$\lambda^m \boldsymbol{\chi}(\mathbf{S}_n) \subset \check{\boldsymbol{\chi}}_k(\mathbf{S}_n) \quad \text{for all} \quad k \in \mathbb{N}_0 \quad \text{and} \quad m \geq \bar{m}.$$

That is, the image of all rings $\boldsymbol{\chi}^m$ of the characteristic spline $\boldsymbol{\chi} = E[v_1, v_2]$ with index $m \geq \bar{m}$ is contained in the image of all characteristic proxy surfaces. In most applications, the inclusion holds already for $\bar{m} = 0$ or $\bar{m} = 1$.

We say that the parameter $\mathrm{s} \in \mathbf{S}_n$ *has ring index* m if $2^m \mathrm{s} \in \mathbf{S}_n^0$ so that $\mathbf{x}(\mathrm{s}) = \mathbf{x}^m(2^m \mathrm{s})$. For a parameter s with ring index $m \geq \bar{m}$, we have $2^{\bar{m}} \mathrm{s} \in \mathbf{S}_n$. Then, depending on the level $k \in \mathbb{N}_0$, there exists $\mathrm{s}_k \in \mathbf{S}_n$ with

$$\check{\boldsymbol{\chi}}_k(\mathrm{s}_k) = \lambda^{\bar{m}} \boldsymbol{\chi}(2^{\bar{m}} \mathrm{s}) = \boldsymbol{\chi}(\mathrm{s}).$$

If $\check{\boldsymbol{\chi}}_k$ is not injective, then s_k is not necessarily unique. However, by arbitrary choice, we can define a (possibly discontinuous) function $\mathbf{r}_k : \mathbf{S}_n \to \mathbf{S}_n$ that maps s to a parameter s_k satisfying the above equation. If s lies in a ring with index $m < \bar{m}$, we simply set $\mathbf{r}_k(\mathrm{s}) = \mathrm{s}$. This yields the identity

$$\bar{\boldsymbol{\chi}}_k(\mathrm{s}) := \check{\boldsymbol{\chi}}_k\big(\mathbf{r}_k(\mathrm{s})\big) = \begin{cases} \boldsymbol{\chi}(\mathrm{s}) & \text{if } 2^{\bar{m}} \mathrm{s} \in \mathbf{S}_n \\ \check{\boldsymbol{\chi}}_k(\mathrm{s}) & \text{if } 2^{\bar{m}} \mathrm{s} \notin \mathbf{S}_n, \end{cases} \tag{8.13}$$

which characterizes $\mathbf{r}_k$. We denote the reparametrized proxies of eigensplines by

$$\bar{E}_k := \check{E}_k \circ \mathbf{r}_k,$$

and define the constants

$$\bar{c}_E := \|E - \bar{E}_0\|_\infty, \quad \bar{c}_F^k := q^{-k} \|E^{\bar{m}} - \bar{E}_k^{\bar{m}}\|_\infty.$$

If $\bar{m} = 0$, as in many applications, then $e_\ell^0 = f_\ell$ are the eigenrings and $\bar{f}_{\ell,k} := \bar{e}_{\ell,0}^0$ their reparametrized level k proxies. Proving the boundedness of the sequence of constants $\bar{c}_F^k$ is the main challenge in the reparametrized setting since it requires verifying the decay of the reparametrized distance $\|E^{\bar{m}} - \bar{E}_k^{\bar{m}}\|_\infty$ by at least q^k even when the parameterization is different from the one used to define q.

Lemma 8.4 (Boundedness of $\bar{c}_F^k$). *If the sequence $\check{B}_k$ of generating proxy splines has uniformly bounded gradient, then the sequence $\bar{c}_F^k$ is bounded, i.e.,*

$$\bar{c}_F := \sup_k \bar{c}_F^k < \infty.$$

Proof. We have

$$\bar{c}_F^k = q^{-k} \|E^{\bar{m}} - \bar{E}_k^{\bar{m}}\|_\infty \leq q^{-k} \|E^{\bar{m}} - \check{E}_k^{\bar{m}}\|_\infty + q^{-k} \|\check{E}_k^{\bar{m}} - \bar{E}_k^{\bar{m}}\|_\infty$$

and may assume $k \geq \bar{m}$ since $\bar{m}$ is fixed. The first summand is bounded since $E^{\bar{m}} - \check{E}_k^{\bar{m}} = (E^0 - \check{E}_{k-\bar{m}}^0) J^{\bar{m}} = (F - \check{F}_{k-\bar{m}}) J^{\bar{m}}$, and hence

$$q^{-k} \|E^{\bar{m}} - \check{E}_k^{\bar{m}}\|_\infty = q^{-k} \|(F - \check{F}_{k-\bar{m}}) J^{\bar{m}}\|_\infty \leq c_F \, \|J^{\bar{m}}\|.$$

To estimate the second summand, we consider an arbitrary parameter $\mathbf{s} \in \mathbf{S}_n$ with ring index $\bar{m}$ and denote the index of the corresponding parameter $\mathbf{s}_k := \mathbf{r}_k(\mathbf{s})$ by m_k so that

$$\mathbf{s}^0 := 2^{\bar{m}}\mathbf{s} \in \mathbf{S}_n^0, \quad \text{and} \quad \mathbf{s}_k^0 := 2^{m_k} \in \mathbf{S}_n^0.$$

We will show that there exist constants m_* and c_L so that for each component of the summand and any $\mathbf{s}^0 \in \mathbf{S}_n^0$,

$$q^{-k}|\check{e}_{\ell,k}^{\bar{m}}(\mathbf{s}^0) - \bar{e}_{\ell,k}^{\bar{m}}(\mathbf{s}^0)| \le c_L c_F\, q^{-m_*}. \tag{8.14}$$

First, we show that for sufficiently large k the index m_k is bounded by some number m_* that is independent of $\mathbf{s}$. By definition of $\mathbf{r}_k$ (8.13/164) and by (8.11/163), we obtain the upper bound

$$\|\boldsymbol{\chi}(\mathbf{s}) - \boldsymbol{\chi}(\mathbf{s}_k)\|_\infty = \|\check{\boldsymbol{\chi}}_k(\mathbf{s}_k) - \boldsymbol{\chi}(\mathbf{s}_k)\|_\infty \le c\max\{\lambda^k, q^k\}.$$

To obtain a lower bound, we observe that, by (8.9/162),

$$\boldsymbol{\chi}(\mathbf{s}) - \boldsymbol{\chi}(\mathbf{s}_k) = \lambda^{\bar{m}}\boldsymbol{\psi}(\mathbf{s}^0) - \lambda^{m_k}\boldsymbol{\psi}(\mathbf{s}_k^0).$$

Also the image of $\boldsymbol{\psi}$ is compact and does not contain the origin, so that there exist constants $r, R > 0$ with $r \le \|\boldsymbol{\psi}(\tilde{\mathbf{s}})\| \le R$ for all $\tilde{\mathbf{s}} \in \mathbf{S}_n^0$. Hence,

$$\|\boldsymbol{\chi}(\mathbf{s}) - \boldsymbol{\chi}(\mathbf{s}_k)\|_\infty \ge \|\lambda^{\bar{m}}\boldsymbol{\psi}(\mathbf{s}^0)\|_\infty - \|\lambda^{m_k}\boldsymbol{\psi}(\mathbf{s}_k^0)\|_\infty \ge r\lambda^{\bar{m}} - R\lambda^{m_k}.$$

Combining the upper and lower bound on $\|\boldsymbol{\chi}(\mathbf{s}) - \boldsymbol{\chi}(\mathbf{s}_k)\|_\infty$, we obtain

$$r\lambda^{\bar{m}} - R\lambda^{m_k} \le c\max\{\lambda^k, q^k\}.$$

There exists $k_* \in \mathbb{N}$ such that $c\max\{\lambda^k, q^k\} \le r\lambda^{\bar{m}}/2$ for all $k \ge k_*$. Hence,

$$R\lambda^{m_k} \ge r\lambda^{\bar{m}}/2, \quad k \ge k_*,$$

shows that m_k cannot be arbitrarily large. More precisely, there exists a constant m_* independent of $\mathbf{s}$ with $m_k \le m_*$ for all $k \ge k_*$.

We now derive the constant c_L of uniform Lipschitz continuity of the functions $\check{E}_k \circ \boldsymbol{\chi}^{-1}$ on a certain part of their domain. Let $\boldsymbol{\chi}_*$ and $\check{E}_{*,k} = [\check{e}_{*,0,k}, \ldots, \check{e}_{*,\bar{\ell},k}]$ denote the restriction of $\boldsymbol{\chi}$ and $\check{E}_k$ to arguments with ring index $m \le m_*$, respectively. For level $k \ge m_*$ and a parameter $\mathbf{s}'$ with ring index $m \le m_*$, we obtain using the second property of (8.3/160)

$$D\check{E}_{*,k}(\mathbf{s}') = D\check{E}_{*,k}^m(2^m\mathbf{s}') = D\check{B}_{*,k}^0(2^m\mathbf{s}')A^mV.$$

By assumption, the functions in $D\check{B}_{*,k}^0$ are uniformly bounded, and A^mV is one out of finitely many matrices since $m \le m_*$. So we conclude that $\check{E}_{*,k}$ has uniformly bounded gradient. Further, $\boldsymbol{\chi}_*^{-1}$ is a continuous and piecewise differentiable function with bounded Jacobian. From the latter two observations it follows that the functions $\breve{E}_{*,k} \circ \boldsymbol{\chi}_*^{-1}$ are uniformly Lipschitz continuous in the sense that there exists a constant c_L independent of ℓ and k with

$$\left|\breve{e}_{*,\ell,k}(\boldsymbol{\chi}_*^{-1}(\boldsymbol{\sigma})) - \breve{e}_{*,\ell,k}(\boldsymbol{\chi}_*^{-1}(\boldsymbol{\sigma}'))\right| \le c_L \, \|\boldsymbol{\sigma} - \boldsymbol{\sigma}'\|_\infty$$

for all $\boldsymbol{\sigma}, \boldsymbol{\sigma}'$ in the image of $\boldsymbol{\chi}_*$.

We now have the constants needed to establish (8.14/165). For a single component, we find

$$\begin{aligned}
|\breve{e}^{\bar{m}}_{\ell,k}(\mathbf{s}^0) - \bar{e}^{\bar{m}}_{\ell,k}(\mathbf{s}^0)| &= |\breve{e}_{*,\ell,k}(\mathbf{s}) - \breve{e}_{*,\ell,k}(\mathbf{s}_k)| \\
&= |\breve{e}_{*,\ell,k}(\boldsymbol{\chi}_*^{-1}(\boldsymbol{\chi}_*(\mathbf{s}))) - \breve{e}_{*,\ell,k}(\boldsymbol{\chi}_*^{-1}(\boldsymbol{\chi}_*(\mathbf{s}_k)))| \\
&\le c_L \, \|\boldsymbol{\chi}_*(\mathbf{s}) - \boldsymbol{\chi}_*(\mathbf{s}_k)\|_\infty \\
&= c_L \, \|\boldsymbol{\chi}(\mathbf{s}) - \boldsymbol{\chi}(\mathbf{s}_k)\|_\infty .
\end{aligned}$$

By (8.13/164), we can replace $\boldsymbol{\chi}(\mathbf{s})$ by $\breve{\boldsymbol{\chi}}_k(\mathbf{s}_k)$, and the estimate (8.12/163) yields

$$\|\boldsymbol{\chi}(\mathbf{s}) - \boldsymbol{\chi}(\mathbf{s}_k)\|_\infty = \|\breve{\boldsymbol{\chi}}_k(\mathbf{s}_k) - \boldsymbol{\chi}(\mathbf{s}_k)\|_\infty \le c_F q^{k-m_k} \le c_F q^{k-m_*}.$$

Together, we have (8.14/165):

$$q^{-k} |\breve{e}^{\bar{m}}_{\ell,k}(\mathbf{s}^0) - \bar{e}^{\bar{m}}_{\ell,k}(\mathbf{s}^0)| \le c_L c_F \, q^{-m_*}$$

for all $\mathbf{s}^0 \in \mathbf{S}^0_n$. Summation of the respective maxima over ℓ yields the desired bound

$$q^{-k} \|\breve{E}^{\bar{m}}_k - \bar{E}^{\bar{m}}_k\|_\infty \le (\bar{\ell} + 1) \, c_L c_F \, q^{-m_*}. \qquad \square$$

Now, we are prepared to estimate

$$\bar{\mathbf{d}}_k = (B - \breve{B}_k \circ \mathbf{r}_k)\mathbf{Q} = (E - \bar{E}_k)\mathbf{P},$$

which, in turn, is an upper bound on the geometric deviation $d(\mathbf{x}, \breve{\mathbf{x}}_k)$.

Theorem 8.5 (Geometric distance to proxy spline). *The geometric distance* $d(\breve{\mathbf{x}}_k, \mathbf{x})$ *between the spline surface* $\mathbf{x} = B\mathbf{Q}$ *and the proxy surface* $\breve{\mathbf{x}}_k = \breve{B}_k\mathbf{Q}$ *is bounded by*

$$d(\mathbf{x}, \breve{\mathbf{x}}_k) \le \min\{\|\mathbf{d}_k\|_\infty, \|\bar{\mathbf{d}}_k\|_\infty\},$$

where

$$\begin{aligned}
\|\bar{\mathbf{d}}_k\|_\infty \le \max\Big\{ &\max_{0 \le m < \bar{m}} c_F^{k-m} q^{k-m} \, |J^m \mathbf{P}|_1, \\
&\max_{\bar{m} \le m \le k+\bar{m}} \bar{c}_F^{k-m+\bar{m}} q^{k-m+\bar{m}} \, |J^{m-\bar{m}} \mathbf{P}|_3, \; \bar{c}_E \, |J^k \mathbf{P}|_3, \Big\}.
\end{aligned} \tag{8.15}$$

In particular, if the sequence $\breve{B}_k$ of generating proxy splines has uniformly bounded gradient, there exists a constant $\bar{c}$ such that

$$\|\bar{\mathbf{d}}_k\|_\infty \le \bar{c}\,\max\{\mu^k, q^k\}\,|\mathbf{P}|_1. \tag{8.16}$$

Proof. With $\bar{\mathbf{d}}_k^m = (B^m - \bar{B}_k^m)\mathbf{Q}$ denoting the rings composing $\bar{\mathbf{d}}_k$, it is

$$\|\bar{\mathbf{d}}_k\|_\infty = \sup_{m\in\mathbb{N}_0} \|\bar{\mathbf{d}}_k^m\|_\infty.$$

The three maximands in (8.15/166) refer to bounds on $\bar{\mathbf{d}}_k^m$ for m varying in three different intervals.

For $m < \bar{m}$, $\mathbf{r}$ is the identity so that $\bar{\mathbf{d}}_k^m$ and the parametric difference $\mathbf{d}_k^m$ coincide. The estimates of Theorem 8.2/161 apply and yield the first maximand. With the estimate $\|J^m\mathbf{P}\|_\infty \le c_J^1\lambda^m|\mathbf{P}|_1$, derived in the proof of Theorem 8.2/161, we obtain

$$\begin{aligned}\max_{m\le\bar{m}} c_F^{k-m}q^{k-m}\,|J^m\mathbf{P}|_1 &\le c_F c_J^1 \max_{m\le\bar{m}}(q^{k-m}\lambda^m)\,|\mathbf{P}|_1\\ &\le (c_F c_J^1 q^{-\bar{m}})q^k\,|\mathbf{P}|_1.\end{aligned}$$

To estimate the other two cases, we note that, as before, the first component of E and $\bar{E}_k$ coincides,

$$\bar{e}_{0,k} = \breve{e}_{0,k}\circ\mathbf{r}_k = 1\circ\mathbf{r}_k = 1 = e_0.$$

But moreover, the reparametrization (8.13/164) is designed so that the next two components coincide as well for rings with index $m \ge \bar{m}$,

$$[\bar{e}_{1,k}^m, \bar{e}_{2,k}^m] = \big([\breve{e}_{1,k}, \breve{e}_{2,k}]\circ\mathbf{r}_k\big)^m = (\breve{\chi}_k\circ\mathbf{r}_k)^m = \chi^m = [e_1^m, e_2^m].$$

In other words, the first *three* components of $E^m - \bar{E}_k^m$ vanish. The remaining part of the proof follows exactly the pattern developed in the parametric case. If $\bar{m} \le m \le k+\bar{m}$ then, abbreviating $m' := m - \bar{m}$,

$$\|\bar{\mathbf{d}}_k^m\|_\infty = \|(E^{\bar{m}} - \bar{E}_{k-m'}^{\bar{m}})J^{m'}\mathbf{P}\|_\infty \le \bar{c}_F^{k-m'}q^{k-m'}\,|J^{m'}\mathbf{P}|_3,$$

which is the second maximand in (8.15/166). The Jordan blocks $J_3,\dots,J_{\bar{\ell}}$ have either modulus μ and dimension 1, or they have modulus $<\mu$. Hence, there exists a constant c_J^3 with $\|J_r^m\| \le c_J^3\mu^m$ for all $r \ge 3$, and we find

$$\|\bar{\mathbf{d}}_k^m\|_\infty \le \bar{c}_F c_J^3\, q^{k-m'}\mu^{m'}\,|\mathbf{P}|_3 \le \bar{c}_F c_J^3\,\max\{\mu^k, q^k\}\,|\mathbf{P}|_3.$$

If $m \ge k+\bar{m}$ then

$$\|\bar{\mathbf{d}}_k^m\|_\infty = \|(E^{m-k} - \bar{E}_0^{m-k})J^k\mathbf{P}\|_\infty \le \bar{c}_E\,|J^k\mathbf{P}|_3$$

yields the third maximand in (8.15/166). Further, in this case,

$$\|\bar{\mathbf{d}}_k^m\|_\infty \le \bar{c}_E c_J^3\,\mu^k\,|\mathbf{P}|_3.$$

Combining the results of the three cases, we obtain (8.15/166) and, with

$$\bar{c} := \max\{c_F c_J^1 q^{-\bar{m}}, c_F c_J^3, \bar{c}_E c_J^3\}$$

and with $|\mathbf{P}|_3 \leq |\mathbf{P}|_1$ also (8.16/167). □

The first two maximands in (8.15/166) show that convergence on rings with fixed index m is governed by the factor q^k, as before. However, due the appropriate choice of the reparametrization, the decay near the central point is now of order μ^k as $k \to \infty$. Typically, for binary refinement, $q = \mu = 1/4$ so that quadratic convergence of the regular setting also applies for proxies of subdivision surfaces. If, as in many applications, $\bar{m} = 0$, then the bound given by the first maximand does not apply, and the factor $|\mathbf{P}|_1$ in (8.16/167) can be replaced by $|\mathbf{P}|_3$.

By means of a spline surface with eigencoefficients $\mathbf{p}_1 = [1,0,0]$, $\mathbf{p}_2 = [0,1,0]$, $\mathbf{p}_3 = [0,0,1]$ and $\mathbf{p}_\ell = [0,0,0]$ else, one can show that, in general, $\|\bar{\mathbf{d}}_k\|$ cannot decay faster than μ^k. The example below confirms that the same is true even for the Hausdorff distance.

*Example 8.6 (Hausdorff distance of Catmull–Clark control net).*Let us consider the Catmull–Clark algorithm according to Sect. 6.1/109 for $n = 3$ with modified weights

$$\alpha = 5/8,\ \beta = 1/2,\ \gamma = -1/8.$$

The leading eigenvalues are

$$\lambda_0 = 1,\ \lambda = \lambda_1^1 = \lambda_1^2 = \frac{9+\sqrt{17}}{32} \approx 0.41,\ \mu = \lambda_1^0 = \frac{3}{8} = 0.375.$$

In a vicinity of the center, the proxy surface $\check{\mathbf{x}}_k$ is given by the piecewise bilinear interpolant of the control points $\mathbf{Q}^k$ at level k. Let us consider a spline surface $\mathbf{x} = [\chi, z] = B[v_1, v_2, q_3]$, where the first two coordinates coincide with the characteristic spline, and the third coordinate z corresponds do initial data with the central control point set to 1, and all other 0. For the technical reasons described earlier, the central control point appears in threefold, i.e., $\mathbf{q}_{0,1} = \mathbf{q}_{1,1} = \mathbf{q}_{2,1} = [0,0,1]$ (see Fig. 6.3/111). The submatrix $\hat{A}_0^{0,0}$ has the eigenvalues $\lambda_0 = 0, \mu = 3/8$ and 0. Using the Jordan form, one can compute its powers and show that the central control point $\tilde{\mathbf{q}}^k$ at level $k \geq 1$ is

$$\mathbf{q}_{0,1}^k = \mathbf{q}_{1,1}^k = \mathbf{q}_{2,1}^k := \tilde{\mathbf{q}}^k, \quad \tilde{\mathbf{q}}^k := [0,0,1+\mu^k/3].$$

These points are converging to the central point

$$\mathbf{x}^c = \lim_{k\to\infty} \tilde{\mathbf{q}}^k = [0,0,1].$$

To determine a lower bound on the Hausdorff distance $d(\mathbf{x}, \check{\mathbf{x}}_k)$, we consider the point $\tilde{\mathbf{q}}^k = \check{\mathbf{x}}_k(\mathbf{0})$ on $\check{\mathbf{x}}_k$. We have

$$d(\mathbf{x}, \check{\mathbf{x}}_k) \geq \min_{\mathbf{s}\in\mathbf{S}_n} \|\mathbf{x}(\mathbf{s}) - \tilde{\mathbf{q}}^k\|,$$

and claim that the minimum is attained for $\mathbf{s} = \mathbf{0}$, i.e.,

$$\min_{\mathbf{s}\in\mathbf{S}_n} \|\mathbf{x}(\mathbf{s}) - \tilde{\mathbf{q}}^k\| = \|\mathbf{x}^{\mathrm{c}} - \tilde{\mathbf{q}}^k\| = \mu^k/3.$$

To show this, we note that the third component z of $\mathbf{x}$ attains its maximum at the center, and this value is less than the third component $1 + \mu^k$ of $\tilde{\mathbf{q}}^k$. Hence, the moduli of all three components of $\mathbf{x}^{\mathrm{c}} - \tilde{\mathbf{q}}^k$ are minimized simultaneously for $\mathbf{s} = \mathbf{0}$. Thus,

$$d(\mathbf{x}, \check{\mathbf{x}}_k) \geq \mu^k/3$$

shows that the order of convergence of the Hausdorff distance cannot exceed the bound given in the theorem. □

8.2 Local and Global Linear Independence

In this section, we take a closer look at the independence of generating splines. This topic is less important for geometric properties of subdivision surfaces, but is crucial for the existence and uniqueness of solutions of interpolation or approximation problems in subdivision spaces. So far, we know by Lemma 4.25/78 that the eigensplines corresponding to non-zero eigenvalues are linearly independent on the domain $\mathbf{S}_n$ if there are no ineffective eigenvectors. Beyond that observation, a local and a global aspect of the problem are relevant:

First, we discuss *local* linear independence. Here, we consider subsets $\bar{\mathbf{S}} \subset \mathbf{S}_n$ with an non-empty interior and ask whether all generating splines with support on $\bar{\mathbf{S}}$ are linearly independent when restricted to that subset. Surprisingly, the answer may differ from that for B-splines, even when the subdivision surfaces are derived from them.

Second, we briefly comment on the situation when stationary subdivision, represented by the powers of A, is preceded by a few initial steps which differ from that rule to account for, say, neighboring extraordinary vertices. When starting from a *global* control net, corresponding to a global subdivision surface defined on a superset $\mathbf{S} \supset \mathbf{S}_n$ of the n-cell domain $\mathbf{S}_n$, the initial steps split the configuration until the local, stationary rules are applicable. While the finitely many initial steps are typically irrelevant for smoothness properties, they are essential for shape properties, and they can also lead to a globally non-injective relation between the initial control points and the resulting surfaces.

Let us start with defining local linear dependence:

Definition 8.7 (Local linear dependence). Let B be the sequence of generating splines of a subdivision algorithm. Given a subset $\bar{\mathbf{S}} \subset \mathbf{S}_n$, we denote by

$$B_{|\bar{\mathbf{S}}} := \{b_{\ell|\bar{\mathbf{S}}} : b_\ell \neq 0 \text{ on } \bar{\mathbf{S}}\}$$

the restriction of all non-vanishing generating splines to this subset. The generating splines B are called *locally linearly dependent*, if there exists a subset $\bar{\mathbf{S}} \subset \mathbf{S}_n$ with non-empty interior such that the functions $B_{\bar{\mathbf{S}}}$ are linearly dependent, i.e.,

$$\#B_{\bar{\mathbf{S}}} > \operatorname{rank} B_{\bar{\mathbf{S}}}.$$

If no such subset $\bar{\mathbf{S}}$ exists, then B is called *locally linearly independent.*

Local linear independence is neither a stronger nor a weaker property than linear dependence in the usual sense. As a simple univariate example, consider the system $[u, |u|], u \in [-1, 1]$. It is linearly independent, but locally linearly dependent since we may choose $\bar{\mathbf{S}} = [0, 1]$. By contrast, the system $[0, u]$ is linearly dependent, but locally linearly independent since the 0-function is disregarded.

If the generating rings G are linearly dependent, then the generating splines B are locally linearly dependent. To show this, we may simply choose $\bar{\mathbf{S}} := \mathbf{S}_n^0$. However, if G is linearly independent, B can still be locally linearly dependent. To discuss this point, we introduce the following terms:

- A generating spline b_ℓ *covers the central knot* $\mathbf{0}$, if $\mathbf{0}$ is an interior point of its support. The set of all such b_ℓ is denoted by

$$B^{\mathbf{0}} := \{b_\ell : b_\ell \text{ covers } \mathbf{0}\}.$$

- The *local rank* $\operatorname{loc} G$ of G is defined as the minimal rank that can be obtained when restricting G to an open subset $\bar{\mathbf{S}} \subset \mathbf{S}_n^0$,

$$\operatorname{loc} G := \min\{\operatorname{rank} G_{|\bar{\mathbf{S}}} : \bar{\mathbf{S}} \text{ is open in } \mathbf{S}_n^0\}.$$

For instance, for the Catmull–Clark algorithm, the number of generating splines covering the central knot is $\#B^{\mathbf{0}} = 2n + 1$. The generating rings are piecewise bicubic so that we have the local rank $\operatorname{loc} G = 4 \times 4 = 16$. For $n > 7$, the first number exceeds the second one so that, as shown below, local linear independence is lost. This situation is typical for many families of algorithms: The number of generating splines covering the central knot *grows* with n, while the local rank does *not change* with n. For symmetric algorithms, this inevitably leads to local linear dependence for sufficiently large n.

Theorem 8.8 (Local linear dependence). *Consider a subdivision algorithm* (A, G) *with generating splines* B *according to Definition 4.12/68. If the number of generating splines covering* $\mathbf{0}$ *exceeds the local rank,*

$$\#B^{\mathbf{0}} > \operatorname{loc} G,$$

then B *is locally linearly dependent. In particular, for a shift invariant standard algorithm according to Definition 5.13/96,* B *is locally linearly dependent if*

$$n > \operatorname{loc} G.$$

Proof. The intersection of the finitely many generating splines covering the central knot contains a neighborhood of the central knot. Hence, there exists $m \in \mathbb{N}$ such that

$$\mathbf{S}_n^m \subset \bigcap_{b_\ell \in B^\mathbf{o}} \operatorname{supp} b_\ell.$$

Let $\bar{\mathbf{S}} \subset \mathbf{S}_n^0$ be a subset with $\operatorname{loc} G = \operatorname{rank} G_{|\bar{\mathbf{S}}}$, and define $\bar{\mathbf{S}}^m := 2^{-m}\bar{\mathbf{S}}$. We consider the set $B_{\bar{\mathbf{S}}^m}$ of generating splines that do not vanish on $\bar{\mathbf{S}}^m$. On one hand, because $b_\ell(2^{-m}\mathbf{s}) = g_\ell(\mathbf{s})$ for $\mathbf{s} \in \bar{\mathbf{S}}$, we have $\operatorname{rank} B_{\bar{\mathbf{S}}^m} = \operatorname{loc} G$. On the other hand, because $\bar{\mathbf{S}}^m \subset \mathbf{S}_n^m$, none of the functions in $B^\mathbf{o}$ vanishes on $\bar{\mathbf{S}}^m$. Hence, $B_{\bar{\mathbf{S}}^m}$ contains at least as many functions as $B^\mathbf{o}$, and

$$\#B_{\bar{\mathbf{S}}^m} \geq \#B^\mathbf{o} > \operatorname{loc} G = \operatorname{rank} B_{\bar{\mathbf{S}}^m}$$

shows that $B_{\bar{\mathbf{S}}^m}$ is linearly dependent.

To prove the second assertion, we consider initial data $\mathbf{Q}[\ell] := [v_1, v_2, \delta_\ell]$, where v_1, v_2 are the subdominant eigenvectors of A, and δ_ℓ is the ℓth unit vector. The corresponding subdivision surface $\mathbf{x}[\ell] = B\mathbf{Q}[\ell]$ is normal continuous with central normal $\mathbf{n}^c[\ell]$, and its third component is $b_\ell = B\delta_\ell$. Let us assume that none of these generating splines is covering the central knot. Then $\mathbf{n}^c[\ell] = [0, 0, \pm 1]$ for all ℓ. Hence, for the initial data $\mathbf{Q} = [v_1, v_2, v_z]$ and any choice of the third component v_z, we equally have the central normal $\mathbf{n}^c = [0, 0, \pm 1]$ for the subdivision surface $\mathbf{x} = B\mathbf{Q}$. In particular, $\mathbf{n}^c = [0, 0, \pm 1]$ for $\mathbf{Q} = [v_1, v_2, v_1]$. This is a contradiction because here, the subdominant eigencoefficients are $\mathbf{p}_1 = [1, 0, 1], \mathbf{p}_2 = [0, 1, 0]$ so that, by Theorem 5.6/87, the central normal is $\mathbf{n}^c = [-1, 0, 1]/\sqrt{2}$. Hence, there must exist a generating spline b_ℓ covering the central knot. Following Sect. 5.4/95, $\delta_{\ell'} := S^i \delta_\ell$ is also a unit vector for all powers $i \in \mathbb{Z}_n$ of the shift matrix S. The corresponding generating spline $b_{\ell'}$ has segments

$$b_{\ell'}(\cdot, j) = b_\ell(\cdot, j - i), \quad i, j \in \mathbb{Z}_n.$$

Hence, $b_{\ell'}$ is covering the central knot, too. That way, for i running from 0 to $n-1$, we obtain n different generating splines which belong to $B^\mathbf{o}$, implying $\#B^\mathbf{o} \geq n$.
□

According to the theorem, families of subdivision algorithms generating piecewise polynomial surfaces of fixed degree reveal local linear dependence of the generating splines for large n. More precisely, for $C_r^{k,q}$-algorithms as introduced in Definition 7.17/143, local linear independence is certainly lost if $n > \operatorname{loc} G = (q+1)^2$. The actual bound, however, is typically much lower. As an example, we examine the Catmull–Clark algorithm and derive a sequence of sharper and sharper results using the arguments provided above.

Example 8.9 (Local linear dependence for the Catmull–Clark algorithm). We consider the Catmull–Clark algorithm with control points labelled as shown in Fig. 6.3/111. Here, we have $q = 3$ and $\operatorname{loc} G = 16$.

- According to the second condition of Theorem 8.8/170, we have local linear dependence for $n > 16$.

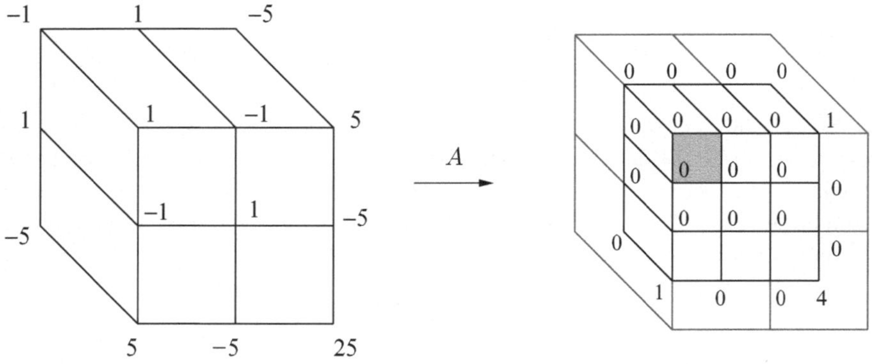

Fig. 8.2 Illustration of Example 8.9/171, fourth item: (*left*) The nonzero input to a Catmull–Clark mesh with isolated node of valence $n = 3$ results (*right*) in the zero function on the shaded area.

- The generating splines corresponding to the central control point $\tilde{\mathbf{q}}_0$ and the innermost ring $\mathbf{q}_{j,1}, \mathbf{q}_{j,2}$ cover the central knot. Hence, according to the second condition of Theorem 8.8/170, we have local linear dependence for $\#B^{\mathbf{0}} = 2n + 1 > 16$, i.e., for $n > 7$.
- When considering the subset $\bar{\mathbf{S}} := [2^{-m-1}, 2^{-m}]^2 \times \{j\} \subset \mathbf{S}_n^m \subset \mathbf{S}_n$ for a sufficiently large value of m, the set $B_{\bar{\mathbf{S}}}$ consists of the the functions in $B^{\mathbf{0}}$ and seven further generating splines. Because $\operatorname{rank} B_{\bar{\mathbf{S}}} = \operatorname{loc} G = 16$, we have local linear dependence for $\#B_{\bar{\mathbf{S}}} = 2n + 8 > 16$, i.e., for $n > 4$.
- For the regular case $n = 4$, the generating splines coincide with standard bicubic tensor product B-splines, which are known to be locally linearly independent. Hence, it remains to consider the case $n = 3$. Here, the counting arguments used above yield no indication for local linear dependence. However, as can be verified by explicit computation, the generating splines fail to be locally linearly independent even here. In Fig. 8.2/172, we see an arrangement of control points which yields vanishing segments near the central point.

Summarizing, the generating splines of the Catmull–Clark algorithm are locally linearly dependent unless $n = 4$. □

The example shows that for low valences the problem of local linear dependence may be subtle and can be settled only by an analysis which is specific to the given algorithm, [PW06].

We conclude with some remarks on global linear independence. So far, our analysis was geared to the case that the same subdivision rules apply at each step. However, for a given *global* mesh of control points, it might be necessary to apply a few special initial steps with different rules. Without going into the details, we are facing the situation that the initial data $\mathbf{Q}$, as used for stationary subdivision, are computed from some vector $\mathbf{Q}_\star$ of *primal data* forming the input mesh. For these primal data, no special connectivity, or a block structure, or the like is assumed. For a linear algorithm, $\mathbf{Q}$ and $\mathbf{Q}_\star$ are related by some matrix $A_\star$, that is not necessarily square,

$$\mathbf{Q} = A_\star \mathbf{Q}_\star.$$

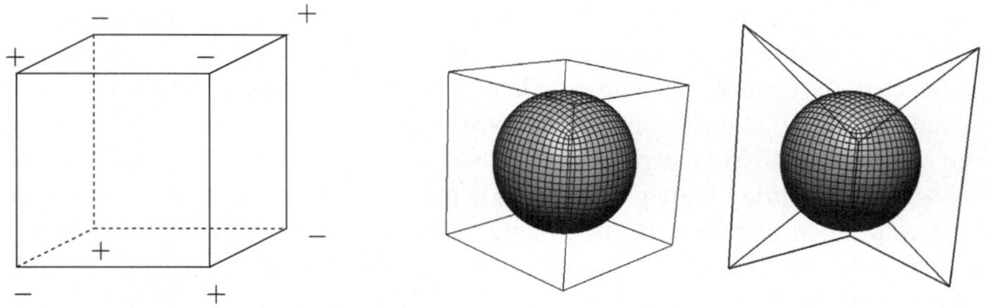

Fig. 8.3 Illustration of Example 8.10/173: Choosing any nonzero number for $+$ and its negative for $-$ results in the zero function. By the refinement stencils of Fig. 6.2/110, new face and edge nodes have value 0. For a vertex, $\alpha - 3\beta + 3\gamma = (5-6+1)/12 = 0$, too, so that all coefficients are zero after one step. *(right)* The two control nets generate the same Catmull–Clark limit surface.

Accordingly, we define the primal eigensplines $B_\star$ via their rings

$$B_\star^m = G A^m A_\star.$$

Then, the subdivision surface $\mathbf{x}_\star$ corresponding to $\mathbf{Q}_\star$ is $\mathbf{x}_\star = B_\star \mathbf{Q}_\star$. Linear independence of the functions $B_\star$, is referred to as *global linear independence*. Of course, this property relies not only on the matrix A, but also on $A_\star$. One can certainly say that $B_\star$ is linearly dependent if $A_\star$ has a non-trivial kernel. However, beyond that elementary statement, it is hardly possible to derive universal results. Instead, we exemplarily consider the Catmull–Clark algorithm, again:

Example 8.10 (Global linear dependence for the Catmull–Clark algorithm). We consider the Catmull–Clark algorithm in its original form for a primal control mesh with the connectivity of a cube. Here, two initial steps are requested before stationary subdivision can be launched. As a matter of fact, the primal generating splines are globally linearly dependent. More precisely, when assigning the values ± 1 to the primal control points in an alternating way then anything is annihilated by the two initial steps. Exploiting this fact, Fig. 8.3/173 shows two different primal control meshes which yield the same resulting subdivision surface. □

Bibliographic Notes

1. For a polynomial spline, the parametric distance to the control polyhedron is proportional to 4^{-k} in k uniform refinement steps. Constants of proportionality can be found in [PK94, NPL99, Rei00] and, for the more general class of box-splines, in [dHR93, Thm. 30].

2. There are a number of *a posteriori* estimates both for regular control polyhedra and for specific subdivision algorithms [FMM86, LP01, Kob98a, GS01]. References [WP04, WP05] provide software for tight and safe *a posteriori* bounds and Peters

and Wu [PW07] derive tight estimates on the minimal number of subdivision steps necessary to ensure that every point of the limit subdivision surface is within a given tolerance of the control polyhedron. In lieu of applying the bound of Theorem 8.5/166 directly to the initial control points, the best bounds are obtained by first applying one or more subdivision steps to each neighborhood. Predicting the deviation, as opposed to measuring it after refinement, is practically relevant since it helps with (a) preallocation of resources, (b) guiding adaptive refinement, and (c) optimizing rendering. Estimates derived from control-net differences predict considerably more subdivision steps than necessary [ZC06, CC06, CY06, CCY06, HW07b, HW07a].

3. The distance of the control polyhedron to the limit surface of Loop subdivision is analyzed in [PW07]. That paper also shows that the easily-computed heuristic prediction of the distance, as the distance between control points and their limit, is in many cases a good estimate, but is unsafe since the maximal distance between the subdivision surface and the control polyhedron is in general not taken on at iterates of the control points.

4. In [GPU07], Ginkel et al. show that the normals of the control polyhedron can be misleading when estimating the limit surface normal.

5. Piecewise polynomial proxy surface constructions with one piece per facet of the control polyhedron have been derived for evaluation on graphics processing units [LS07, NYM$^+$07, MYP07].

6. Local and global linear independence of Catmull–Clark and Loop subdivision is considered in detail in [PW06]. In particular, the amazing phenomena for $n = 3$ discussed in Examples 8.9/171 and 8.10/173 are observed there for the first time. An early version of the Catmull–Clark subdivision algorithm, quoted in [DS78], computes the vertex node as $(Q + R + 2S)/4$. This results in linearly independent generating splines also for $n = 3$.

Chapter 9
Conclusion

The analysis of stationary linear subdivision algorithms presented in this book summarizes and enhances the results of three decades of intense research and it combines them into a full framework. While our understanding of C^1-algorithms is now almost complete, the generation and the analysis of algorithms of higher regularity still offers some challenges. Guided subdivision and the PTER-framework pave a path towards algorithms of higher regularity, that, for a long time, were considered not constructible. Various aspects of these new ideas have to be investigated, and the development is in full swing, at the time of writing.

In focusing on analytical aspects of subdivision surfaces from a differential geometric point of view, we left out a number of other interesting and important topics, such as the following.

- *Implementation issues:* In many applications, subdivision is considered a recipe for mesh refinement, rather than a recursion for generating sequences of rings. The availability of simple, efficient strategies for implementation [ZS00, SAUK04], even in the confines of the Graphics Processing Unit [BS02, SP03, SJP05a], evaluation of refinable functions at arbitrary rational parameters [CDM91] and, for polynomial subdivision, at arbitrary parameters [Sta98a, Sta98c] and their inclusion into the graphics pipeline [DeR98, DKT98] largely account for the overwhelming success of subdivision in Computer Graphics.
- *Sharp(er) features:* By using modified weights for specially 'tagged' vertices or edges, it is possible to blend subdivision of space curves and subdivision of surfaces to deliberately sharpen features and even reduce the smoothness to represent creases or cusps [DKT98, Sch96]. In a similar way, subdivision algorithms can be adapted to match curves and boundaries [Nas91, Lev99c, Lev00, Nas03].
- *Multiresolution:* Based on the inherent hierarchy of finer and finer spaces, one can develop strategies for multiresolution editing of subdivision surfaces [LDW97]. There are close relations to the study of wavelets, but this development is still in its infancy.
- *Applications in scientific computing:* Beyond the world of Computer Graphics, subdivision surfaces can be employed for the simulation of thin shells and

plates [COS00, GKS02, Gri03, GTS02], and possibly also in the boundary element method.

Many of these and other application-oriented issues are discussed in the book of Warren and Weimer [WW02].

Necessarily, the material presented here is a compromise between generality and specificity. Therefore, to conclude, we want to review the basic assumptions of our analysis framework, check applicability to the rich 'zoo' of subdivision algorithms in current use, and discuss possible generalizations. We consider in turn function spaces, types of recursion, and the underlying combinatorial structure.

9.1 Function Spaces

The spaces of splines and rings that we considered throughout the book cover a very wide range of algorithms. Many popular algorithms are generalizations of subdivision schemes for B-splines or box-splines [Vel01b, Vel01a, VZ01, ZS01, CC78, DS78, PR97]. The classical definition of the term 'spline' implies that the surface rings are piecewise polynomial then. However, except were explicitly mentioned, e.g. in the degree estimate in Theorem 7.19/144, the concepts and proofs of this book never relied on the particular type of functions used to build the generating rings g_ℓ. In fact, Definitions 3.1/43 and 4.9/65 are very general. The only requirements on the function spaces are the following:

- The segments are C^1 and join *parametrically* smooth.
- All rings $\mathbf{x}^m$ can be represented by using *one and the same finite-dimensional* system G of generating rings.
- The generating rings g_ℓ form a partition of unity.

Therefore, the framework covers for instance all schemes where the g_ℓ are derived from tensoring univariate subdivision, such as the interpolatory algorithms discussed in [DGL87, DL99, HIDS02, Leb94, Kob96a]. Moreover, not even the tensor product structure is crucial. Genuine bivariate subdivision operating on a regular square grid, such as Simplest subdivision or non-polynomial variants thereof, is equally covered by the framework. By contrast, subdivision algorithms generating patches that join *geometrically smooth* are not considered by our analysis. In the terminology of this book, geometric continuity can be defined using non-rigid embeddings. The first algorithm of that type was suggest by Prautzsch for subdividing Freeform splines [Pra97]. Here, the limit consists of *finitely many* polynomial patches so that no limit analysis is required. Other algorithms exploiting geometric continuity can be found in [KP07b] and in [ZLLT06].

The last bullet above, requiring that generating rings form a partition of unity, and, similarly, that the rows of the subdivision matrix to sum to 1, are not strictly necessary but a relict of the fact that most algorithms were formulated by forming local affine combinations of 'control points'. Only the following is actually needed. The ring with constant value 1 must be contained in the space spanned by G and it is mapped to itself by the recursion: for some vector e

$$Ge = 1, \quad Ae = e.$$

The changes required to adapt our analysis to this more general setting are marginal.

Not stipulating *linear independence* of the generating system G is a crucial for generality. Example 4.14/70 gives a first indication that the class of overcomplete generating systems provides a much richer source of algorithms. In fact, by not assuming linear independence of G, our analysis applies *unchanged* to most kinds of *vector-valued or matrix-valued, Hermite or jet subdivision* algorithms, see for instance [XYD06, KMP06, CJ06]. The mapping to the framework of the book is as follows. All components of the vectors, matrices, or jets in question are collected in the single vector $\mathbf{Q}$, and all subdivision rules are encoded in a single subdivision matrix A. Since some of the coefficients, representing for instance derivative information, are not needed for the parametrization of the ring, (although they are involved in determining the coefficients of subsequent rings), the vector G of generating rings is simply padded with zeros at the corresponding places. The iteration $\mathbf{x}^m = GA^m\mathbf{Q}$ is then perfectly able to model the vector-valued or matrix-valued, Hermite or jet subdivision. All that remains to be done is to remove ineffective eigenvectors from A, if there are any, by the procedure of Theorem 4.20/77.

9.2 Recursion

Throughout, we assume that the recursion $\mathbf{Q}^m \to \mathbf{Q}^{m+1}$ is

- stationary and
- linear

That is, there exists a square matrix A such that we can write

$$\mathbf{Q}^m = A\mathbf{Q}, \quad m \in \mathbb{N}_0.$$

Due to its simplicity, this class of algorithms is predominant in applications. *Non-stationary* linear algorithms may use a different rule $A(m)$ for each step,

$$\mathbf{Q}^{m+1} = A(m)\mathbf{Q}^m.$$

In the univariate case, such schemes arise in a natural way when subdividing L-splines. In particular, subdivision for exponential and trigonometric splines falls into that class. An important feature of special trigonometric splines is their ability to reproduce conic section, and in particular circles [MWW01, Sab04, CJar, ANM06]. Such schemes are *not* covered by our approach. Typically, the analysis of non-stationary algorithms is complicated, compared to our setup, by the need to characterize the joint spectrum of a sequence of subdivision matrices, i.e., the spectrum of arbitrary products of members of these families [Rio92, DL02].

Even less is known about *non-linear* algorithms. Here, we have

$$\mathbf{Q}^{m+1} = A(m, \mathbf{Q}^m)$$

for some non-linear function A. For instance, positions of new control points could be defined by geometric algorithms like the intersection of planes associated with the given points. Such an algorithm, which provably generates smooth surfaces, was suggested by Dyn, Levin, and Liu [DLL92]. Defining $\mathbf{Q}^{m+1}$ as the coefficients of the solution of a non-linear fairing problem for the corresponding ring yields another rich source of non-linear subdivision schemes. However, the analysis is far from well understood. Further, non-linear schemes are encountered when generalizing subdivision to operate on manifolds or Lie groups. In the univariate case, there are some results in that direction due to Wallner and Dyn [WD05], while the regular bivariate case is currently studied by Grohs [Gro07].

Concerning the recursion, the framework is implicitly based on a further assumption. The algorithms are assumed to be *local* in the sense that only control points in the vicinity of an extraordinary point influence the recursion. This need not hold. There are algorithms, known as *variational subdivision*, where new control points are computed as the solution of a global, linear or non-linear fairing problem [HKD93, Kob95b, Kob96b]. The visual results of these algorithms are impressive, but more or less nothing is known about the underlying theoretical properties.

9.3 Combinatorial Structure

In discussing the topological and combinatorial structure of the space of the parameters, we have to, first of all, distinguish subdivision in the univariate, the bivariate, and the general multivariate setting. As summarized in Sect. 1.7/8, there exists an extensive literature covering the univariate case. Refinable shift-invariant constructions can be obtained in the multivariate setting by tensoring univariate subdivision, or by subdividing box splines [dHR93]. The non-shift invariant setting, in three variables, has been studied, for instance in [BMDS02, SHW04]. In this book, we focus on the bivariate case.

For local algorithms, it is admissible to reduce the analysis to a vicinity of an isolated singularity. In Chap. 4/57 we chose

- the local domain $\mathbf{S}_n := [0,1]^2 \times \mathbb{Z}_n$ to consist of n copies of the unit *square*;
- a *binary refinement* of cells. That is, each square is subdivided into four new ones.

The algorithms of Catmull–Clark and Doo–Sabin, but also Simplest subdivision or $4-8$-subdivision fit that pattern. However, many other popular algorithms, like Loop's subdivision [Loo87], the Butterfly algorithm [DGL90, ZSS96] or the three-direction box splines of Sabin's thesis [Sab77] are based on triangular patches. Here,

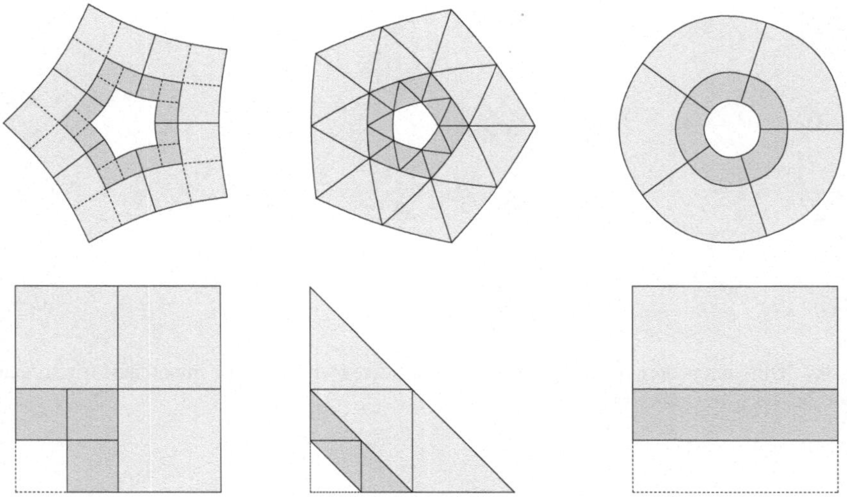

Fig. 9.1 Combinatorial layout. (*left*) □-sprocket (Catmull–Clark subdivision) layout; (*middle*) △-sprocket (Loop subdivision) layout; (*right*) polar layout (9.1/180). (*top*) Two nested characteristic rings. (*bottom*) $\mathbf{S}$ and $\mathbf{S}/2$ of the prolongation $\varphi(\mathbf{S}/2) := \lambda\varphi(\mathbf{S})$.

the appropriate structure is obtained when replacing the unit square $\boldsymbol{\Sigma} = [0,1]^2$ by an equilateral triangle. Conceptually, the framework for such *triangular subdivision algorithms* is exactly the same. However, certain technical details concerning the neighbor relation and the contact conditions have to be adapted. The regular case for triangular subdivision is $n = 6$, corresponding to a lattice spanned by three directions. As in the quadrilateral case, the regular part $\boldsymbol{\Sigma}^0$ of the cells is obtained by subtracting one half of $\boldsymbol{\Sigma}$ from itself,

$$\boldsymbol{\Sigma}^0 := \boldsymbol{\Sigma}\backslash(\boldsymbol{\Sigma}/2).$$

As shown in Fig. 9.1/179, $\boldsymbol{\Sigma}^0$ is a trapezoid rather than L-shaped.

Another generalization is to modify the number of new cells into which a given cell is subdivided. For binary refinement, each quadrilateral or triangular cell is split into $2 \times 2 = 4$ new ones. It is equally possible to use a $(\nu \times \nu)$-partition. In such a *ν-ary refinement*, we set

$$\boldsymbol{\Sigma}^0 := \boldsymbol{\Sigma}\backslash(\boldsymbol{\Sigma}/\nu).$$

For example, Kobbelt's $\sqrt{3}$-subdivision [Kob00] is based on *ternary refinement*, i.e., $\nu = 3$, and several other algorithms use the same or even finer tessellations [DIS03, OS03, Loo02b, NNP07, IDS05, LG00]. The analysis can proceed in exactly the same way as in the binary setting except for replacing, by $\nu^{\pm m}$, the factors $2^{\pm m}$ that are ubiquitous in the binary setup. The partition of the domain in ν-ary subdivision is uniform in that the domain is split evenly into subdomains. Goldman and Warren [GW93] extend uniform subdivision of curves to knot intervals that are

Fig. 9.2 Different patterns of control nets to be considered when proving smoothness of a tri/quad algorithm for *(from top)* $n = 3, 4, 6, 12$.

in a globally geometric progression. Sederberg et al. [SSS98] propose Non-uniform Recursive Subdivision Surfaces where every edge of the initial domain is allowed to be scaled differently. This presents challenges, currently not met by the state of the analysis. By contrast, the *non-uniform* subdivision algorithm [KP07a] has been fully analyzed within the framework presented in this book. Here the edges near a singularity are recursively subdivided in a ratio $\sigma : (1 - \sigma)$ where $\sigma \in (0, 1)$ is chosen to adjust the relative width of the rings and the 'speed of convergence'.

For high valence n, the rings of subdivision surfaces considered so far typically feature n sharp corners corresponding to the corners of the domain $\mathbf{S}_n$. Referring to Fig. 6.4/114 *right*, we call this arrangement the *sprocket layout*. To remove related shape artifacts, the so-called *polar layout*, as described below, offers an interesting alternative. Recall that, in (3.13/48), we connect the cells $\boldsymbol{\Sigma}$ using the neighborhood relations $(\varepsilon_4, j) \sim (\varepsilon_1, j + 1), j \in \mathbb{Z}_n$. As illustrated in Fig. 9.1/179, *right*, the cells can be glued together differently to form a ring structure. In *polar subdivision* [KP07c, KMP06, KP07d]

$$(\varepsilon_4, j) \sim (\varepsilon_2, j + 1), \quad j \in \mathbb{Z}_n. \tag{9.1}$$

That is, *opposite* edges of each cell are identified with the edges of its two neighbors. A ring with polar structure is therefore bounded on either side by a closed smooth curve. The papers on Guided subdivision [KP05, KP07b, KP08] juxtapose polar and sprocket layout. It turns out that polar subdivision nicely models features of high valence, especially rotationally extruded features. Again, for the analysis of polar subdivision, the neighbor relations and contact conditions have to be adapted, while the core methodology applies with minor modifications.

A further generalization is obtained when permitting a mix of *quadrilateral and triangular cells* to enclose an extraordinary knot [LL03, SL03, PS04, SW05]. Basically, one has to cope with two technical difficulties not encountered in the standard, symmetric setting. First, due to a lack of shift invariance, we cannot apply the Discrete Fourier Transform to simplify the analysis of spectral properties of the subdivision matrix. Rather, to obtain general results, a multitude of different patterns has to be analyzed individually. Figure 9.2/180 gives an impression of the combinatorial complexity of the problem. Second, the smooth transition between neighboring patches of different type may be more difficult to establish than between triangles only or squares only. This problem, which is very interesting in its own right, is considered in [Rio92, DL02, LL03, PS04]. For example, [PS04] proves C^1-continuity and bounded curvature for a quad-tri algorithm.

References

[ACES01] Akleman, E., Chen, J., Eryoldas, F., Srinivasan, V.: Handle and hole improvement by using new corner cutting subdivision scheme with tension. In: SMI 2001 International Conference on Shape Modeling and Applications, IEEE (2001)

[ADS06] Augsdörfer, U.H., Dodgson, N.A., Sabin, M.A.: Tuning subdivision by minimising Gaussian curvature variation near extraordinary vertices. Comput. Graph. Forum (Proc. Eurograph.) **25**(3), 263–272 (2006)

[Ale02] Alexa, M.: Refinement operators for triangle meshes. Comput. Aided Geom. Des. **19**(4), 169–172 (2002)

[Alk05] Alkalai, N.: Optimising 3d triangulations: improving the initial triangulation for the Butterfly subdivision scheme. In: Dodgson, N.A., Floater, M.S., Sabin, M.A. (eds.) Advances in Multiresolution for Geometric Modelling, pp. 231–244. Springer, Berlin Heidelberg New York (2005)

[ANM06] Abbas, A., Nasri, A., Ma, W.: An approach for embedding regular analytic shapes within subdivision surfaces. In: Seidel, H.P., Nishita, T., Peng, Q. (eds.) Proceedings of the Computer Graphics International 2006, Springer Lecture Notes in Computer Science, vol. 4035, pp. 417–429 (2006)

[AS03] Akleman, E., Srinivasan, V.: Honeycomb subdivision. (2003, preprint)

[BAD+01] Bóo, M., Amor, M., Doggett, M., Hirche, J., Strasser, W.: Hardware support for adaptive subdivision surface rendering. In: HWWS '01: Proceedings of the ACM SIGGRAPH/EUROGRAPHICS Workshop on Graphics Hardware, pp. 33–40. ACM Press, New York, NY (2001)

[BGSK05] Barthe, L., Gérot, C., Sabin, M., Kobbelt, L.: Simple computation of the eigencomponents of a subdivision matrix in the Fourier domain. In: Dodgson, N.A., Floater, M.S., Sabin, M.A. (eds.) Advances in Multiresolution for Geometric Modelling, pp. 245–257. Springer, Berlin Heidelberg New York (2005)

[BK04] Barthe, L., Kobbelt, L.: Subdivision scheme tuning around extraordinary vertices. Comput. Aided Geom. Des. **21**(6), 561–583 (2004)

[BKS00] Bischoff, S., Kobbelt, L., Seidel, H.-P.: Towards hardware implementation of Loop subdivision. In: HWWS '00: Proceedings of the ACM SIGGRAPH/EUROGRAPHICS Workshop on Graphics Hardware, pp. 41–50. ACM Press, New York, NY (2000)

[BLZ00] Biermann, H., Levin, A., Zorin, D.: Piecewise smooth subdivision surfaces with normal control. In: SIGGRAPH 2000 Conference Proceedings: Computer Graphics Annual Conference Series, pp. 113–120 (2000)

[BMDS02] Barthe, L., Mora, B., Dodgson, N., Sabin, M.: Triquadratic reconstruction for interactive modelling of potential fields. In: Proceeding: Shape Modeling and Applications 2002, pp. 145–153 (2002)

[BMZ04] Boier-Martin, I., Zorin, D.: Differentiable parameterization of Catmull–Clark subdivision surfaces. In: SGP '04: Proceedings of the 2004 EUROGRAPHICS/ACM SIGGRAPH Symposium on Geometry Processing, pp. 155–164. ACM Press, New York, NY (2004)

[Böh83] Böhm, W.: Subdividing multivariate splines. Comput. Aided Des. **15**(6), 345–352 (1983)

[BR97] Bohl, H., Reif, U.: Degenerate Beźier patches with continuous curvature. Comput. Aided Geom. Des. **14**(8), 749–761 (1997)

[BS] Bolz, J., Schröder, P.: Evaluation of subdivision surfaces on programmable graphics hardware. http://www.multires.caltech.edu/pubs/GPUSubD.pdf

[BS84] Ball, A., Storry, D.: Recursively generated B-spline surfaces. In: Proceedings of CAD84, pp. 112–119 (1984)

[BS86] Ball, A.A., Storry, D.J.T.: A matrix approach to the analysis of recursively generated B-spline surfaces. Comput. Aided Des. **18**, 437–442 (1986)

[BS88] Ball, A.A., Storry, D.J.T.: Conditions for tangent plane continuity over recursively generated B-spline surfaces. ACM Trans. Graph. **7**, 83–102 (1988)

[BS90] Ball, A.A., Storry, D.J.T.: An investigation of curvature variations over recursively generated B-spline surfaces. Technical Report, Loughborough University of Technology (1990)

[BS01] Bartels, R., Samavati, F.: Reversing subdivision rules: local linear conditions and observations on inner products. J. Comput. Appl. Math. **119**(1–2), 29–67 (2001)

[BS02] Bolz, J., Schröder, P.: Rapid evaluation of Catmull–Clark subdivision surfaces. In: Web3D '02: Proceeding of the 7th International Conference on 3D Web Technology, pp. 11–17. ACM Press, New York, NY (2002)

[BS07] Boubekeur, T., Schlick, C.: QAS: Real-time quadratic approximation of subdivision surfaces. In: Proceedings of Pacific Graphics 2007 (2007)

[BSWX02] Bajaj, C., Schaefer, S., Warren, J., Xu, G.: A subdivision scheme for hexahedral meshes. Visual Comput. **18**(5–6), 343–356 (2002)

[Bun05] Bunnell, M.: Adaptive tessellation of subdivision surfaces with displacement mapping. In: GPU Gems 2: Programming Techniques for High-Performance Graphics and General-Purpose Computation. Addison-Wesley, Reading, MA (2005)

[BXHN02] Bajaj, C., Xu, G., Holt, R., Netravali, A.: Hierarchical multiresolution reconstruction of shell surfaces. Comput. Aided Geom. Des. **19**(2), 89–112 (2002)

[CBvR+01] Claes, J., Beets, K., van Reeth, F., Iones, I., Krupkin, A.: Turning the approximating Catmull–Clark subdivision scheme into an locally interpolating surface modelling tool. In: SMI 2001 International Conference on Shape Modeling and Applications, pp. 42–48 (2001)

[CC78] Catmull, E., Clark, J.: Recursively generated B-spline surfaces on arbitrary topological meshes. Comput. Aided Des. **10**, 350–355 (1978)

[CC06] Chen, G., Cheng, F.: Matrix based subdivision depth computation for extraordinary Catmull–Clark subdivision surface patches. In: Proceedings of GMP 2006, LNCS, vol. 4077, pp. 545–552 (2006)

[CCS05] Charina, M., Conti, C., Sauer, T.: Regularity of multivariate vector subdivision schemes. Numer. Algorithms **39**, 97–113 (2005)

[CCY06] Cheng, F., Chen, G., Yong, J.: Subdivision depth computation for extra-ordinary Catmull–Clark subdivision surface patches. In: Proceedings of GMP 2006, LNCS, vol. 4077, pp. 545–552 (2006)

[CDKP00] Cohen, A., Dyn, N., Kaber, K., Postel, M.: Multiresolution schemes on triangles for scalar conservation laws. J. Comput. Phys. **161**, 264–286 (2000)

[CDL96] Cohen, A., Dyn, N., Levin, D.: Stability and inter-dependence of matrix subdivision schemes. In: Advanced Topics in Multivariate Approximation, Montecatini Terme, 1995, Series in Approximations and Decompositions, vol. 8, pp. 33–45. World Scientific Publishing, River Edge, NJ (1996)

[CDM91] Cavaretta, A.S., Dahmen, W., Micchelli, C.A.: Stationary subdivision. Mem. Am. Math. Soc. **93**(453), 1–186 (1991)

[CDM04] Cohen, A., Dyn, N., Matei, B.: Quasilinear subdivision schemes with applications to ENO interpolation. Appl. Comput. Harmonic Anal. (2004)

[Cha74] Chaikin, G.: An algorithm for high speed curve generation. Comput. Graph. Image Process. **3**, 346–349 (1974)

[CJ06] Chui, C.K., Jiang, Q.: Matrix-valued subdivision schemes for generating surfaces with extraordinary vertices. Comput. Aided Geom. Des. **23**(5), 419–438 (2006)

[CJar] Chalmoviansky, P., Juetler, B.: A non-linear circle-preserving subdivision scheme. Adv. Comput. Math. (to appear)

[CLR84] Cohen, E., Lyche, T., Riesenfeld, R.F.: Discrete box splines and refinement algorithms. Comput. Aided Geom. Des. **1**(2), 131–148 (1984)

[CMQ02] Chang, Y., McDonnell, K., Qin, H.: A new solid subdivision scheme based on box splines. In: Proceedings of the 7th ACM Symposium on Solid Modeling and Applications, pp. 226–233 (2002)

[CMS03] Cohen, A., Merrian, J.-L., Schumaker, L.L. (eds.): Curve and Surface Fitting, Nashboro Press, Brentwood, TN (2003) ISBN 0-9728482-1-5

[COS00] Cirak, F., Ortiz, M., Schröder, P.: Subdivision surfaces: a new paradigm for thin-shell finite element analysis. Int. Numer. J. Meth. Eng. **47**(12), 2039–2072 (2000)

[CSA+02] Cirak, F., Scott, M.J., Antonsson, E.K., Ortiz, M., Schröder, P.: Integrated modeling, finite element analysis and engineering design for thin-shell structures using subdivision. Comput. Aided Des. **34**, 137–148 (2002)

[CY06] Cheng, F., Yong, J.: Subdivision depth computation for Catmull–Clark subdivision surfaces. Comput. Aided Des. Appl. **3**, 485–494 (2006)

[CZ04] Conti, C., Zimmermann, G.: Interpolatory rank-1 vector subdivision schemes. Comput. Aided Geom. Des. **21**, 341–351 (2004)

[Dau88] Daubechies, I.: Orthonormal bases of compactly supported wavelets. Comm. Pure. Appl. Math. **41**, 909–996 (1988)

[Dau92] Daubechies, I.: Ten Lectures on Wavelets. SIAM, Philadelphia (1992)

[dC76] do Carmo, M.P.: Differential Geometry of Curves and Surfaces. Prentice-Hall, Englewood Cliffs, NJ (1976) Translated from the Portuguese

[DD89] Deslauriers, G., Dubuc, S.: Symmetric iterative interpolation processes. Constr. Approx. **5**, 49–68 (1989)

[DD92] Deslauriers, G., Dubuc, S.: Erratum: symmetric iterative interpolation processes. Const. Approx. **8**, 125–126 (1992)

[DDL85] Dahmen, W., Dyn, N., Levin, D.: On the convergence rates of subdivision algorithms for box splines. Constr. Approx. **1**, 305–322 (1985)

[DeR98] DeRose, T., et al.: Texture mapping and other uses of scalar fields on subdivision surfaces in computer graphics and animation US Patent 6,037,949, 1998

[DF00] Dyn, N., Farkhi, E.: Spline subdivision schemes for convex compact sets. J. Comput. Appl. Math. **119**, 133–144 (2000)

[DFS05] Dodgson, N., Floater, M., Sabin, M. (eds.): Advances in Multiresolution for Geometric Modelling. Springer, Berlin Heidelberg New York (2005)

[DGL87] Dyn, N., Gregory, J.A., Levin, D.: A four-point interpolatory subdivision scheme for curve design. Comput. Aided Geom. Des. **4**, 257–268 (1987)

[DGL90] Dyn, N., Gregory, J.A., Levin, D.: A Butterfly subdivision scheme for surface interpolation with tension control. ACM Trans. Graph. **9**(2), 160–169 (1990)

[DGL91] Dyn, N., Gregory, J.A., Levin, D.: Analysis of uniform binary subdivision schemes for curve design. Constr. Approx. **7**(2), 127–148 (1991)

[DGL95] Dyn, N., Gregory, J., Levin, D.: Piecewise uniform subdivision schemes. In: Lyche, T., Dahlen, M., Schumaker, L.L. (eds.) Mathematical Methods for Curves and Surfaces: Ulvik 1996, pp. 111–119. Vanderbilt University Press, Nashville, TN (1995)

[DGS99] Daubechies, I., Guskov, I., Sweldens, W.: Regularity of irregular subdivision. Constr. Approx. **15**(3), 381–426 (1999)

[DGS00] Dreger, A., Gross, M., Schlegel, J.: Construction of multiresolution triangular B-spline surfaces using hexagonal filters. Visual Comput. 16(6) (2000)

[DGS01] Daubechies, I., Guskov, I., Sweldens, W.: Commutation for irregular subdivision. Constr. Approx. **17**(4), 479–514 (2001)

[DGSS99] Daubechies, I., Guskov, I., Schröder, P., Sweldens, W.: Wavelets on irregular point sets. Philos. Trans. Roy. Soc. Lond. A **357**(1760), 2397–2413 (1999)

[dHR93] de Boor, C., Höllig, K., Riemenschneider, S.: Box Splines: Applied Mathematical Sciences, vol. 98, Springer, Berlin Heidelberg New York (1993)

[DIS03] Dodgson, N., Ivrissimtzis, I., Sabin, M.: Characteristics of dual triangular $\sqrt{3}$ subdivision. In: Curve and Surface Fitting: St Malo 2002, pp. 119–128 (2003)

[DKT98] DeRose, T., Kass, M., Truong, T.: Subdivision surfaces in character animation. In: SIGGRAPH '98: Proceedings of the 25th Annual Conference on Computer Graphics and Interactive Techniques, pp. 85–94. ACM Press, New York, NY (1998)

[DL92a] Daubechies, I., Lagarias, J.C.: Two-scale difference equations. I. Existence and global regularity of solutions. SIAM J. Math. Anal. **22**, 1388–1410 (1992)

[DL92b] Daubechies, I., Lagarias, J.C.: Two-scale difference equations. II. Local regularity, infinite products of matrices and fractals. SIAM J. Math. Anal. **23**, 1031–1079 (1992)

[DL92c] Dyn, N., Levin, D.: Stationary and non-stationary binary subdivision schemes. In: Lyche, T., Schumaker, L.L. (eds.) Mathematical Methods in Computer Aided Geometric Design II, pp. 209–216. Academic Press, New York (1992)

[DL95] Dyn, N., Levin, D.: Analysis of asymptotically equivalent binary subdivision schemes. J. Math. Anal. Appl. **193**, 594–621 (1995)

[DL99] Dyn, N., Levin, D.: Analysis of Hermite interpolatory subdivision schemes. In: Dubuc, S. (ed.) Spline Functions and the Theory of Wavelets, AMS series CRM Proceedings and Lecture Notes, pp. 105–113 (1999)

[DL02] Dyn, N., Levin, D.: Subdivision schemes in geometric modelling. Acta Numer **11**, 73–144 (2002)

[DLL92] Dyn, N., Levin, D., Liu, D.: Interpolatory convexity-preserving subdivision schemes for curves and surfaces. Comput. Aided Des. **24**, 211–216 (1992)

[DLL03] Dyn, N., Levin, D., Luzzatto, A.: Non-stationary interpolatory subdivision schemes reproducing spaces of exponential polynomials. Found. Comput. Math. 187–206 (2003)

[DLM90] Dyn, N., Levin, D., Micchelli, C.A.: Using parameters to increase smoothness of curves and surfaces generated by subdivision. Comput. Aided Geom. Des. **7**, 129–140 (1990)

[DLS03] Dyn, N., Levin, D., Simoens, J.: Face value subdivision schemes on triangulations by repeated averaging. In: Merrien, J.-L., Cohen, A., Schumaker, L.L. (eds.) Curve and Surface Fitting: St Malo 2002, pp. 129–138 (2003)

[DM84] Dahmen, W., Micchelli, C.: Subdivision algorithms for the generation of box spline surfaces. Comput. Aided Geom. Des. **1**(2), 115–129 (1984)

[DM86] Dahmen, W., Micchelli, C.: On the piecewise structure of discrete box-splines. Comput. Aided Geom. Des. **3**, 185–191 (1986)

[Doo78] Doo, D.: A subdivision algorithm for smoothing down irregularly shaped polyhedrons. In: International Conference on Ineractive Techniques in Computer Aided Design, pp. 157–165. IEEE Computer Society, Bologna, Italy (1978)

[dR47] de Rham, G.: Un peu de mathématiques à propos d'une courbe plane. Elemente der Math. **2**, 73–76, 89–97 (1947)

[dR56] de Rham, G.: Sur une courbe plane. J. Math. Pure. Appl. **35**(9), 25–42 (1956)

[DS78] Doo, D., Sabin, M.: Behaviour of recursive division surfaces near extraordinary points. Comput. Aided Des. **10**(6), 356–360 (1978)

[DSBH02] Dodgson, N.A., Sabin, M.A., Barthe, L., Hassan, M.F.: Towards a ternary interpolating subdivision scheme for the triangular mesh. Technical Report UCAM-CL-TR-539, University of Cambridge, Computer Laboratory (2002)

[Dub86] Dubuc, S.: Interpolation through an iterative scheme. J. Math. Anal. Appl. **114**, 185–204 (1986)

[DY00] Donoho, D., Yu, T.: Nonlinear pyramidal transforms based on median-interpolation. SIAM J. Math. Anal. **31**, 1030–1061 (2000)

[Dyn92] Dyn, N.: Subdivision schemes in Computer Aided Geometric Design. In: Light, W. (ed.) Advances in Numerical Analysis – Volume II, Wavelets, Subdivision Algorithms and Radial Basis Functions, pp. 36–104. Clarendon Press, Oxford (1992)

[Dyn02a] Dyn, N.: Analysis of convergence and smoothness by the formalism of Laurent polynomials. In: Primus Workshop Proceedings, pp. 51–68 (2002)

[Dyn02b] Dyn, N.: Interpolatory subdivision schemes. In: Primus Workshop Proceedings, pp. 25–50 (2002)

[Far97] Farin, G.: Curves and Surfaces for Computer-Aided Geometric Design: A Practical Guide, 4th edn. Academic Press, New York, pub-ACADEMIC:adr (1997)

[FM98] Floater, M.S., Micchelli, C.A.: Nonlinear Stationary Subdivision. Marcel Dekker, New York (1998)

[FMM86] Filip, D., Magedson, R., Markot, R.: Surface algorithms using bounds on derivatives. Comput. Aided Geom. Des. **3**(4), 295–311 (1986)

[FMS03] Friedel, I., Mullen, P., Schröder, P.: Data-dependent fairing of subdivision surfaces. In: Proceedings of the 8th ACM Symposium on Solid Modeling and Applications, pp. 185–195. ACM Press, New York, NY (2003)

[For74] Forrest, A.R.: Notes on Chaikin's algorithm. Technical Report CGP74/1, University of East Anglia, Computational Geometry Project Memo (1974)

[Ful95] Fulton, W.: Algebraic Topology. Graduate Texts in Mathematics. Springer, Berlin Heidelberg New York (1995)

[G0́3a] Gérot, C.: Analyzing a subdivision scheme with the eigenanalysis of a new subdivision matrix. Technical Report, Laboratoire des Images et des Signaux (2003)

[G0́3b] Gérot, C.: Rereading a subdivision scheme as the computation of C^p surfaces series samplings. Technical Report, Laboratoire des Images et des Signaux (2003)

[GBDS05] Gérot, C., Barthe, L., Dodgson, N., Sabin, M.: Subdivision as a sequence of sampled C^p surfaces. In: Dodgson, N.A., Floater, M.S., Sabin, M.A. (eds.) Advances in Multiresolution for Geometric Modelling, pp. 259–270. Springer, Berlin Heidelberg New York (2005)

[GH89] Gregory, J.N., Hahn, J.M.: A C^2 polygonal surface patch. Comput. Aided Geom. Des. **6**(1), 69–75 (1989)

[GKS02] Grinspun, E., Krysl, P., Schröder, P.: CHARMS: a simple framework for adaptive simulation. In: Hughes, J. (ed.) SIGGRAPH 2002 Conference Proceedings, Annual Conference Series, pp. 281–290. ACM Press/ACM SIGGRAPH, New York, NY (2002)

[GPU07] Ginkel, I., Peters, J., Umlauf, G.: Normals of subdivision surfaces and their control polyhedra. Comput. Aided Geom. Des. **24**(2), 112–116 (2007)

[GQ96a] Gregory, J., Qu, R.: Non-uniform corner cutting. Comput. Aided Geom. Des. **13**, 763–772 (1996)

[GQ96b] Gregory, J.A., Qu, R.: Nonuniform corner cutting. Comput. Aided Geom. Des. **13**(8), 763–772 (1996)

[Gri03] Grinspun, E.: The basis refinement method. Ph.D. thesis, California Institute of Technology (2003) p. 39

[Gro07] Grohs, P.: Smoothness analysis of nonlinear subdivision schemes on regular grids. Ph.D. thesis, TU-Wien (2007)

[GS01] Grinspun, E., Schröder, P.: Normal bounds for subdivision-surface interference detection. In: Ertl, T., Joy, K., Varshney, A. (eds.) Proc Visualization, pp. 333–340. IEEE (2001)

[GTS02] Green, S., Turkiyyah, G., Storti, D.: Subdivision-based multilevel methods for large scale engineering simulation of thin shells. In: Lee, K., Patrikalakis, N.M. (eds.) Proceedings of the 7th ACM Symposium on Modeling and Application (SM-02), pp. 265–272. ACM Press, New York, NY (2002)

[GU96] Goodman, T.N.T., Unsworth, K.: Injective bivariate maps. Ann. Numer. Math. **3**, 91–104 (1996)

[GU06a] Ginkel, I., Umlauf, G.: Controlling a subdivision tuning method. In: Schumaker, L.L., Cohen, A., Merrien, J.L. (eds.) Curve and Surface Fitting, pp. 170–179. Nashboro Press, Brentwood, TN (2006)

[GU06b] Ginkel, I., Umlauf, G.: Loop subdivision with curvature control. In: Scheffer, A., Polthier, K. (eds.) Proceedings of a Symposium on Geometry Processing, Cagliari, Italy, 26–28 June 2006. pp. 163–172. ACM Press, New York, NY (2006)

[GU07a] Ginkel, I., Umlauf, G.: Analyzing a generalized loop subdivision scheme. Computing **79**(2–4), 353–363 (2007)

[GU07b] Ginkel, I., Umlauf, G.: Tuning subdivision algorithms using constrained energy minimization. In: Winkler, J. (ed.) Mathematics of Surfaces XII, pp. 166–176. Springer, Berlin Heidelberg New York (2007)

[GU08] Ginkel, I., Umlauf, G.: Symmetry of shape charts. Comput. Aided Geom. Des. (2008, to appear)

[GW93] Goldman, R., Warren, J.: An extension of Chaikin's algorithm to B-spline curves with knots in geometric progression. CVGIP-Graph. Model. Image Process. **55**(1), 58–62 (1993)

[Han03a] Han, B.: Classification and construction of bivariate subdivision schemes. In: Curve and Surface Fitting: St Malo 2002, pp. 187–198 (2003)

[Han03b] Han, B.: Computing the smoothness exponent of a symmetric multivariate refinable function. SIAM J. Matrix Anal. Appl. **24**, 693–714 (2003)

[Har05] Hartmann, R.: Formeigenschaften von Subdivisionsalgorithmen. Master's thesis, Technische Universität Darmstadt (2005)

[HD03] Hassan, M., Dodgson, N.: Ternary and three-point univariate subdivision schemes. In: Curve and Surface Fitting: St Malo 2002, pp. 199–208 (2003)

[Hed90] Hed, S.: Analysis of subdivision schemes for surfaces. Master's thesis, Tel-Aviv University (1990)

[HIDS02] Hassan, M., Ivrissimtzis, I.P., Dodgson, N., Sabin, M.: An interpolating 4-point C^2 ternary stationary subdivision scheme. Comput. Aided Geom. Des. **19**, 1–18 (2002)

[HJ98] Han, B., Jia, R.-Q.: Multivariate refinement equations and convergence of subdivision schemes. SIAM J. Math. Anal. **29**(5), 1177–1199 (1998)

[HKD93] Halstead, M., Kass, M., DeRose, T.: Efficient, fair interpolation using Catmull–Clark surfaces. In: Proceedings of SIGGRAPH 93, pp. 35–44 (1993)

[HM90] Höllig, K., Mögerle, H.: G-splines. Comput. Aided Geom. Des. **7**(1–4), 197–208 (1990)

[Hol96] Holt, F.: Towards a curvature-continuous stationary subdivision algorithm. Z. Angew. Math. Mech. **76**(1), 423–424 (1996)

[HW99] Habib, A.W., Warren, J.: Edge and vertex insertion for a class of c^1-subdivision surfaces. Comput. Aided Geom. Des. **16**(4), 223–247 (1999)

[HW07a] Huang, Z., Wang, G.: Distance between a Catmull–Clark subdivision surface and its limit mesh. In: Proceedings of the 2007 ACM Symposium on Solid and Physical Modeling, pp. 233–240 (2007)

[HW07b] Huang, Z., Wang, G.: Improved error estimate for extraordinary Catmull–Clark subdivision surface patches. In: Proceedings of CGI 2007 (2007)

[HYX05] Han, B., Yu, T., Xue, Y.-G.: Non-interpolatory Hermite subdivision schemes. Math. Comput. **74**, 1345–1367 (2005)

[IDH02] Ivrissimtzis, I.P., Dodgson, N., Hassan, M.: On the geometry of recursive subdivision. Int. J. Shape Model. **8**(1), 23–42 (2002)

[IDHS02] Ivrissimtzis, I.P., Dodgson, N., Hassan, M., Sabin, M.: The refinability of the four point scheme. Comput. Aided Geom. Des. **19**(4), 235–238 (2002)

[IDS02] Ivrissimtzis, I.P., Dodgson, N.A., Sabin, M.A.: A generative classification of mesh refinement rules with lattice transformations. Technical Report UCAM-CL-TR-542, University of Cambridge, Computer Laboratory (2002)

[IDS05] Ivrissimtzis, I.P., Dodgson, N., Sabin, M.: $\sqrt{5}$ Subdivision. In: Dodgson, N.A., Floater, M.S., Sabin, M.A. (eds.) Advances in Multiresolution for Geometric Modelling, pp. 285–300. Springer, Berlin Heidelberg New York (2005)

[ISD02] Ivrissimtzis, I.P., Sabin, M., Dodgson, N.: On the support of recursive subdivision surfaces. Technical Report UCAM-CL-TR-544, University of Cambridge, Computer Laboratory (2002)

[ISS03] Ivrissimtzis, I.P., Shrivastava, K., Seidel, H.-P.: Subdivision rules for general meshes. In: St Malo 2002, pp. 229–237 (2003)

[JS02] Jüttler, B., Schwanecke, U.: Analysis and design of Hermite subdivision schemes. Visual Comput. **18**(5–6), 326–342 (2002)

[JSD02] Jena, M., Shunmugaraj, P., Das, P.: A subdivision algorithm for trigonometric spline curves. CAGD **19**, 71–88 (2002)

[JSD03] Jena, M., Shunmugaraj, P., Das, P.: A non-stationary subdivision scheme for generalizing trigonometric spline surfaces to arbitrary meshes. CAGD **20**, 61–77 (2003)

[Jur64] Jury, E.I.: Theory and Applications of the z-Transform Method. Wiley, New York (1964)

[KMP06] Karčiauskas, K., Myles, A., Peters, J.: A C^2 polar jet subdivision. In: Scheffer, A., Polthier, K. (eds.) Proceedings of a Symposium on Geometry Processing, Cagliari, Italy, 26–28 June 2006. pp. 173–180. ACM Press, New York, NY (2006)

[Kob95a] Kobbelt, L.: Interpolatory refinement as a low pass filter. In: Dæhlen, M., Lyche, T., Schumaker, L.L. (eds.) Proceedings of the 1st Conference on Mathematical Methods for Curves and Surfaces (MMCS-94), pp. 281–290. Vanderbilt University Press, Nashville, USA (1995)

[Kob95b] Kobbelt, L.: Interpolatory refinement by variational methods. In: Chui, C., Schumaker, L.L. (eds.) Approximation Theory VIII, Wavelets and Multilevel Approximation, vol. 2, pp. 217–224. World Scientific Publishing, River Edge, NJ (1995)

[Kob96a] Kobbelt, L.: Interpolatory subdivision on open quadrilateral nets with arbitrary topology. Comput. Graph. Forum (EUROGRAPHICS '96 Conference Issue) **15**(3), 409–420 (1996)

[Kob96b] Kobbelt, L.: A variational approach to subdivision. Comput. Aided Geom. Des. **13**, 743–761 (1996)

[Kob98a] Kobbelt, L.: Tight bounding volumes for subdivision surfaces. In: Bob Werner (ed.) Pacific-Graphics'98, pp. 17–26. IEEE (1998)

[Kob98b] Kobbelt, L.: Using the discrete Fourier-transform to analyze the convergence of subdivision schemes. Appl. Comput. Harmonic Anal. **5**, 68–91 (1998)

[Kob00] Kobbelt, L.: $\sqrt{3}$-Subdivision. In: SIGGRAPH '00: Proceedings of the 27th Annual Conference on Computer Graphics and Interactive Techniques, pp. 103–112. ACM Press/Addison-Wesley Publishing, New York, NY (2000)

[KP05] Karčiauskas, K., Peters, J.: Guided subdivision. http://www.cise.ufl.edu/research/SurfLab/papers.shtml (2005)

[KP07a] Karčiauskas, K., Peters, J.: Adjustable speed subdivision surfaces. Technical Report TR-2007-430, University of Florida, Department of CISE (2007)

[KP07b] Karčiauskas, K., Peters, J.: Concentric tessellation maps and curvature continuous guided surfaces. Comput. Aided Geom. Des. **24**(2), 99–111 (2007)

[KP07c] Karčiauskas, K., Peters, J.: Surfaces with polar structure. Computing **79**, 309–315 (2007)

[KP07d] Karčiauskas, K., Peters, J.: Bicubic polar subdivision. ACM Trans. Graph. **26**(4), 14 (2007)

[KP08] Karčiauskas, K., Peters, J.: On the curvature of guided surfaces. Comput. Aided Geom. Des. **25**(2), 69–79 (2008)

[KPR04] Karčiauskas, K., Peters, J., Reif, U.: Shape characterization of subdivision surfaces – case studies. Comput. Aided Geom. Des. **21**(6), 601–614 (2004)

[KS98] Kobbelt, L., Schröder, P.: A multiresolution framework for variational subdivision. ACM Trans. Graph. **17**(4), 209–237 (1998)

[KS99] Khodakovsky, A., Schröder, P.: Fine level feature editing for subdivision surfaces. In: ACM Solid Modeling 1999 (1999)

[KSH01] Karbacher, S., Seeger, S., Häusler, G.: Refining triangle meshes by non-linear subdivision. In: Third International Conference on 3-D Digital Imaging and Modeling, pp. xiii+408 (2001)

[KvD98] Kuijt, F., van Damme, R.: Convexity preserving interpolatory subdivision schemes. Constr. Approx. **14**(14), 609–630 (1998)

[LDW97] Lounsbery, M., DeRose, T.D., Warren, J.: Multiresolution analysis for surfaces of arbitrary topological type. ACM Trans. Graph. **16**(1), 34–73 (1997)

[Leb94] Leber, M.: Interpolierende Unterteilungsalgorithmen. Master's thesis, Universität Stuttgart (1994)

[Lev99a] Levin, A.: Combined subdivision schemes. Ph.D. thesis, Tel-Aviv University (1999)

[Lev99b] Levin, A.: Combined subdivision schemes for the design of surfaces satisfying boundary conditions. CAGD **16**, 345–354 (1999)

[Lev99c] Levin, A.: Interpolating nets of curves by smooth subdivision surfaces. In: Proceedings of the ACM SIGGRAPH 1999, pp. 57–64 (1999)

[Lev99d] Levin, D.: Using Laurent polynomial representation for the analysis of non-uniform binary subdivision schemes. Adv. Comput. Math. **11**, 41–54 (1999)

[Lev00] Levin, A.: Surface design using locally interpolating subdivision schemes. J. Approx. Theory **104**, 98–120 (2000)

[Lev03] Levin, A.: Polynomial generation and quasi-interpolation in stationary non-uniform subdivision. Comput. Aided Geom. Des. **20**(1), 41–60 (2003)

[Lev06] Levin, A.: Modified subdivision surfaces with continuous curvature. ACM Trans. Graph. **25**(3), 1035–1040 (2006)

[LG00] Labsik, U., Greiner, G.: Interpolatory $\sqrt{3}$-subdivision. Comput. Graph. Forum (EUROGRAPHICS 2000) **19**(3), 131–138 (2000)

[LL03] Levin, A., Levin, D.: Analysis of quasi uniform subdivision. Appl. Comput. Harmonic Anal. **15**(1), 18–32 (2003)

[LLS01a] Litke, N., Levin, A., Schröder, P.: Fitting subdivision surfaces. In: Ertl, T., Joy, K., Varshney, A. (eds.) Proceedings of the Conference on Visualization 2001 (VIS-01), pp. 319–324. IEEE Computer Society, Piscataway, NJ (2001)

[LLS01b] Litke, N., Levin, A., Schröder, P.: Trimming for subdivision surfaces. CAGD **18**, 463–481 (2001)

[LMH00] Lee, A., Moreton, H., Hoppe, H.: Displaced subdivision surfaces. In: SIGGRAPH 2000, pp. 85–94 (2000)

[Loo87] Loop, C.: Smooth subdivision surfaces based on triangles. Master's thesis, Department of Mathematics, University of Utah (1987)

[Loo02a] Loop, C.: Bounded curvature triangle mesh subdivision with the convex hull property. Visual Comput. **18**(5–6), 316–325 (2002)

[Loo02b] Loop, C.: Smooth ternary subdivision of triangle meshes. In: Proceedings of Curve and Surface Fitting, pp. 295–302. Saint-Malo, France (2002)

[LP01] Lutterkort, D., Peters, J.: Tight linear bounds on the distance between a spline and its B-spline control polygon. Numer. Math. **89**, 735–748 (2001)

[LR80] Lane, J.M., Riesenfeld, R.F.: A theoretical development for computer generation and display of piecewise polynomial surfaces. IEEE Trans. Pattern Anal. Mach. Intell. **2**(1), 35–46 (1980)

[LS07] Loop, C., Schaefer, S.: Approximating Catmull–Clark subdivision surfaces with bicubic patches. Technical Report MSR-TR-2007-44, Microsoft Research (2007)

[MDL05] Marinov, M., Dyn, N., Levin, D.: Geometrically controlled 4-point interpolatory schemes. In: Sabin, M.A., Dodgson, N.A., Floater, M.S. (eds.) Advances in Multiresolution for Geometric Modelling, pp. 301–318. Springer, Berlin Heidelberg New York (2005)

[Mer92] Merrien, J.-L.: A family of Hermite interpolants by bisection algorithms. Numer. Algorithms **2**, 187–200 (1992)

[Mer94a] Merrien, J.-L.: Convexity-preserving interpolatory subdivision. Comput. Aided Geom. Des. **11**, 17–37 (1994)

[Mer94b] Merrien, J.-L.: Dyadic Hermite interpolation on a triangulation. Numer. Algorithms **7**, 391–410 (1994)

[MFD05] Mustafa, G., Falai, C., Deng, J.: Estimating error bounds for binary subdivision curves/surfaces. J. Comput. Appl. Math. **193**, 596–613 (2005)

[MG00] Morin, G., Goldman, R.: A subdivision scheme for Poisson curves and surfaces. CAGD **17**, 813–833 (2000)

[MJ96] MacCracken, R., Joy, K.: Free-form deformations with lattices of arbitrary topology. In: Proceedings of the ACM/SIGGRAPH 1996, pp. 181–188 (1996)

[MKP07] Myles, A., Karčiauskas, K., Peters, J.: Extending Catmull–Clark subdivision and PCCM with polar structures. In: Alexa, M., Gortler, S., Ju, T. (eds.) Proceedings of Pacific Graphics, Hawaii, pp. 313–320. ACM Press, New York, NY (2007)

[MMTP02] Ma, W., Ma, X., Tso, S.-K., Pan, Z.: Subdivision surface fitting from a dense triangle mesh. In: Proceedings of the Geometric Modeling and Processing 2002, pp. 94–103 (2002)

[MMTP04] Ma, W., Ma, X., Tso, S.K., Pan, Z.: A direct approach for subdivision surface fitting from a dense triangle mesh. Comput. Aided Des. **36**(6), 525–536 (2004)

[MP87a] Micchelli, C., Prautzsch, H.: Computing surfaces invariant under subdivision. CAGD **4**, 321–328 (1987)

[MP87b] Micchelli, C.A., Prautzsch, H.: Computing curves invariant under halving. Comput. Aided Geom. Des. **4**, 133–140 (1987)

[MP89] Micchelli, C.A., Prautzsch, H.: Uniform refinement of curves. Linear Algebra Appl. **114/115**, 841–870 (1989)

[MS01] Maillot, J., Stam, J.: A unified subdivision scheme for polygonal modeling. Comput. Graph. Forum **20**(3), 471–479 (2001)

[MS04] Möller, H., Sauer, T.: Multivariate refinable functions of high approximation order via quotient ideals of Laurent polynomials. Adv. Comput. Math. **20**, 205–228 (2004)

[MWW01] Morin, G., Warren, J., Weimer, H.: A subdivision scheme for surfaces of revolution. Comput. Aided Geom. Des. **18**(5), 483–502 (2001)

[MYP07] Myles, A., Yeo, Y., Peters, J.: Bi-quintic C^1 surfaces via perturbation. Technical Report TR-2007-423, University of Florida, Department of CISE (2007)

[MZ00] Ma, W., Zhao, N.: Catmull–Clark surface fitting for reverse engineering applications. In: GMP 2000, IEEE, pp. 274–283 (2000)

[Nas87] Nasri, A.H.: Polyhedral subdivision methods on free-form surfaces. ACM Trans. Graph. **6**(1), 29–73 (1987)

[Nas91] Nasri, A.: Boundary-corner control in recursive subdivision surfaces. Comput. Aided Des. **23**(6), 405–410 (1991)

[Nas03] Nasri, A.: Interpolating an unlimited number of curves meeting at extraordinary points on subdivision surfaces. Comput. Graph. Forum **22**(1), 87–97 (2003)

[NF01] Nasri, A., Farin, G.: A subdivision algorithm for generating rational curves. J. Graph. Tools **6**(1), 35–47 (2001)

[NKL01] Nasri, A., Kim, T.-W., Lee, K.: Fairing recursive subdivision surfaces with curve interpolation constraints. In: 3rd International Conference on Shape Modeling and Applications (SMI 2001), pp. 49–59 (2001)

[NNP07] Ni, T., Nasri, A.H., Peters, J.: Ternary subdivision for quadrilateral meshes. Comput. Aided Geom. Des. **24**(6), 361–370 (2007)

[NP94] Neamtu, M., Pfluger, P.R.: Degenerate polynomial patches of degree 4 and 5 used for geometrically smooth interpolation in R^3. Comput. Aided Geom. Des. **11**(4), 451–474 (1994)

[NPL99] Nairn, D., Peters, J., Lutterkort, D.: Sharp, quantitative bounds on the distance between a polynomial piece and its Bézier control polygon. Comput. Aided Geom. Des. **16**(7), 613–633 (1999)

[NS02a] Nasri, A., Sabin, M.: Taxonomy of interpolation constraints on recursive subdivision curves. Visual Comput. **18**, 259–272 (2002)

[NS02b] Nasri, A., Sabin, M.: A taxonomy of interpolation constraints on recursive subdivision surfaces. Visual Comput. **18**, 382–403 (2002)

[NSZ+06] Nasri, A., Sabin, M., Abu Zaki, R., Nassiri, N., Santina, R.: Feature curves with cross curvature control on Catmull–Clark subdivision surfaces. In: Seidel, H.P., Nishita, T., Peng, Q. (eds.) Proceedings of the Computer Graphics International 2006, Springer Lecture Notes in Computer Science, vol. 4035, pp. 761–768 (2006)

[NvOW01] Nasri, A., van Overfeld, K., Wyvill, B.: A recursive subdivision algorithm for piecewise circular spline. Comput. Graph. Forum **20**(1), 35–45 (2001)

[NYM+07] Ni, T.L., Yeo, Y., Myles, A., Goel, V., Peters, J.: Smooth surfaces from 4-sided facets. Technical Report TR-2007-427, University of Florida, Department of CISE (2007)

[OS03] Oswald, P., Schröder, P.: Composite primal/dual $\sqrt{3}$ subdivision schemes. CAGD **20**, 135–164 (2003)

[Osw03] Oswald, P.: Smoothness of a nonlinear subdivision scheme. In: St Malo 2002, pp. 323–332 (2003)

[PBP02] Prautzsch, H., Boehm, W., Paluszny, M.: Bézier and B-spline Techniques. Mathematics and Visualization. Springer, Berlin Heidelberg New York (2002)

[Pet91] Peters, J.: Parametrizing singularly to enclose vertices by a smooth parametric surface. In: MacKay, S., Kidd, E.M. (eds.) Graphics Interface '91, pp. 1–7. Canadian Information Processing Society, Calgary, Alberta (1991)

[Pet96] Peters, J.: C^2 surfaces built from zero sets of the 7-direction box spline. In: Glen Mullineux (ed.) The Mathematics of Surfaces VI, pp. 463–474. Clarendon Press, NY (1996)

[Pet98] Peters, J.: Some properties of subdivision derived from splines. In: Subdivision for Modeling and Animation, Number 36 in Course Notes of 25th International conference on Computer Graphics and Interactive Techniques, pp. 83–90. ACM SIGGRAPH, New York, NY (1998)

[Pet00] Peters, J.: Patching Catmull–Clark meshes. In: Kurt Akeley (ed.) Siggraph 2000, Computer Graphics Proceedings, Annual Conference Series, pp. 255–258. ACM Press/ACM SIGGRAPH/Addison Wesley Longman, New York, NY (2000)

[Pet02a] Peters, J.: C^2 free-form surfaces of degree (3,5). Comput. Aided Geom. Des. **19**(2), 113–126 (2002)

[Pet02b] Peters, J.: Geometric continuity. In: Handbook of Computer Aided Geometric Design, pp. 193–229. Elsevier, Amsterdam (2002)

[PK94] Prautzsch, H., Kobbelt, L.: Convergence of subdivision and degree elevation. Adv. Comput. Math. **2**, 143–154 (1994)

[Plo97] Plonka, G.: Approximation order provided by refinable function vectors. Constr. Approx. **13**, 221–244 (1997)

[PN97] Peters, J., Nasri, A.: Computing volumes of solids enclosed by recursive subdivision surfaces. Comput. Graph. Forum **16**(3):C89–C94 (1997)

[PR97] Peters, J., Reif, U.: The simplest subdivision scheme for smoothing polyhedra. ACM Trans. Graph. **16**(4), 420–431 (1997)

[PR98] Peters, J., Reif, U.: Analysis of algorithms generalizing B-spline subdivision. SIAM J. Numer. Anal. **35**(2), 728–748 (1998)

[PR99] Prautzsch, H., Reif, U.: Degree estimates for C^k-piecewise polynomial subdivision surfaces. Adv. Comput. Math. **10**(2), 209–217 (1999)

[PR04] Peters, J., Reif, U.: Shape characterization of subdivision surfaces – basic principles. Comput. Aided Geom. Des. **21**(6), 585–599 (2004)

[Pra84] Prautzsch, H.: Unterteilungsalgorithmen für multivariate Splines, Ein geometrischer Zugang. Ph.D. thesis, Dissertation, Technische Universitat Braunschweig (1984)

[Pra97] Prautzsch, H.: Freeform splines. Comput. Aided Geom. Des. **14**(3), 201–206 (1997)

[Pra98] Prautzsch, H.: Smoothness of subdivision surfaces at extraordinary points. Adv. Comput. Math. **9**, 377–389 (1998)

[PS04] Peters, J., Shiue, L.: Combining 4- and 3-direction subdivision. ACM Trans. Graph. **23**(4), 980–1003 (2004)

[PSSK03] Prusinkiewicz, P., Samavati, F., Smith, C., Karwowski, R.: L-system description of subdivision curves. Int. J. Shape Model. **9**, 41–59 (2003)

[PU98a] Prautzsch, H., Umlauf, G.: A G^2 subdivision algorithm. Computing **13**, 217–224 (1998)

[PU98b] Prautzsch, H., Umlauf, G.: Improved triangular subdivision schemes. In: CGI '98: Proceedings of the Computer Graphics International 1998, pp. 626–632. IEEE Computer Society, Washington, DC, USA (1998)

[PU00a] Peters, J., Umlauf, G.: Gaussian and mean curvature of subdivision surfaces. In: Proceedings of the 9th IMA Conference on the Mathematics of Surfaces, pp. 59–69. Springer, London, UK (2000)

[PU00b] Prautzsch, H., Umlauf, G.: A G^1 and a G^2 subdivision scheme for triangular nets. Int. J. Shape Model. **6**(1), 21–35 (2000)

[PU01] Peters, J., Umlauf, G.: Computing curvature bounds for bounded curvature subdivision. Comput. Aided Geom. Des. **18**, 455–462 (2001)

[PU06] Prautzsch, H., Umlauf, G.: Parametrizations for triangular G^k spline surfaces of low degree. ACM Trans. Graph. **25**(4), 1281–1293 (2006)

[PW97] Peters, J., Wittman, M.: Smooth blending of basic surfaces using trivariate box splines. In: Mathematics of Surfaces VII, pp. 409–426 (1997)

[PW06] Peters, J., Wu, X.: On the local linear independence of generalized subdivision functions. SIAM J. Numer. Anal. **44**(6), 2389–2407 (2006)

[PW07] Peters, J., Wu, X.: Net-to-surface distance of subdivision functions. JAT (2007, in revision)

[PX04] Pan, Q., Xu, G.: Fast evaluation of the improved loop's subdivision surfaces. In: Proceedings of the Geometric Modeling and Processing 2004. IEEE Computer Society (2004)

[QG92] Qu, R., Gregory, J.: A subdivision algorithm for non-uniform B-splines. In: Singh (ed.) Approximation Theory, Spline Functions and Applications, pp. 423–436 (1992)

[QW99] Qin, K., Wang, H.: Eigenanalysis and continuity of non-uniform Doo-Sabin surfaces. In: Proceedings of Pacific Graphics 99, pp. 179–186 (1999)

[RDS+05] Rahman, I., Drori, I., Stodden, V., Donoho, D., Schröder, P.: Multiscale representations for manifold valued data. SIAM J. Multiscale Model. Simulat. **4**(4), 1201–1232 (2005)

[Rei93] Reif, U.: Neue Aspekte in der Theorie der Freiformflächen beliebiger Topologie. Ph.D. thesis, Universität Stuttgart (1993)

[Rei95a] Reif, U.: A note on degenerate triangular Beźier patches. Comput. Aided Geom. Des. **12**(5), 547–550 (1995)

[Rei95b] Reif, U.: Some new results on subdivision algorithms for meshes of arbitrary topology. In: Chui, C.K., Schumaker, L.L. (eds.) Wavelets and Multilevel Approximation, Series in Approximations and Decompositions, vol. 2, pp. 367–374. World Scientific Publishing, River Edge, NJ (1995)

[Rei95c] Reif, U.: A unified approach to subdivision algorithms near extraordinary vertices. Comput. Aided Geom. Des. **12**(2), 153–174 (1995)

[Rei96a] Reif, U.: A degree estimate for subdivision surfaces of higher regularity. Proc. AMS **124**(7), 2167–2174 (1996)

[Rei96b] Reif, U.: On constructing topologically unrestricted B-splines. Z. Angew. Math. Mech. **76**(1), 73–74 (1996)

[Rei97] Reif, U.: A refinable space of smooth spline surfaces of arbitrary topological genus. J. Approx. Theory **90**(2), 174–199 (1997)

[Rei98] Reif, U.: TURBS – topologically unrestricted rational B-splines. Constr. Approx. **14**(1), 57–77 (1998)

[Rei99] Reif, U.: Analyse und Konstruktion von Subdivisionsalgorithmen für Freiformflächen beliebiger Topologie. Shaker Verlag (1999) Habilitationsschrift

[Rei00] Reif, U.: Best bounds on the approximation of polynomials and splines by their control structure. Comput. Aided Geom. Des. **17**(6), 579–589 (2000)

[Rei07] Reif, U.: An appropriate geometric invariant for the C^2-analysis of subdivision surfaces. In: Martin, R., Sabin, M., Winkler, J. (eds.) Mathematics of Surfaces XII, pp. 364–377. Springer, Berlin Heidelberg New York (2007)

[Rie75] Riesenfeld, R.F.: On Chaikin's algorithm. IEEE Comput. Graph. Appl. **4**, 304–310 (1975)

[Rio92] Rioul, O.: Simple regularity criteria for subdivision schemes. SIAM J. Math. Anal. **23**(6), 1544–1576 (1992)

[RP05] Reif, U., Peters, J.: Structural analysis of subdivision surfaces – a summary. In: Jetter, K., et al. (eds.) Topics in Multivariate Approximation and Interpolation, vol. 12, pp. 149–190 (2005)

[RS01] Reif, U., Schröder, P.: Curvature smoothness of subdivision surfaces. Adv. Comput. Math. **14**(2), 157–174 (2001)

[Rus96] Rushmeier, H. (ed.): SIGGRAPH Proceedings 1996. Addison Wesley, New York, NY (1996)

[Sab77] Sabin, M.A.: The use of piecewise forms for the numerical representation of shape. Ph.D. thesis, MTA, Budapest (1977)

[Sab86] Sabin, M.: Recursive division. In: Gregory, J. (ed.) The Mathematics of Surfaces, pp. 269–282. Clarendon Press, Oxford (1986)

[Sab91a] Sabin, M.: Cubic recursive division with bounded curvature. In: Schumaker, L.L., Laurent, J., LeMehaute, A. (eds.) Curves and Surfaces, pp. 411–414. Academic Press, New York (1991)

[Sab91b] Sabin, M.: ω-Convergence, a criterion for linear approximation. In: Schumaker, L.L., Laurent, J., LeMehaute, A. (eds.) Curves and Surfaces, pp. 415–420. Academic Press, New York (1991)

[Sab02a] Sabin, M.: Eigenanalysis and artifacts of subdivision curves and surfaces. In: Primus Workshop Proceedings, pp. 69–92 (2002)

[Sab02b] Sabin, M.: Interrogation of subdivision surfaces. In: Handbook of Computer Aided Geometric Design, pp. 327–341. Elsevier, Amsterdam (2002)

[Sab02c] Sabin, M.: Subdivision of box-splines. In: Primus Workshop Proceedings, pp. 3–23 (2002)

[Sab02d] Sabin, M.: Subdivision surfaces. In: Handbook of Computer Aided Geometric Design, pp. 309–326. Elsevier, Amsterdam (2002)

[Sab04] Sabin, M.: A circle-preserving interpolatory subdivision scheme. (2004, in preparation)

[SAP01] Skaria, S., Akleman, E., Parke, F.: Modeling subdivision control meshes for creating cartoon faces. In: Proceedings of Shape Modeling and Applications 2001, pp. 216–225 (2001)

[SAUK04] Shiue, L.-J., Alliez, P., Ursu, R., Kettner, L.: A tutorial on CGAL polyhedron for subdivision algorithms. In: 2nd CGAL User Workshop (2004). http://www.cgal.org/Tutorials/Polyhedron/

[SB99] Samavati, F., Bartels, R.: Multiresolution curve and surface representation: reversing subdivision rules by least-squares data fitting. Comput. Graph. Forum **18**, 97–119 (1999)

[SB01] Samavati, F., Bartels, R.: Reversing subdivision using local linear conditions: generating multiresolutions on regular triangular meshes. Technical Report http://www.cgl.uwaterloo.ca/rhbartel/Papers/TriMesh.pdf, University of Waterloo (2001)

[SB02] Sabin, M., Barthe, L.: Analysis and tuning of subdivision algorithms. In: Curves and Surfaces. Vanderbilt University Press, Nashville, TN (2002)

[SB03a] Sabin, M., Barthe, L.: Artifacts in recursive subdivision surfaces. In: St Malo 2002, pp. 353–362 (2003)

[SB03b] Sabin, M., Bejancu, A.: Boundary conditions for the 3-direction box-spline. In: Maths of Surfaces X (2003)

[Sch81] Schumaker, L.L.: Spline Functions. Wiley, New York (1981)

[Sch96] Schweitzer, J.: Analysis and applications of subdivision surfaces. Ph.D. thesis, Department of Computer Science and Engineering, University of Washington, Seattle (1996)

[Sch02] Schröder, P.: Subdivision as a fundamental building block of digital geometry processing algorithms. J. Comput. Appl. Math. **149**(1), 207–219 (2002)

[SDHI03] Sabin, M., Dodgson, N., Hassan, M., Ivrissimtzis, I.: Curvature behaviours at extraordinary points of subdivision surfaces. Comput. Aided Des. **35**(11), 1047–1051 (2003)

[SDL99] Shenkman, P., Dyn, N., Levin, D.: Normals of the butterfly subdivision scheme surfaces and their applications. J. Comput. Appl. Math. (Special Issue: Computational Methods in Computer Graphics) **102**(1), 157–180 (1999)

[SHW04] Schaefer, S., Hakenberg, J., Warren, J.: Smooth subdivision of tetrahedral meshes. In: Fellner, D., Spencer, S. (eds.) Proceedings of the 2004 EUROGRAPHICS/ACM SIGGRAPH Symposium on Geometry Processing (SGP-04), pp. 151–158. Eurographics Association, Aire-la-Ville, Switzerland (2004)

[SJP05a] Shiue, L.-J., Jones, I., Peters, J.: A realtime GPU subdivision kernel. In: Marcus Gross (ed.) Siggraph 2005, Computer Graphics Proceedings, Annual Conference Series, pp. 1010–1015. ACM Press/ACM SIGGRAPH/Addison Wesley Longman, New York, NY (2005)

[SJP05b] Shiue, L.-J., Jones, I., Peters, J.: A realtime GPU subdivision kernel. ACM Trans. Graph. **24**(3), 1010–1015 (2005)

[SL03] Stam, J., Loop, C.T.: Quad/triangle subdivision. Comput. Graph. Forum **22**(1), 79–86 (2003)

[SNB02] Samavati, F., Nezam, M.-A., Bartels, R.: Multiresolution surfaces having arbitrary topologies by a reverse Doo subdivision method. Comput. Graph. Forum **21**(2), 121–134 (2002)

[SP03] Shiue, L.J., Peters, J.: Mesh mutation in programmable graphics hardware. In: EUROGRAPHICS/SIGGRAPH Hardware Workshop, pp. 1–10 (2003)

[SP05] Shiue, L.-J., Peters, J.: Mesh refinement based on Euler encoding. In: Proceedings of the International Conference on Shape Modeling and Applications 2005, pp. 343–348 (2005)

[SSS98] Sederberg, T., Sewell, D., Sabin, M.: Non-uniform recursive subdivision surfaces. In: Proceedings of SIGGRAPH 1998, pp. 387–394 (1998)

[Sta98a] Stam, J.: Evaluation of Loop subdivision surfaces. In: SIGGRAPH '98 CDROM Proceedings (1998)

[Sta98b] Stam, J.: Evaluation of Loop subdivision surfaces (1998)

[Sta98c] Stam, J.: Exact evaluation of Catmull–Clark subdivision surfaces at arbitrary parameter values. In: Cohen, M. (ed.) SIGGRAPH 98 Proceedings, pp. 395–404. Addison Wesley, New York, NY (1998)

[Sta01] Stam, J.: On subdivision schemes generalizing uniform B-spline surfaces of arbitrary degree. CAGD **18**, 383–396 (2001)

[STK99] Suzuki, H., Takeuchi, S., Kanai, T.: Subdivision surface fitting to a range of points. In: Proceedings of Pacific Graphics 99, pp. 158–167 (1999)

[Sto85] Storry, D.J.T.: B-Spline surfaces over an irregular topology by recursive subdivision. Ph.D. thesis, Leicestershire (1985)

[Str96] Strela, V.: Multiwavelets: theory and applications. Ph.D. thesis, Massachusetts Institute of Technology (1996)

[SW03] Schaefer, S., Warren, J.: A factored interpolatory subdivision scheme for quadrilateral surfaces. In: St Malo 2002, pp. 373–382 (2003)

[SW05] Schaefer, S., Warren, J.: On C^2 triangle/quad subdivision. ACM Trans. Graph. **24**(1), 28–36 (2005)

[SW07a] Schaefer, S., Warren, J.: Exact evaluation of non-polynomial subdivision schemes at rational parameter values. In: PG '07: 15th Pacific Conference on Computer Graphics and Applications, pp. 321–330. IEEE Computer Society, Los Alamitos, CA, USA (2007)

[SW07b] Schaefer, S., Warren, J.: Exact evaluation of non-polynomial subdivision schemes at rational parameter values. In: Proceedings of Pacific Graphics 2007 (2007)

[SWY04] Shi, X., Wanga, T., Yu, P.: A practical construction of G^1 smooth biquintic B-spline surfaces over arbitrary topology. Comput. Aided Des. **36**(5), 413–424 (2004)

[SZBN03] Sederberg, T.W., Zheng, J., Bakenov, A., Nasri, A.: T-splines and T-nurccs. ACM Trans. Graph. **22**(3), 477–484 (2003)

[Tak99] Takahashi, S.: Multiresolution constraints for designing subdivision surfaces via local smoothing. In: Proceedings of Pacific Graphics 99. IEEE, pp. 168–178, 323 (1999)

[Tau02] Taubin, G.: Detecting and reconstructing subdivision connectivity. Visual Comput. **18**(5–6), 357–367 (2002)

[Uml99] Umlauf, G.: Glatte Freiformflächen und optimierte Unterteilungsalgorithmen. Ph.D. thesis, Informatik (1999)

[Uml00] Umlauf, G.: Analyzing the characteristic map of triangular subdivision schemes. Constr. Approx. **16**(1), 145–155 (2000)

[Uml04] Umlauf, G.: A technique for verifying the smoothness of subdivision schemes. In: Neamtu, M., Lucian, M.L. (eds.) Geometric Modeling and Computing: Seattle 2003. Nashboro Press, Brentwood, TN (2004)

[Uml05] Umlauf, G.: Analysis and tuning of subdivision algorithms. In: SCCG '05: Proceedings of the 21st Spring Conference on Computer Graphics, pp. 33–40. ACM Press, New York, NY (2005)

[vD97] van Damme, R.: Bivariate Hermite subdivision. Comput. Aided Geom. Des. **14**, 847–875 (1997)

[Vel01a] Velho, L.: Quasi 4–8 subdivision. Comput. Aided Geom. Des. **18**(4), 345–358 (2001)

[Vel01b] Velho, L.: Using semiregular 4–8 meshes for subdivision surfaces. J. Graph. Tools **5**(3), 35–47 (2001)

[VZ01] Velho, L., Zorin, D.: 4–8 subdivision. Comput. Aided Geom. Des. **18**(5), 397–427 (2001)

[War95] Warren, J.: Binary subdivision schemes for functions of irregular knot sequences. In: Ulvik 1996, pp. 543–562 (1995)

[War97] Warren, J.: Sparse filter banks for binary subdivision schemes. In: Mathematics of Surfaces VII, pp. 427–438 (1997)

[WD05] Wallner, J., Dyn, N.: Convergence and C^1 analysis of subdivision schemes on manifolds by proximity. Comput. Aided Geom. Des. **22**(7), 593–622 (2005)

[WP04] Wu, X., Peters, J.: Interference detection for subdivision surfaces. Comput. Graph. Forum (EUROGRAPHICS 2004) **23**(3), 577–585 (2004)

[WP05] Wu, X., Peters, J.: An accurate error measure for adaptive subdivision surfaces. In: Proceedings of the International Conference on Shape Modeling and Applications 2005, pp. 51–57 (2005)

[WP06] Wallner, J., Pottmann, H.: Intrinsic subdivision with smooth limits for graphics and animation. ACM Trans. Graph. **25**(2), 356–374 (2006)

[WQ99] Wang, Y.-P., Qu, R.: Fast implementation of scale-space by interpolatory subdivision scheme. IEEE Trans. Pattern Anal. Mach. Intell. **21**(9), 933–939 (1999)

[WS04] Warren, J., Schaefer, S.: A factored approach to subdivision surfaces. Comput. Graph. Appl. IEEE **24**, 74–81 (2004)

[WW98] Weimer, H., Warren, J.: Subdivision schemes for thin plate splines. Comput. Graph. Forum **17**(3), 303–313 (1998)

[WW99] Weimer, H., Warren, J.: Subdivision schemes for fluid flow. In: Proceedings of ACM SIGGRAPH 99, pp. 111–120 (1999)

[WW02] Warren, J., Weimer, H.: Subdivision Methods for Geometric Design. Morgan Kaufmann, New York (2002)

[XY04] Xie, G., Yu, T.P.: Smoothness analysis of nonlinear subdivision schemes of homogeneous and affine invariant type. Constr. Approx. (2004, to appear). Preprint available at http://www.rpi.edu/~yut/Papers/quasilinear.pdf

[XYD06] Xue, Y., Yu, T., Duchamp, T.: Jet subdivision schemes on the k-regular complex. Comput. Aided Geom. Des. **23**(4), 361–396 (2006)

[YZ04] Ying, L., Zorin, D.: A simple manifold-based construction of surfaces of arbitrary smoothness. ACM Trans. Graph. **23**(3), 271–275 (2004)

[ZC06] Zeng, X.-M., Chen, X.J.: Computational formula of depth for Catmull–Clark subdivion surfaces. J. Comput. Appl. Math. **195**(1), 252–262 (2006)

[ZK02] Zorin, D., Kristjansson, D.: Evaluation of piecewise smooth subdivision surfaces. Visual Comput. **18**(5–6), 299–315 (2002)

[ZLLT06] Zulti, A., Levin, A., Levin, D., Teicher, M.: C^2 subdivision over triangulations with one extraordinary point. Comput. Aided Geom. Des. **23**(2), 157–178 (2006)

[Zor97] Zorin, D.: Subdivision and multiresolution surface representations. Ph.D. thesis, Caltech, Pasadena (1997)

[Zor99] Zorin, D.: Implementing subdivision and multiresolution meshes. In: Chapter 6 of Course Notes 37 of SIGGRAPH 99 (1999)

[Zor00a] Zorin, D.: A method for analysis of C^1-continuity of subdivision surfaces. SIAM J. Numer. Anal. **37**(5), 1677–1708 (2000)

[Zor00b] Zorin, D.: Smoothness of stationary subdivision on irregular meshes. Constr. Approx. **16**(3), 359–397 (2000)

[Zor06] Zorin, D.: Constructing curvature-continuous surfaces by blending. In: Sheffer, A., Polthier, K. (eds.) Eurographics Symposium on Geometry Processing, pp. 31–40. Eurographics Association, Aire-la-Ville, Switzerland (2006)

[ZS00] Zorin, D., Schröder, P. (eds.): Subdivision for Modeling and Animation, Course Notes. ACM SIGGRAPH, New York, NY (2000)

[ZS01] Zorin, D., Schröder, P.: A unified framework for primal/dual quadrilateral subdivision schemes. CAGD **18**, 429–454 (2001)

[ZSS96] Zorin, D., Schröder, P., Sweldens, W.: Interpolatory subdivision for meshes with arbitrary topology. In: Proceedings of ACM SIGGRAPH 1996, pp. 189–192 (1996)

[ZW02] Zhang, H.-X., Wang, G.-J.: Honeycomb subdivision. J. Softw. **13**(4), 1–9 (2002)

[Zwa73] Zwart, P.B.: Multivariate splines with nondegenerate partitions. SIAM J. Numer. Anal. **10**(4), 665–673 (1973)

Index